MANGROVE ECOSYSTEM : STRUCTURE AND FUNCTION

V.P. Singh
Prof. & Head,
IEMPS, Vikram University,
Ujjain

Prof. Kazuo Odaki
Director, ERL, Chiba,
Japan

SCIENTIFIC PUBLISHERS (INDIA)
P.O. Box 91 JODHPUR

Published by:
PAWAN KUMAR
Scientific Publishers (India)
5-A, New Pali Road, P.O. Box 91
JODHPUR – 342 001
Tel.: +91-291-2433323 & 2624154
Fax: + 91-291-2512580
E-mail: scienti@sancharnet.in
artrans@tantramail.com

ISBN: 81-7233-353-6

Lasertypeset: Rajesh Ojha
Printed in India

PREFACE

Probably, no other distinct plant community has attracted as much curiously and scientific attention for as long as have the mangrove forests. Rollets (1981) annotated bibliography lists 5,608 published titles through 1975 with one of the first being the written account from the chronicle of Nearchus, dating back to the Greek mariners of 325 BC.

Mangroves are the rainforests by the sea. The majority of the sub-tropical and tropical coastline is dominated by mangroves, estimated to cover an area of 22 million hectares. However, over the past several decades, the global area in mangroves has increasingly diminished as a result of a variety of human activities, such as over harvesting, freshwater diversion and conversion to other uses.

Mangroves forests are comprised of taxonomically diverse, salt–tolerant trees and other plant species which thrive in inter–tidal zones of sheltered tropical shores, "overwash" islands, and estuaries. Mangrove trees have specially adapted aerial and salt-filtering-roots and salt–excreting-leaves that enable them to occupy the saline wetlands where other plant life cannot survive.

Mangrove forests are vital for healthy coastal ecosystems. The forest detritus, consisting mainly of fallen leaves and branches from the mangroves, provide nutrients for the marine environment and supports immense varieties of sea-life in intricate foodwebs associated directly through detritus or indirectly through the planktonic and epiphytic algal food chains.

These complex ecosystems are found between the latitudes of 32° North and 38° South, along the tropical coast of Africa, Australia, Asia and Americas. These are varying scientific classifications of what constitutes a mangrove plant. According to two reputable scientific studies, mangroves include approximately 16–24 families and 54–75 species. The greatest diversity of mangroves species exist in South-east Asia.

Mangrove forests literally live in two worlds at once, acting as the interface between land and sea. Mangroves help to protect coastlines from erosion, storm damage and wave action. The stability, mangroves provide is of immense importance. They prevent shoreline erosion by acting as buffers and catch alluvial materials, thus stabilizing land elevation by sediment accretion that balance sediment loss. Vital coral reefs and sea grass beds are also protected from damaging situation.

A primary factor of the natural environment that affects mangrove over the long term is sea level and its fluctuations. Other short–term factors are air, temperature, salinity, ocean currents, storms, shore slope and soil substrate. Most mangroves live on muddy soils, but they also grow on sand, peat and coral rock. If tidal conditions are optimal, mangrove can flourish far inland, along the upper reaches of coastal estuaries.

Mangrove ecosystems have traditionally been sustainably managed by local populations for the production of food, medicines, tannins, fuel wood and construction materials. For millions of indigenous coastal residents, mangrove forest offer dependable, basic livelihoods and sustain their traditional cultures.

Naturally resilient, mangrove forests have withstood severe storms and changing tides for many millennia, but they are now being devastated by modern encroachments. Today, mangrove forests are among the most threatened habitats in the world – disappearing at an accelerating rate, yet with little public notice. Lenticles in the exposed portions of mangrove roots are highly susceptible to clogging by crude oil and other pollutants, attacks by parasites and prolonged flooding from artificial dikes or causeways. Over time, environmental stress can kill large numbers of mangroves trees. In addition, the charcoal and timber industries have also severely impacted mangrove forests, as well as tourism and other coastal developments. The rapidly expanding shrimp aquaculture industry poses the greatest threat to the world's remaining mangroves. Literally, thousands of hectares of lush mangroves forests have been cleared to make room for the artificial shrimp ponds of this boom and bust industry. For the last many years, countless scientists and NGO groups involved in protection of mangrove forests at regional levels and global levels. The damage caused by the excessive intervention of plants and animal life and the irretrievable loss of natural resources are the dramatic results of global significance. In connection with global destruction of the tropical forests, increasing attention has recently been drawn to the ecosystem of the mangroves forests, which are being threatened to an even greater extent.

In the past, mangrove areas were generally regarded as useless, hostile territory. They were not considered to be valuable and were therefore destroyed in varying ways. Only with the help of effective management which takes the various conditions of this highly delicate ecosystem into consideration will it be possible to preserve the genetic multiplicity for future generations. In order to ensure an adequate level of protection, it is essential to recognize the value of these ecosystem on a global scale.

This book is a outcome of the work which we have done separately on various mangrove forests of the world – specifically mangrove forest of

Andaman and Nicobar Islands (nearly 5 years) and mangroves of South Japan (25 years) and worked jointly on mangrove forests of Amami-Oshima Islands, Okinawa Island, Iriomote Island and Ishigaki-Jima Islands of South Japan.

We hope that this book will provide to readers concise, sufficient and recent informations on mangrove ecosystems. The authors wish to thank JSPS-INSA for granting CEP to study the mangrove forests of South Japan and Ecological Research Laboratory, Chiba for other support, pertaining to visit to these Islands.

One of the authors (VPS) thankful to Department of Environment, Govt. of India for financial support for granting MAB project in year 1982.

Our thanks are also preferred to our friends and colleagues who helped us in various ways.

We thank Mr. Suresh Chiplunkar, Silicon Computers, Rishi Nagar, Ujjain for his meticulous typing work.

Finally, we are grateful to Shri Pawan Kumar, Publisher, Scientific Publishers (India), Jodhpur for the printing of this book.

V.P. Singh
Prof. & Head,
IEMPS, Vikram University, Ujjain

Prof. Kazuo Odaki
Director,
ERL, Chiba, Japan

Plate - 1

Kandelia candel forest at Amami-ohshima Is

Inside the forest of *Kandelia candel*

Prof. Odaki at site

Acanthus ebracteatus

Acanthus ilicifolius

Rhizophora apiculata

Plate - 2

Rhizophora stylosa

Rhizophora mucronata

Rhizophora apiculata

Bruguiera cylindrica

Bruguiera parviflora

Lumnitzera littorea

Plate - 3

Aegiceras corniculatum

Schyphiphora hydrophyllacea

Excoecaria agallocha

Avicennia marina

Avicennia officinalis

Xylocarpus granatum

Xylocarpus moluccensis

Plate - 4

Bruguiera gymnorhiza

Ceriops tagal

Bruguiera gymnorhiza & B. sexangula

Sonneratia alba

Heritiera littoralis

Avicennia marina Pneumatophores

Plate - 5

Knee roots of *Bruguiera gymnorhiza*

Nypa fruticans

Phoenix paludosa

Pandanus tectorius

Dischidia nummularia

Acrostichum aureum community

Plate - 6

Forest of *Sonneratia alba* at Okinawa Is. Japan

Root system of *Rhizophora mucronata*

Avicennia marina Forest at Goa

Seedling of *Bruguiera gymnorhiza*

Kandelia candel forest at Amami-ohshima Is.

Inside the forest of *Kandelia candel*

Plate - 7

Mangrove forest is place of recreation

General view of mangrove forest at Okinawa Isl

Rhizophora community

Rhizophora species showing seedling emergence

Kandelia candel community at proximal zone

Zonational pattern in mangrove of Amami ohsima Is.

Plate - 8

Rhizophora showing pattern of root systems

Ceriops tagal showing knee roots

Cutting of mangrove forest at Andaman Isl

Regeneration of species in disturbed area

Kandelia candel along Minato River Japan

Viviparous seeds of *Kandelia candel*

Plate - 9

Warty lenticillate bark in *Bruguira gymnorhiza*

Kandelia candel stem horbour lichens, mosses and ferns

Regeneration of *Kandelia* species *In situ*

Plate - 10

Pneumatophores of *Sonneratia caseolaris*

Old knee roots of *Bruguiera gymnorhiza*

Close-up of the Pneumatophores [2m height] of *Sonneratia caseolaris*

A small tree of *Lumnitzera littorea*

Plate - 11

Fruits of *Aegiceras corniculatum*

Germination of seedling of *Aegiceras corniculatum*

Plate - 12

Seeds of *Xylocarpus granatum*

Fruits and flower of *Nypa fruticans*

CONTENTS

Chapter – 1

INTRODUCTION

1.1. HISTORY

In recent years, mangrove forest have been attracted much scientific attention and main curiosity due to strange morphological, anatomical adaptations, special physiology (high osmotic potential of cell sap, reaction to salinity, desalination, viviparity etc.), seed and seed dispersal and survival, species succession in time and zonation in space (vertical and horizontal) paleo–history of the shores and biological assemblages. Besides, mangrove ecosystems are commercially very significant and provide many direct and indirect services to man. Consequently, the literature on these tropical forest now exceeds 10,000 times. This characteristic, the littoral plant formations of tropical and subtropical sheltered coastlines have been described variously as "coastal woodland", "littoral swamp forest", "tidal forest" and "mangrove forest". Basically the word mangrove is formed by two words the Portuguese "Mangue" and the English "grove". The 'mangue' is used for individual species of mangrove habitat and thus the word mangrove can be said to be a 'grove' of mangrove species (trees and shrubs).

Macnae (1968) coined a new term to the mangroves viz. 'mangal' for mangrove community and retained the 'mangrove' for only individual species. The correspondence French word are manglier and Paletuvier respectively (Chapman, 1976).

Firstly, Nearchus (325 BC) during his voyage from Indus delta to Persian Gulf described about *Rhizophora* trees along Arabian shore of Persian Gulf. Afterward, Theophrastus (AD 305) described these trees around 45 ft (14m) tall in blossom with white flowers, growing along sea upto their middle. Pliny in his book 'Historia naturalis' in AD 97 said about similar trees which occurred along coastlines. Aboul Abass En–Neboty (1230), Moorish botanist living in Spain gave an account of perhaps *Rhizophora* seedling under the name *Kandelia*. Oviedo (1926), firstly described the West Indies mangroves and wrote that native used the hypocotyle of *Rhizophora* seedling as food in famine. Van Reede in his 'Hortus Malabaricus' gave an account of Indian ocean mangrove in 1678. First detailed account about this was given by Schimper (1903) who observed that mangroves due to saline habitat show xerophytic nature. Among the workers are warming (1909), Watson (1928), Chapman (1944 & 1975), Davis (1940), Van Steenis (1962), Macnae (1968), Lugo and Snedaker (1974), Blasco (1977), Walsh

(1974) and Woodroffe (1982) and others. The progress of studies on mangrove had been phenomenal increased in 20th century. Because prior to the year, 1600 only references were there, 25 references in 17th century, upto 18th century the number was only 48 which rose 427 in 19th century and increased 5,000 references in 1975, mostly after 1939. Upto 1981 the figures would be above 7000 (Rollet, 1981).

Today, a total of some 10,000 studies on mangroves exist, of which approximately 3000 have appeared since 1978 [cf. UNESCO, 1981; ASFA, 1993). At present 150–200 studies on mangrove ecology are published in International Journals and collection of research funding annually.

Mangrove areas are a natural features of the tropical forests. Not only since the conference in Rio de Janerio in June 1992 but for several decades how, countless scientists and groups involved in environmental protection have been warning about the increasing global destruction of the various ecosystem, in particular those found in tropics. In connection with global destruction of tropical forests, increasing attention has recently been drawn to the ecosystem of the mangrove forests, which are being threatened to an even greater extent of have already been completely destroyed. In contrast to tropical rainforests, however, the dramatic devastation of these saline tolerant tidal forests, which represent an open ecosystem, has not been the subject of major discussion.

In the past mangrove areas were general regarded as useless, hostile territory. They were not considered to be valuable and were therefore destroyed in varying ways. Nearly 80 percent of the previously existing mangrove forests have been devastated or have been subjected to degradation as result of land deposits, construction of salinity landfills or simply due to overuse, whether this may be through charcoal extraction on the setting up of aqua–culture systems.

For the last two decades the important of mangrove ecosystems have been recognized at state and Government levels in all country and attention has been paid towards this fragile ecosystems to conserved and managed.

In 1978, the scientific committee on oceanic research (SCOR), in collaboration with UNESCO's division of marine sciences, created SCOR working group 60 on mangrove ecology. At the first meeting of this group in San Jose, Costa Rica, decided to undertake a biosphere inventory of mangrove lands and to determine their current status, managing institutions and research directions. In 1980s, the commission on ecology (COE) of the International Union for conservation of nature and natural ecology (IUCN) created the working group on mangroves to collate existing information on the global status of mangroves for International use in guiding the management and conservation of this natural resources. Interest in mangrove research has increased of late, as documented by the bibliographies gathered by Chapman (1974), Walsh (1974), Van Steenis and Snedaker (1975), Frith (1977) and Untawale (1980). In the last quarter of the 20 century, fundamental research focussed mainly on plant physiology, plant ecology and biodiversity conservation of mangrove ecosystem, where Australians and Americans played a prominent role. In the last 25 years, some new fields have been investigated in plant and animal ecology–physiology, geomorphology, sedimentology, marine toxicology and biodiversity conservation and reclamation. Some countries were in the past or have recently become important centres of mangrove research – USA (Florida, Puerto Rico, Hawaii) Senegal, Madagascar, South Africa, Malaya, Australia, New Zealand, India and Pakistan etc.

1.2. MANGROVES

Mangroves are trees or bushes that usually grow between high spring tide and mean sea level. The term mangrove here refers to individual trees or species, while the mangrove plant community is called a mangal (Macnae, 1968). Mangroves belong to many different families which may not be closely related in a phylogenetic sense but may have in common a number of special characteristics such as pneumatophores and a viviparous seeding habit.

Mangroves are one of the dominant features of coastal ecosystem of the tropics while ecologically they have been considered an important component of coastal watersheds, they continually exploited for aquaculture, agriculture, charcoal, urban development and other economic activities.

Throughout the world, there are some 60–70 species distributed in over 20 families. They have common, ecological, morphological and physiological characters which does not found in other groups of plant kingdom. The mangrove community is best developed where basic requirement are met (Macnae, 1968; Walsh, 1974).

1. *Protected shores* – Wave action and strong tides erode established mangals and hinder the establishment of seedling if follows, therefore that mangals are best developed along coasts such as :

(i) Coast of small bay
(ii) Estuaries
(iii) Lagoons
(iv) Creeks
(v) Coral reef
(vi) On the edge of continental shelf
(vii) Narrow sea channels

2. *Warm climate* : Mangals are most extensively and luxuriantly developed on tropical shores, where average temperature of the coldest months is above 20°C.

3. *Salt Water* – Mangroves can more readily compete with terrestrial plants in saline environments and thus saline conditions become one factor that enables mangals to developed successfully. The spread and survival of these tidal forest species depends on their special adaptation to variation in the salinity of the water, frequent anoxia of the soil and the general instability of the diurnally inundated tidal ground that they colonize so successfully. The salt may be exerted via specialized glands in the leaves, or largely excluded by apoplastic or membrane control in the roots (Tomlinson, 1986). Soil anoxia is tolerated by the formation of specialized root structures; some may be lateral pneumatophores with negative geotrophic growth and highly aerenchymatous tissue that are exposed to the air at low tide, as in genus *Avicennia* or, as in *Bruguiera* spp.

Aerial knee root are formed from an upward and subsequent downward growth progression of the otherwise subterranean horizontal roots. Even in the seedling stage before pneumatophores developed, the whole plant body of *Avicennia marine* is aerated through a continum of gas space, from stomata of the leaf and lenticels of the shoot surfaces, to the root system. Stability in the shifting mud is maintained by yet other root adaptations, including buttresses or substantial adventitious 'prop' roots from the trunk, which also ensure that a substantial portion of the root system is above the anoxic mud. Vegetative reproduction does not occur, so survival of mangrove stands and their colonization of new areas depends entirely upon the production of seeds with a high potential for rapid establishment in the harsh and challenging environment.

In the mangrove ecosystem, where tides and coastal currents bring unremitting variation to the forest plants and animals adopt continuously to changing chemical, physical and biological characteristics of their environment. Many species use the environment dominated by mangrove trees for food and shelter during part or all of their life cycles. There is constant movement of living and non–living matter into and out of the mangrove swamp, and the effects of such movement may be felt miles away (Heald, 1971; Odum, 1971). Of course, not all tropical coasts are lined with mangrove forests, often a mangrove stand is small, or only an occasional tree dots the shoreline. The factors which determine development of coastal forests, the ecological role of mangroves in estuaries, and their utilization by man have been studied at length.

Within these ecosystems, the individual plants and animals, the soil microbial populations and the physical environment are linked by processes by which a continuous exchange and assimilation of energy occurs. These 'internal' processes of energy fixation, accumulation of biomass, decomposition of dead organic materials and mineral cycling, are most strongly influenced by small set of critical external process. This set of processes, governing available of water, the pool of available nutrient and the stability of the habitats, is beyond the control of the ecosystem. These internal and external processes govern the biological diversity of the ecosystem, and its functional nature, and are often significant in determining its sensitivity to the various forms of exploitation to which it may be subjected. Providing these process processes are maintained, the ecosystem will persist. Disturbances or over exploitation of any one may lead ultimately to the destruction of the system.

The most critical processes involved in the perpetuation of mangrove ecosystem, depend upon –

(i) an adequate supply of water,
(ii) an adequate supply of nutrient and
(iii) stability of the substrate.

Some of the activities most destructive of mangroves – are those which take place out side the system, yet interfere severely with these processes. Provided the system continues intact, an appreciable quantity of organic litter produced by the plants as they grow (leaves, fruits twigs) makes its way into estuaries. Here biological degradation by bacteria, fungi and lower animals results in detrital complexes which appear to be the primary energy source for tropical marine coastal ecosystems. Consequently, mangroves are thought to play a vitally important role in maintaining estuaries fisheries.

1.3. DISTRIBUTION OF MANGROVES – WORLD DISTRIBUTION

Mangrove vegetation occurs characteristically in coastal flooded areas of tropical and sub–tropical protected shores.

Chapman (1986) presents a model for the paleogeography of the mangrove trees and places the origin of genera *Rhizophora* and *Avicennia*, which are the most widespread mangrove trees in the Malayan centre or Indo–Pacific speciation. In his view, these are also the oldest 'pioneer' genera of the mangrove community they originated in the late Cretaceous or Eocenes times and followed westward the yet open Tethys sea and reached the American continent, still much closer to Eurasia than today. Recent data show that the Mediterranean passageway, eventually followed by the dispersal of the mangrove trees lost its contact with the Indian Ocean some 10 million year ago and dried out completely, though only for a short period about 5 million year ago. Other mangrove genera of the big Indo–Pacific pool of the genera would thus appear to be of an earlier age, since they could not utilize the Mediterranean way anymore and, thus were confined to the Indian and Pacific oceans. There is a distinct possibility that the Mediterranean segment of the Tethys sea, was not a mere passageway, but an important centre of tropical marine speciation. This facts is amply documented for such taxa as coelentrates and echinoderms. There would be thus a possibility for a concomitant eastward and westward spread of *Rhizophora* and *Avicennia*. This provides a better explanation for the difference between the eastern and western mangroves, eventually making the tropical west African mangrove the oldest of its sort.

According to the Chapman (1976), in the present time, mangrove vegetation extends north, in the northern hemisphere, on all coasts to latitudes ranging between 24°N and 32°N. In Florida, they extend to 28°N and to Bermuda (32°N) but on the Pacific coast they not extend beyond 24° 38'. The northern extension in all cases is probably set by the number of killing frosts per annum for the individual species. From observation of *Avicennia marina* var. *resinifera*, Chapman (1976) suggested that three to four nights of 2–3°C are sufficient to kill-off the hardiest of the mangrove species.

In the southern hemisphere mangroves are found on the Brazilian east coast to the tropic of Capricorn, but on the west coast they do not extend much further south than 4, which is probably due to the lack of suitable physiographic conditions. On the east coast of Africa mangroves can be found to nearly 32°S, but on the west coast they do not extend beyond 10. In Australia and New Zealand *Avicennia* is found as far south as 37.

The mangrove vegetation of the world fails naturally into two groups –

(i) The Old World mangroves; and

(ii) The New World mangroves.

Which also includes those on the west coast of Africa. The Old World mangroves extends from East Africa, up the Red Sea, across the Indian Ocean to Australia, the northwards to the Philippines and southern Japan and southward to New Zealand, finally reaching their eastern limit in Samoa.

Being specially best developed in Indo–Malayan and Papua New Guinea (Lam, 1934), in Philippines (Dickenson *et al.*, 1928) and in Andaman and Nicobar (Singh and Garge, 1983). The restriction of the mangals to other regions is due to sensitivity of mangrove to frost. As it is evident from the temperatures of the coldest month is 20°C under tropical conditions and the seasonal fluctuations do not exceed 10°C. Only *Avicennia* which is distinguished particularly by high concentration of cell spa, does not perish to easily from cold and occurs even near Auckland in New Zealand ad *Avicennia marina* var. resinifera. Here occasional frost of –2°C are tolerated in each winter. Thus we find that the temperature restricts the mangroves distribution. Temperature also effects the occurrence of many species in a particular region. It is quite obvious the increase in distance from the equator there is decrease in number of species. As Saenger *et al.* (1983) recognized 6 mangroves species zones in all over the world.

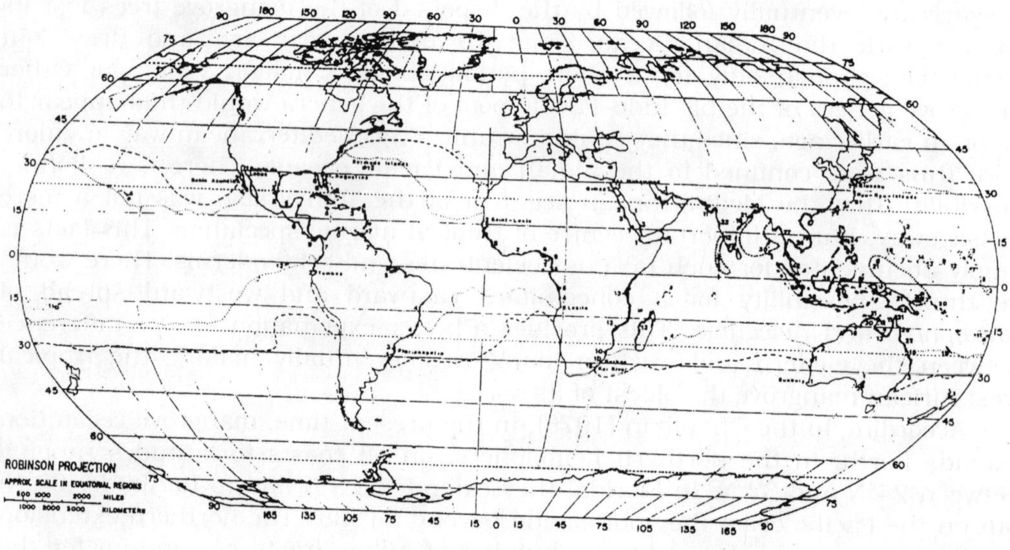

Fig. 1. Mangroves distribution in the world

The zone 1 & 2 have shown greatest abundance of the species (44) along the coasts of Malaya, Indonesia and New Guinea. Further north species are lacking with the failing temperature until in the warm temperature climate of south Japan. There is only one species *Kandelia candel* in the south equator where as *Avicennia marina* var. *resinifera* the only species remain as the least frost – trender in Southern Australia and northern New Zealand. *Rhizophora mucronata* and *Avicennia marina* penetrate into the arid climate zone 3 along Red sea coast and the later species even into Gulf Aguaba. The lack of mangroves on certain areas of west coast of north America as well as Africa results from the low water temperatures of the cold ocean currents which bathe their shores. Mangroves are found in their maximum luxuriance both in number of species and also cover within a few degree of equator in countries where the rainfall is abundant in all seasons. They have been also described to be distributed in two zones. The first zone in Eastern hemisphere excluding West Africa and second in Western hemisphere including West Africa.

Chapman (1984) have recognized 6 major groups some of which have subgroup, as follows –

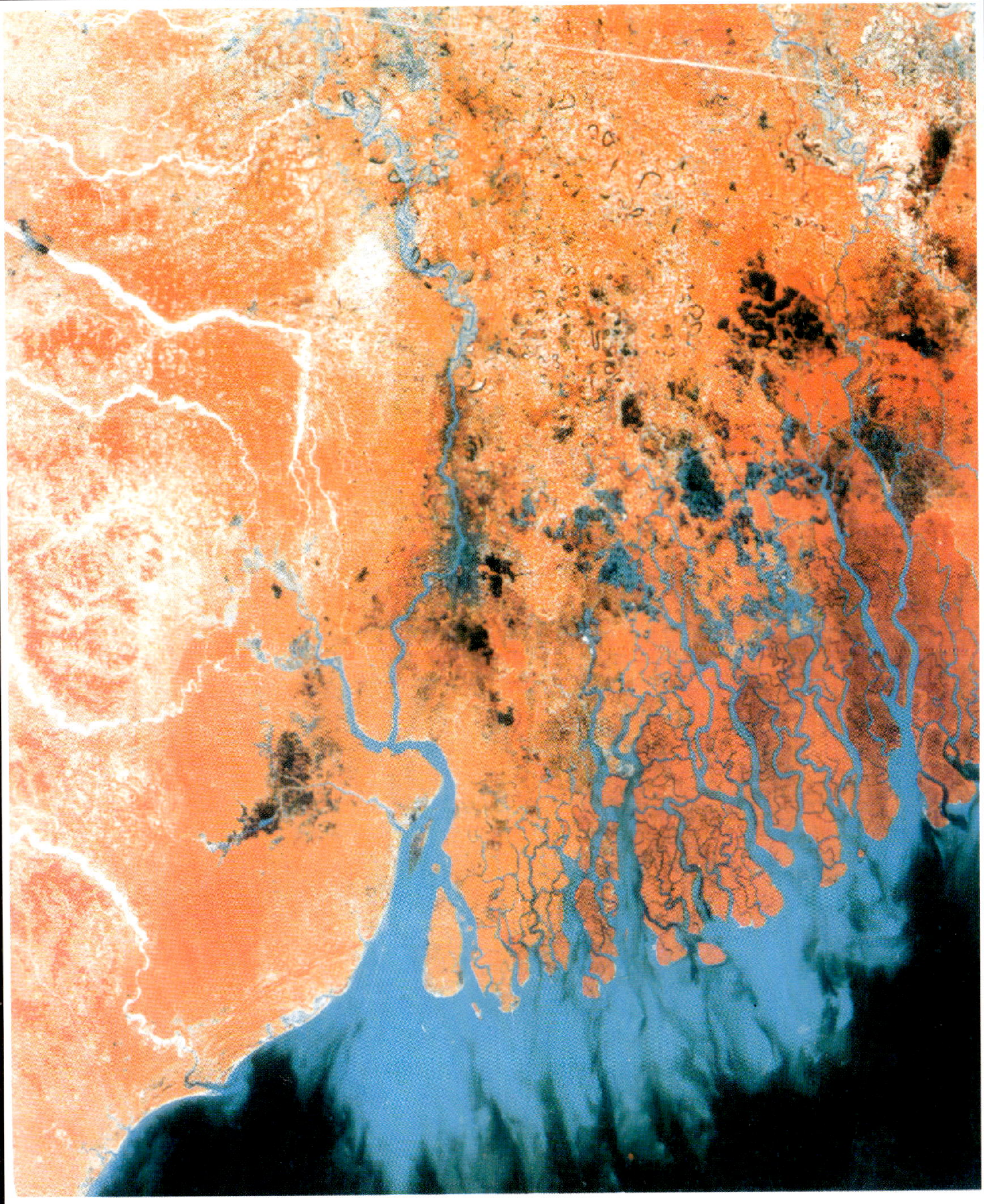

The image shows the East coast of India and Sunderbans. The Bay of Bengal is dark blue, whereas, the waters in the shallow areas near coast are seen in light blue colour. The mangroves in Sunderbans are seen as bright red colour in the wetland areas. The river Hoogly dispersing sediments into the sea can be seen clearly.

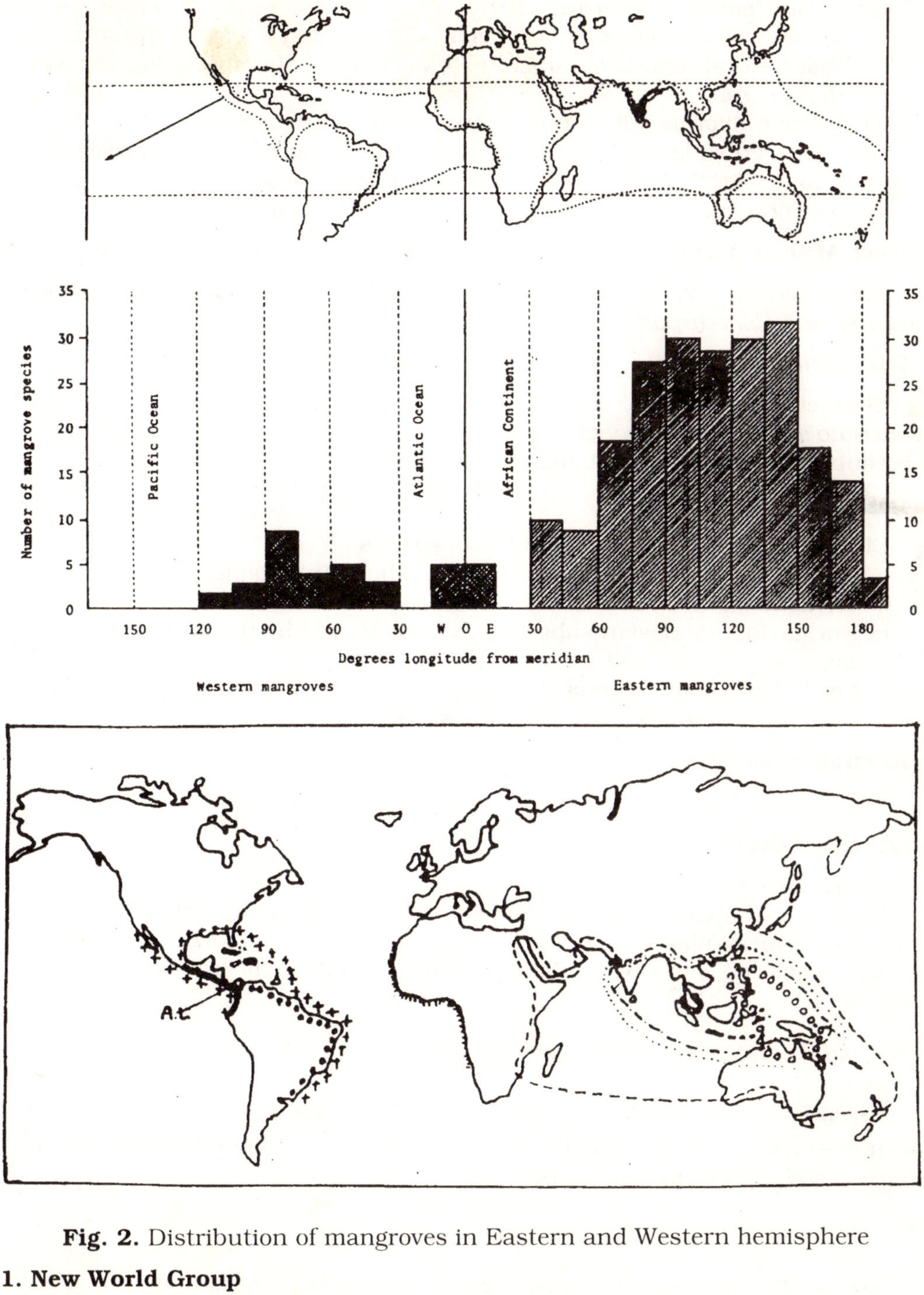

Fig. 2. Distribution of mangroves in Eastern and Western hemisphere

1. New World Group

Characterized by *Rhizophora mangle*, *Avicennia germinans*, *Laguncularia racemosa*, *Conocarpus erectur*.

(a) West Central American subgroup: Characterized by the addition of *Rhizophora racemosa, R. harrisonii, A bicolor,* and *Pelliciera rhizophora.*

(b) East Central American subgroup – same as in 1(a) above but without *P. rhizophorae.*

(c) Gulf of California subgroup – sparse, without *Laguncularia.*

(d) Caribbean subgroup – *A. schaueriana* in addition to 4 basic species.

(e) Atlantic–South American subgroup – Characterised by the addition of *R. racemosa* and *A.schaueriana* and generally lacking *Conocarpus.*

2. West African Group

Characterised by *Rhizophora mangle, Avicennia africanan, Laguncularia racemosa* and *Conocarpus erectus.*

3. East African Group

Characterized by *R. mucronata, A. marina, Bruguiera gymnorhiza, Sonneraria alba, Ceriops tagal, Xylocarpus granatum.* Red Sea : Persian Gulf subgroup – Spares with only *A. marina* and *R. mucronata.*

4. Indo–Malasian Group

Characterised by very large number of species.

a. Indian subgroup : Characterised by abundant *Sonneratia apetala* and *Heriteria minor.*

b. Burma–Indo–Malasian subgroup : Characterized by the largest number of species.

c. North Australian – Papua subgroup : Some species are less predominant. *Xylocarpus moluccensis* (*X. autraliasicus*) is specific to the area.

5. Australian Sub-group

Few species *Avicennia marina* var. *resinifera* is predominant.

6. Oceanian Group

Characterised by *Heritiera littoralis.*

Philippines Subgroup : Characterised by Camplostemon philippinensis. This group was formerly placed in group 4 but is probably better placed here.

The mangrove vegetarian of the world falls into two groups (Tomlinson, 1983).

(i) The Old World Mangroves (Eastern Mangrove Group)

Essentially the Eastern Hemisphere group – including East–Africa, India, South–East Asia, Australia and the Western Pacific. The total number of the true mangrove species in this area is 48.

(ii) The New World Mangroves (Western Mangroves Group)

Essentially the Western Hemisphere – including West Africa, Atlantic, South America, the Caribbean, Florida, Central America and Pacific and South America. The total number of true mangrove species in this area is only 9.

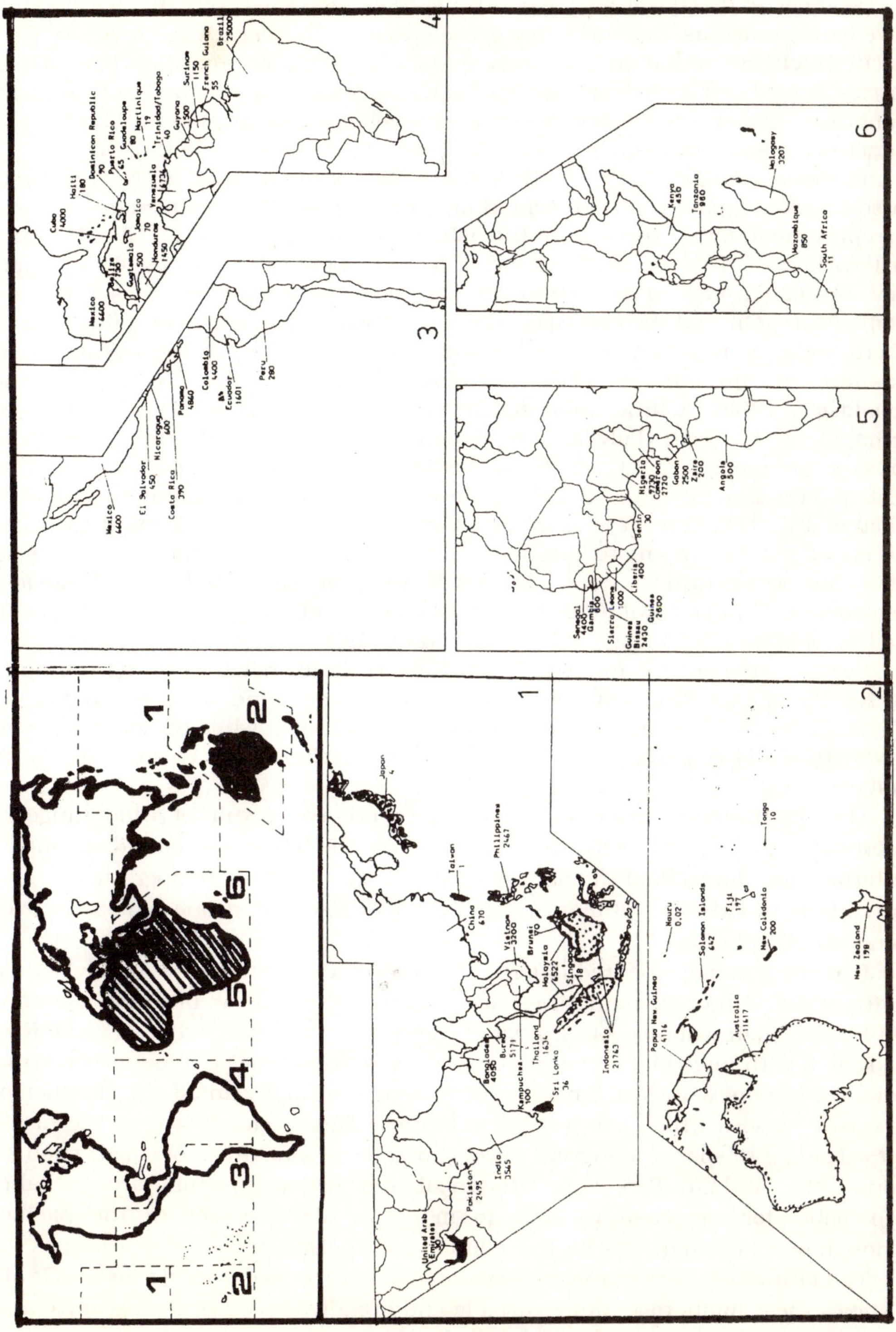

Fig. 3. Geographical zone from 1–6 COE 1963.

Seanger *et al.* (1983) have listed 60 species restricted to mangrove habitats have been named as 'Exclusive mangrove species'. Another 23 species have been listed which are found in mangrove habitat as well as elsewhere too. These species have been called as 'non–exclusive species'. Out of total 60 exclusive mangrove species 48 species occur exclusively in eastern hemisphere and 9 species in western hemisphere with only 3 species common to both.

Eastern mangroves, in north eastern side extend upto 35°N (135°E) to Izuschizuoka (Japan) but represented only one species *Kandelia candel*, while it is represented by 7 species in the island Iriomote, 25°N (125°N). Its south eastern limit is 40°, the north island of New Zealand (Van Steenis, 1962). Here only *Avicennia marina* is abundant and perhaps this limit is set by low temperature. On east African coast the limit is estuary of Kie river mouth (32°S) where only is represented due to low temperature because average daily minimum in July remains 12°C, and absolute minimum 4.4°C (Chapman and Ronaldson, 1958). A little more southward at least London it does not occur perhaps due to mean daily minimum temperature which goes upto 10°C and absolute minimum, 2.8°C. It is surprising that *Avicennia marina* grows in western Port Bay (38°C) near Melbourne where average daily minimum for July remains 4°C. Therefore, it is clear that the Victorian and New Zealand groves of *Avicennia marina* are more low temperature tolerant than the mangroves of other low temperature areas. In north western side (Red sea) *Bruguiera gymnorhiza*, *Ceriops tagal*, *Rhizophora mucronata* and *Avicennia marina* occur in the Bay of Mass (15°N) while only *A. marina* reaches southern tip of Sinai (20°).

Western mangroves, towards North–Eastern limit, reach upto the Bermuda (32°N), though, in American continent mangals are found upto South Florida (27–28°C) North–Western limit is south California (33°N) while South–West limit is North Peru (4°S) perhaps due to dryness of the climate, in Western Africa its limits is Angola.

The distribution patterns of mangroves are the result of wide range of historical and contemporary factors. Perhaps the most obvious distribution patterns, the latitudinal limits, are largely set by low temperatures, and particularly by extremes of temperature. Rainfall also had strong influences over mangrove distribution.

Largely through the reduction of salinity in an otherwise highly saline environment. Although mangroves are adapted to saline or brackish environments the high salinity of sea-water, and the sometimes higher salinities associated with intertidal areas, particularly in arid countries, frequently restrict growth. In areas with low, irregular or limited seasonal rainfall the number of mangrove species which can survive is limited. This is clearly one of the major factor leading to sparse mangrove development over wide areas of coast, such as around the Arabian Peninsula. Historical and tectonic factors are probably responsible for the easterly limit to mangrove development in the pacific, although exact mechanisms for these limits are unclear.

The latitudinal limits of mangroves on each of the major land masses (Table 1) shows these limits to be quite variable and broadly related to temperature and aridity.

Table 1. Latitudinal limits of mangroves on major Land masses

Continental Landmass	Northern Limits	Southern Limits
1. Atlantic America	32° 20'	28° 56'
2. Pacific America	30°15'	05° 32'
3. Atlantic Africa	19° 50'	12° 20'
4. Eastern Africa / Red Sea	27° 40'	32° 59'
5. Western Australia	——	33° 16'
6. Eastern Australia	——	38° 45'
7. Pacific Asia	31° 22'	——

In addition, these latitudinal limits are preceded by a gradual attention of species with increasing latitudes. For example, Saenger and Moverley (1985) have suggested that in the presence of an adequate rainfall (as on the eastern coast of Australia) temperature is the major factor in reducing species abundance with latitude and good correlation between the temperature optima of species and their geographical distribution were found.

The latitudinal limits of mangroves on the west African and South American Pacific coasts coincide with the limits of arid regions. Human activity (such as pollution, water diversion and selective clearing) can significantly change mangrove distributional range.

1.4. AREA — GLOBAL DISTRIBUTION

In many developing countries, the area covered by mangroves is decreasing rapidly due to human intervention. In some countries there has been intensive aforestation of mangroves for purpose of coastal protection. Thus to make estimates for individual countries it is crucial that data be up–to–date, but unfortunately this is seldom the case. The procedure for estimating the area covered with mangroves is also subject to some fundamental methodological problems. Satellite and aerial photographs usually form the basis of such studies.

Table 2. Estimated mangrove coverage area of the 15 countries with the largest mangrove areas.

	Country	Mangrove (1000 ha)	Global Percentage (%)
1	Indonesia	4250	30
2	Brazil	1376	10
3	Australia	1150	08
4	Nigeria	970	07
5	Malaysia	641	05
6	Bangladesh	611	04
7	Myanmar	570	04
8	Vietnam	540	04
9	Cuba	530	04
10	Mexico	525	04
11	Senegal	440	03
12	India	360	03
13	Colombia	358	03
14	Cameroon	350	02
15	Madagascar	327	02
	Total	14200	100

Source : MTO / ISME, 1993 and Hamilton *et al.*, 1989.

Accordingly, the total worldwide area of mangrove forest cited differs from author to author considerably. Reliable estimates range from 10 million ha (Bunt, 1992) to 14.2 million ha (Lacerda *et al.*, 1993) and 17 million ha (Hamilton *et al.*, 1989) to as much as 24 million ha (Twilley *et al.*, 1992). Lacerda *et al.* (1993) have made up-to-date estimate between 15 to 30 million ha presumably the current area covered by mangroves lies between 14 to 15 million ha.

The largest mangrove areas occur in Indonesia, Brazil, Australia and Nigeria. On the whole, tidal forests occur in America and Africa to a considerably smaller extent than in Asia.

Table 3. The another estimate indicate the following areas of mangrove formations in some parts of the tropics (in 10,000 ha).

	Regions	Mangrove areas in 10000 ha
1	Indonesia	4000-6000
2	India	1420
3	Nigeria	700-1000
4	Bangladesh	629
5	Philippines	400
6	Thailand	313-368
7	Vietnam	250-370
8	Papua New Guinea	162–201
9	Malaysia	130–150
10	Srilanka	120
11	UR Tanzania	100
12	Pureto Rico	76
13	Kenya	45
14	Zaire	00
15	Fizi Island	18.20
16	Tampa Bay (Florida)	07

Source : Walsh *et al.*, 1975; Fenco, 1976; Anon, 1977.

Despite having worldwide distribution, most of the coastal areas of the world are lacking the mangroves as it is evident from the fact that total coastal area of the Australia is 11,617 Km^2 and mangroves occupy only 6,064 km. Similarly, in Indian coasts mangals occupy only the 8.0% of the total coastline (Table 4). Table shows the relationship between mangrove areas, length of the mangrove coastline and total coastline of major countries of world.

In India mangrove distribution is not uniform, most of the mangroves are concentrated on the east coast and to the islands of Andaman and Nicobar (Fig. 4).

Table 4. Relationship between mangrove area, length of mangrove coastline and total coastline for those countries where data are available

Country	Mangrove areas (Km^2) (A)	Mangrove Coastline (Km) (B)	Percentage total coastline	Ratio A/B
Australia	11617	6064	22.0	1.92
Mexico	6600	9900	—	0.66
Venezuela	6737	1102	3.34	6.11
Peninsular Malaysia	1136	381	20.0	2.98
Peninsular India	3565	380	8.0	9.38
Andaman & Nicobar	1000	280	40.0	3.8

Courtesy : Peter Senger

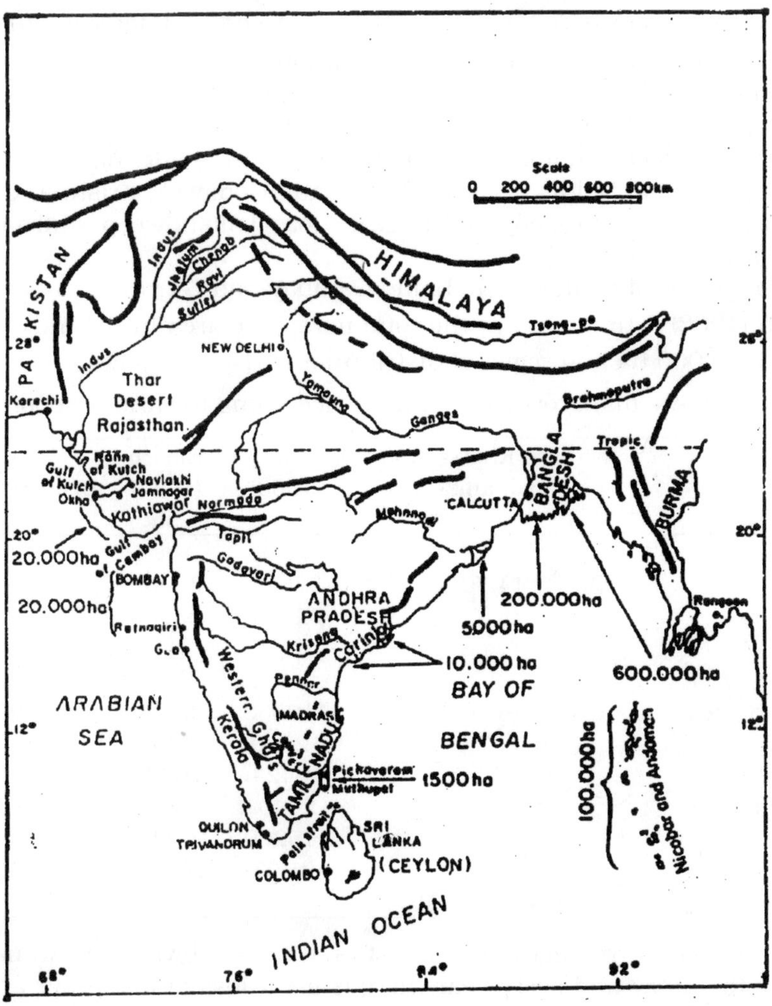

Fig. 4. Mangrove regions in India

Distribution of Mangrove in India

Moreover, mangrove occurrence has also been noticed on the coasts of Bombay, Goa, Pichavaram, Western coast etc (Fig. 4). These coast have very less number of species (13) and all are common to east coast and to the islands.

India is one of the world's top 12 mega–diversity countries. It is very rich in biological diversity. The mangroves of India have some peculiarities they are unknown in other countries:

- Rich in species biodiversity
- Very high complexity index
- Complex community organisation in many regions
- Play decisive role as nursery habitats – India has become one of the major shrimps suppliers in Asia.
- Support major populations needs as firewood and domestic fuel.

On the basis phyto-geographical position and regional situation, the mangroves of India has been grouped into 5 regions.

(1) Mangroves of Sunderbans and Gaya Delta (Fig. 5).
(2) Mangroves of Andaman and Nicobar Archipelgo (Fig. 6).
(3) Mangroves of Deltas of rivers, Mahanadi, Godavari, Krishna, Cauvery.
(4) Mangroves of Bombay, Ratnagiri, Goa, Lakshdweep, Karnataka and Kerala.
(5) Mangroves of Gulfs of Kutch and Cambay and Gujrat.

Siddhu (1963) estimated that the total mangrove area in the Indian territory was about 7,00,000 ha and regional distributions wise as follows –

Table 5. Indicate the mangrove area in India according to various reports

Major Zones	Waheed Khan (1959) sq. km.	Sidhu (1963) ha	Blasco (1977) ha	Blasco and Aizpuru (1999) km^2
West Coast				200
– Gujrat	14.70	52616	20000	
– Maharashtra	96.0	62208	20000	___
– Gulf of Kutch	___	___	___	200
East Coast				
– The Gangetic delta	468.0	418888	200000	2000
– Orissa	1637.0	12000	5000	50
– Andhra Pradesh	74.5	18424	10000	100
– Tamil Nadu	___	2640	15000	10
– Andaman & Nicobar	___	115200	100000	440
Total	2419.5 (5,49,950 ha)	6,81,976	3,56,500	3000

The total area of tree mangrove forest is still unknown, the estimate varies from survey to survey and every year there is great loss of mangrove areas in all part of the country.

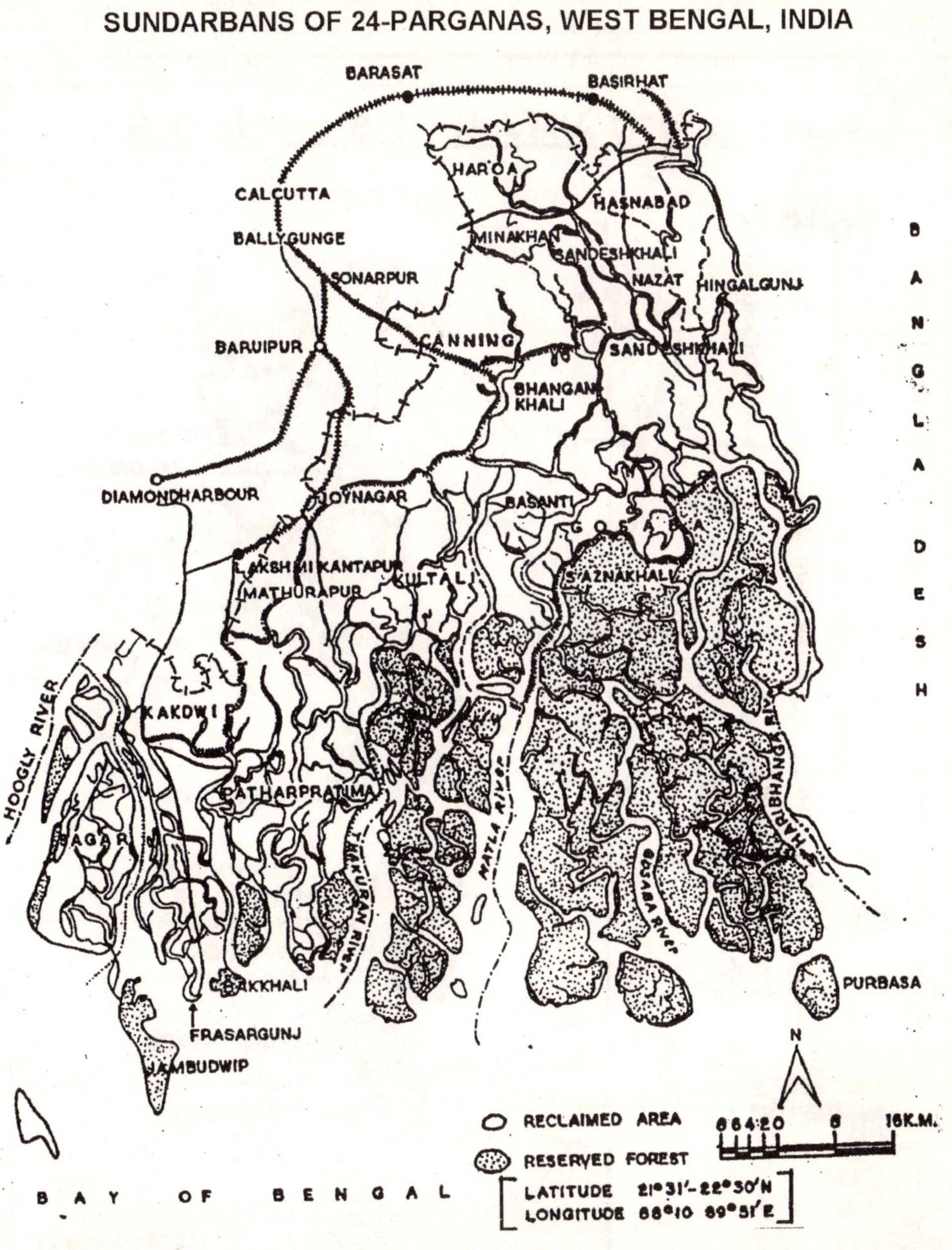

Fig. 5. Mangroves of Sunderbans

Blasco (1977) pointed out that estimate of the total area of the Indian mangals is just one half of the official estimate. The difference may be attributed to the lack of precise teminonology in reference to the coastal formations of the sub-continent. Remote sensing techniques like aerial photographs and satellite data have been largely observed to be precise, economic and rapid tool for the studies of mangrove eco-systems. Upto 17th century mangroves were regarded as dangerous and in general of little value to man and his work.

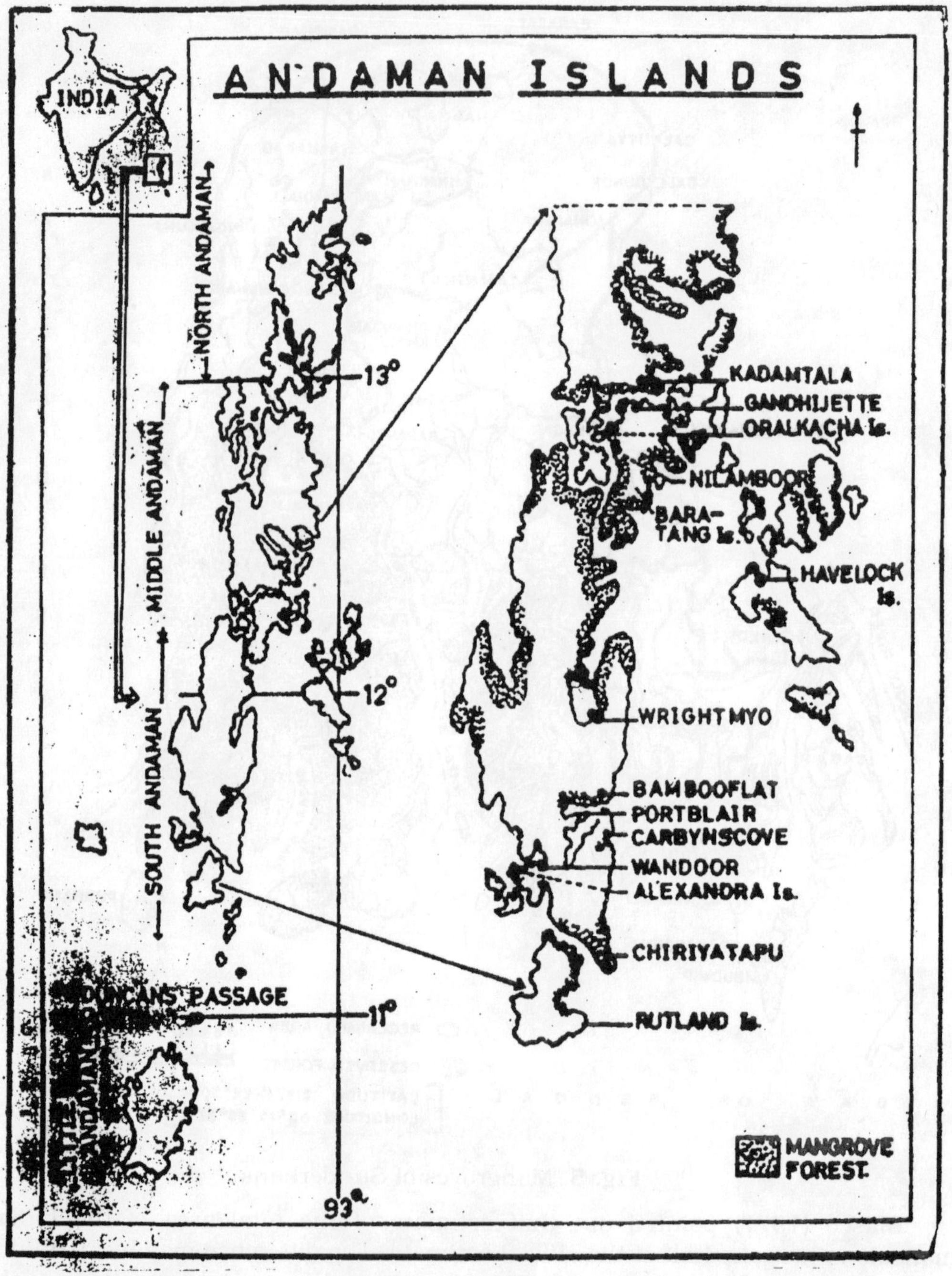

Fig. 6. Sketch map of Andaman Islands, showing their position (insect) mangrove forest (sheding) and study areas mentioned in the text.

Quite early some reports appeared in 'Miami Herald', Nov. 15, 1951, that their roots produce killer gas. Afterwards these ecosystems were often regarded as wastelands of little or no value until they are 'developed'. All too often, developed means conversion of mangroves ecosystems to some other form of use, assumed to be of greater value. In 1950, Commission of Mangroves Land by soil conservation services of USA from classification based on relative suitability for crops, pastures, woodland, wildlife and other uses are highly significant (Lugo and Snedaker, 1974).

It was the studies of Heald (1971) and Odum (1972) that brought in forefront the use of mangroves of Florida in commercial fisheries.

Chapter - 2

ENVIRONMENTAL FACTORS

INTRODUCTION

Mangals are best developed where several basic requirements are met (Macnae, 1968, Walsh, 1974). These requirements are —

(1) *Warm climate* : The average temperature of the coldest month is above 20°C and the seasonal temperature range 5°C or less. Accordingly mangals are most extensively developed on tropical shores. Schimper (1903) believed that mangrove develops in rain forests area with heavy rainfall. It has been confirmed that it growth on tropical shores from even wet to desert.

(2) *Temperature* : Its best growth is near about 24°C.

(3) *Humidity* : It should be high and not lower than 60%.

(4) *Habitats* : Flat land bathed by sea water during tides.

(5) *Soil* : Muddy, sandy or peaty with high salt contents with 3.5% salinity and 1–1.9% chlorinity.

(6) *Protected Shores* : Wave action and strong winds erode established mangals and hinder the establishment of seedlings. Mangals are best developed along coasts such as inlets, lagoons, Creek and gulfs. Where environment is protected from strong currents and waves.

Walsh (1974) suggested that the existence of the extensive Mangal depended upon 5 basic factors, but Chapman (1975) believed these are seven.

(1) *Air Temperature* : The northern limits of mangrove agree closely with the January 16°C isotherm. The southern limits of mangrove in the Atlantic and Indian oceans agree closely with the July 16°C isotherm. In eastern Australia and New Zealand the limits agree closely with the 21°C isotherm.

(2) *Ocean Currents* : The distribution of species flooded with the sea water can be related to ocean currents which can convey seeds, seedlings or plant options.

(3) *Shallow shores* : The shallower the shores, and the more extensive they are, the greater will be the area occupied by mangroves.

(4) *Soil water* : Whilst mangrove species are confined to the coastline, salt water is not necessarily a prerequisite (species *Sonneratia* have been recorded) (Van Sleconis, 1963). However, sea water does enable species

to grow in and tropical areas, e.g. Red sea, Marocco, where they could not survive even if planted inland.

(5) *Tidal range* : This largely controls the vertical condition of some mangrove specics. Thus one tends to find a wider range of communities on coasts with large tidal ranges. Deep tidal penetration up estuaries also enables mangroves to grow for inland.

(6) *Substrate* : Mangroves can be found on sand, mud, peat and coral rock, but the most extensive mangal is always associated with muddy soil and these in turn are commonly found in deltas, lagoons, bays and estuaries.

Those environmental factors act as forces on the mangrove environment. They act at different rates and intensities. The sum of all the forces that act on any ecosystem may be termed, according to Cintron *et al.* (1985), its 'energy signature' and thus 'energy signature' dictates the types of organisms that will survive and the spread of ecological processes.

Since the structure and function of mangroves reflects the nature and intensity of these forces, it should be classified be possible to classify mangroves into a few discrete types that correspond to natural forcing functions. This would allow the conceptualization of ecosystems in relation to the natural forces regulate their behaviour.

2.1. CLIMATE OF MANGROVES

(a) Rainfall

Schimper (1903) believed that mangroves develop in rain forest areas with heavy rainfall, but now it has been confirmed that it grows on tropical shores from even–wet to desert, the later climate is obtained in the middle–east along coastal points of Arabian Peninsula. In semi–dry climate of Western Australia the mangroves are not very much developed and their best development occurs on coastal zone with heavy rainfall where they border between the sea and tropical rain forest. In such situation, there is always a constant supply of fresh water with alluvium rich in minerals from rain forest. In Costa–Rica mangrove forests towards Caribbean coast (annual rainfall 3300 mm) have complexity index as 72.0 and on Pacific coast with lesser and seasonal rainfall (6 months dry season) have complexity index as only 3.6.

As mangroves exist on tropical coastlines, whatever climate; thus a littoral fringe of mangrove is tropical of both arid and equatorial tropical shores. However, these are significant differences between wet and dry coasts in terms of mangrove forest zonation patterns. The difference are clearest in the landward mangrove zones and in the composition of the hinterland vegetation.

Generally speaking the worldwide distribution of mangrove coincides with the 20°C winter isothermal line (Fig. 7) on the Eastern coast of South America, in Australia and in the Northern part of New Zealand, however, the 20°C summer isothermal line marks the boundary within which these plants grow (Duke, 1992). Only in the damp and warm tropics do these plants find ideal condition, however, thus enabling maximal growth and production. Thus, in the South–West Atlantic ocean, the production of mangrove plants and its resulting biomass decreases as one proceeds south (Cintron and Sehaefer Novelli, 1983). These, relatively cold winters with temperature dropping down to as low as 10°C result in dwarf growth of mangrove, so that in South even relatively old plant only reach a height of 1.5 to 2m.

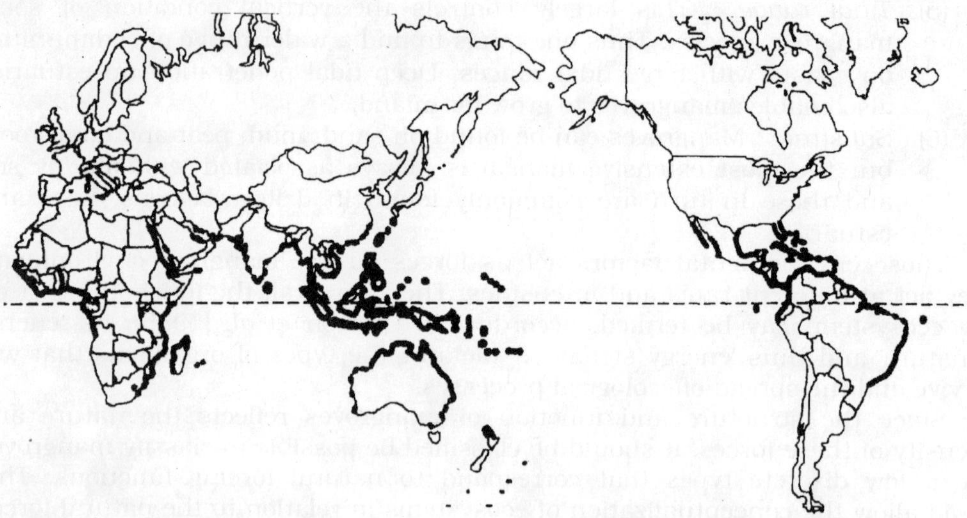

Fig. 7. World-wide distribution of mangroves with winter isothermal line.

Especially in arid regions, a low degree of precipitation and high wind speeds lead to the formation of vegetationless saltflats between the continent and the mangroves. These areas are rarely flooded (for example, only seasonally or when water levels are high), so that due to drying processes extremely high degrees of salinity occur in the pure water.

(b) Temperature

The best growth is near about 24°C. The frost temperature is fatal but it has been reported from areas with winter minimum average as 4.5°C. Well developed mangals are found only along coastal areas where the average temperature of the coldest month is higher than 20°C and the seasonal temperature range does not exceed 5°C (West 1956, Van Steenis 1962).

(c) Humidity

It should be very high and not lower than 60%.

(d) Water

The most critical processes involved in the perpetuation of mangrove ecosystems depend upon —

(i) An adequate supply of water,
(ii) An adequate supply of nutrient and
(iii) Stability of substrate.

The availability of water to mangrove plants and animals depends upon the–

(i) Frequency and volume of tidal exchange,
(ii) Frequency and volume of freshwater supply and
(iii) Evaporative demand of the atmosphere.

When tides occur no less than once per day and there are no other modifying influences the salinity of the surface soil will, in most cases, equilibrate to approximately that of the adjacent water body (estuary or ocean). Mangrove plants and animals are physiologically capable of making use of water of this salinity. However, reduced root aeration or temperature stress may inhibit water uptake. Where there are no diluting factors (Rainfall, groundwater discharge, runoff) and the frequency of tidal influence is less than once per day, the effect of atmospheric evaporation and transpiration of water by the plants causes soil salinities to rise very rapidly. As salinities rise, there is a corresponding increase in the osmotic potential of the interstitial soil water, which makes water uptake by the plant roots more difficult. Under these conditions the exclusion, storage or excretion of excess salt increases the expenditure of energy by the plant. The efficiency with which each species deals with high soil salinities largely determines its position in the intertidal zone.

In the absence of any freshwater addition to the upper intertidal sites by rainfall or seepage, the concentrations of salts in the soil solutions may exceed the physiological tolerances of all the plant species. In these arid conditions mangrove communities are restricted to a narrow band on the coastline, and much of the tidily affected zone is base of vegetation.

The best development of mangrove ecosystems occurs where upper tidal areas are exposed to a continuous supply of freshwater. This occurs either —

(i) where rainfall exceeds evaporation through years,

(ii) where hinterland swamps provide a continual seepage of freshwater,

(iii) where large freshwater catchments allow strong dilution of estuarine waters, reducing salt additions to soil and

(iv) where large freshwater catchments and heavy seasonal rains provide regular and prolonged freshwater flooding of the tidal zone.

The mangroves of south–east Asia are mostly rainfed mangroves hence the productivity and complexity index is very high.

2.2. HABITATS

A well developed mangrove forest appears where the shore shelves gradually resulting into a vast expenses of almost level land bathed by sea water during tides. Chapman (1940) has mentioned four types of habitat in Jamaica, as follows where mangrove were found —

(i) *Muddy substream* – Principally near mouth of rivers : here almost best development of mangroves take place due to aeration of mud from upland.

(ii) *Rocky substratum* : Rocks and boulders of coral reefs and caves are the basic substratum of the mangroves. Here mangroves are not common.

(iii) *Sandy substratum* : Such as sand shoals, sand bars, and sand cays. This is usually along sea coast.

(iv) *Peaty substratum* : Along the shores of sheltered coasts where the rivers do not bring much detritus or else where rivers are absent.

All the four types of substratum have been observed by author in Andaman and Nicobar Islands. Lugo and Snedaker (1974) on the basis of functional and geomorphological aspects, mangrove forest can be differentiated into 5 types –

(i) *Riverine forests* : The most productive and most extensively investigated stands of mangroves are those situated on the banks of estuaries, in areas which are under the direct influence of rivers. These regions are flooded regularly and completely in rhythm with the tide. During the rainy season, they are often permanently submersed.

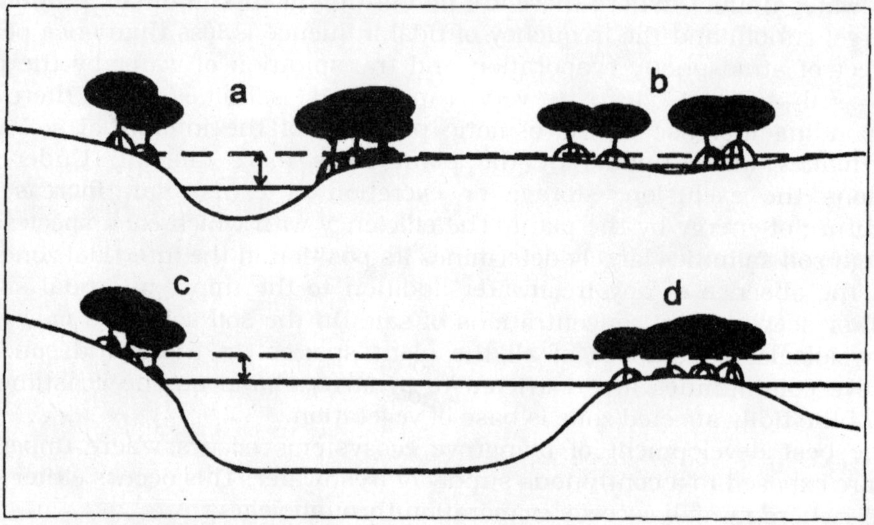

Fig. 8. Schematic view of 4 types of mangrove forests.

(ii) *Basin forests* : Forest located in hollows in the hinterland which tide pools run through and which are rarely effected by the tide during the dry seasons, but which are frequently effected by it during the rainy seasons.

(iii) *Fringe mangroves* : Mangroves on sleep coasts which are flooded regularly.

(iv) *Overwash forests* : This type of forest to be found on flat islands and small peninsulas which are regularly flooded at high tides.

(v) *Scrub mangroves* : Mangroves which lie for above the mean high tide and are only affected by the tide during the rainy season. Here the dwarf growth is to be attributed to the small exchange of water. Dwarf mangrove plants are also to be found in riverine and fringe forests if other parameters (for example temperature, wind stress) are not optimal.

(a) Tidal Inundation

The sea is dominated by waves of many kinds and by tides produced by the pull of moon and sun. Tides are specially important in the shoreward zones where marine life is often especially varied and deuse. Tides are chiefly responsible for the marked periodicities in these communities and entrain the "Lunar day". Since tides have a periodicity of about 12½ hours, high tides occur in most localities. Twice daily, being about 50 minutes later on successive days. Every two weeks when sun and moon are "working together" the amplitude of tide is increased the so called spring tides when high tides are very high and low tides very low – whereas mid–way in the fortnightly periods the range between low and high tide is smallest – the so called neap tides when Sun and Moon tend to cancel one another. The tidal range varies from less than 1 foot in the open sea to 50 feet in certain enclosed bays. There are many factors which modify tides so that tidal patterns vary in different places cover the world.

This has been studied in some details by Watson (1928), Detlaam (1931), Walter and Steiner (1936), Chapman (1944), Chapman and Ronaldson (1958), Macnae (1968) and Baltzer (1969).

Tidal influence is one of the most important factor for proper development and zonation of mangroves. Watson (1928) in his pioneering study of Malaya mangroves characterized 5 classes based on the frequency of inundation (Table 6a, 6b). These are —

(i) Species growing on land flooded at all high tides. No species normally exists under these conditions but *Rhizophora mucronata* will do so exceptionally.
(ii) Species growing on land flooded by medium high tides. Common species of this zone are *Avicennia alba, A. marina* and *Sonnertia griffith.*
(iii) Species growing where they are flooded by normal, high tides, majority of mangroves thrive under these conditions but *Rhizophora* seems to be dominant.
(iv) Species growing on land flooded by spring tides only. Such area are a little too dry for *Rhizophora* and suit *Brugueira gymnorhiza* and *Bruguiera cylindrica* better.
(v) Species on land flooded by equinoctial or other exceptional tides only : *Brugueira gymnorhiza* growing but dominated by *Rhizophora apiculata* and *Xylocarpus granatum.* The most important aspect of the tidal phenomenon is the number of consecutive days at any one when no tide covers the area.

Table 6a. Watson's Inundation classes (Port Swettenham = Port Klang)

Inundation class	Flooded by	Height above datum line (ft))		Times flooded per month	
		From	To	From	To
1	All high tides	00	08	56	62
2	Medium high tides	08	11	45	69
3	Normal high tides	11	13	20	45
4	Spring high tides	13	15	02	20
5	Abnormal (equatorial tides)	15	—	—	02

Table 6b. Inundation classes for Malyan mangroves

Flooded by	Height above admiralty datum (feet)	Times flooded per month (Watson)	Days flooding per month (feet)
All high tides	0 – 8	56–62	20 +
Medium high tides	8 – 11	45–59	10–19
Normal high tides	11–13	20–45	4 – 9
Spring high tides	13–15	2 – 20	2 – 4
Abnormal or equatorial tides	15	2	2
Seasonal in wet seaon only		Seasonal flood	

Tidal movement and inundations are clearly responsible for distribution and establishment of viviparous seedlings of mangroves. It has been commonly thought that the torpedo like shape of *Rhizophora* and *Bruguiera* seedlings favour vertical penetration into mud beneath the parent tree (Krause, 1985).

The tidal range determine tree exposure of different heights in the tidal zone to air and in turn influences the timing of many biological activities. For example, the marine spring tides when the intertidal zone is covered by water to the greatest extent. The famous grunion runs on southern California beaches, during which millions of these small fish come into shallow water to spawn, similarly occur during the highest spring tides (Fig. 9).

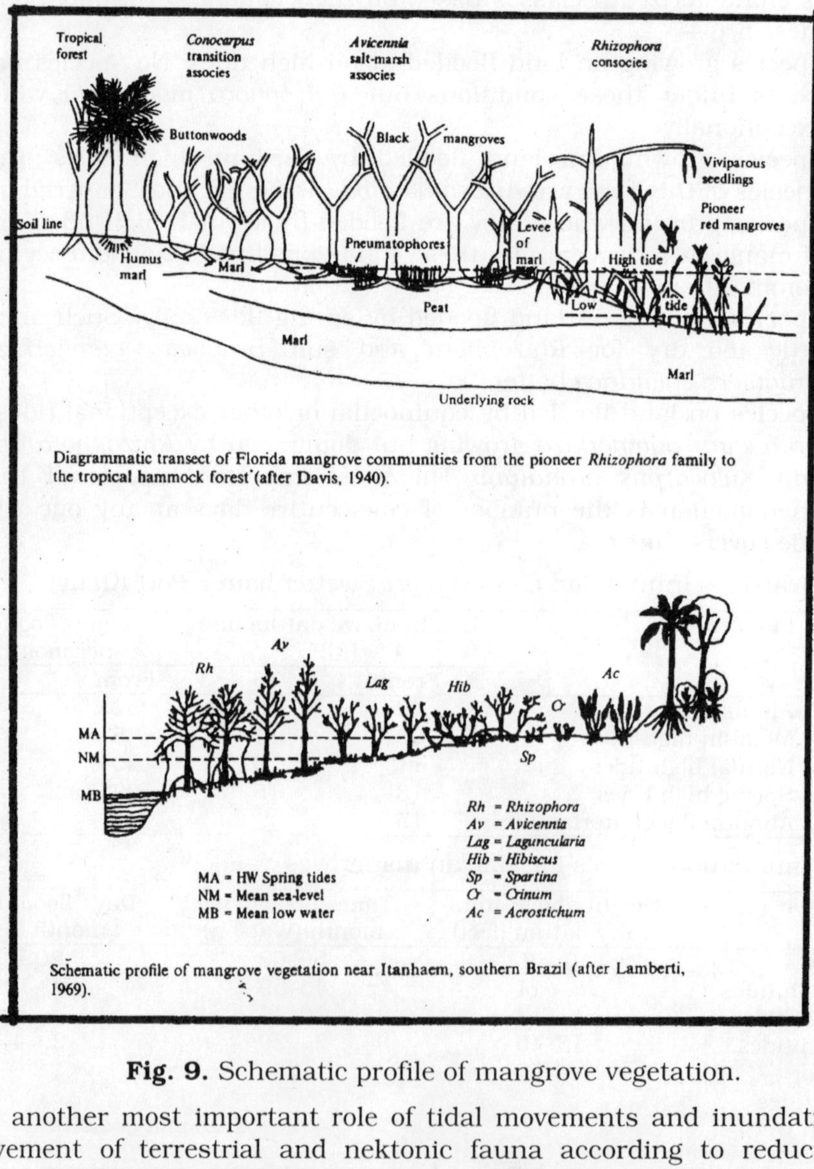

Diagrammatic transect of Florida mangrove communities from the pioneer *Rhizophora* family to the tropical hammock forest (after Davis, 1940).

Schematic profile of mangrove vegetation near Itanhaem, southern Brazil (after Lamberti, 1969).

Fig. 9. Schematic profile of mangrove vegetation.

The another most important role of tidal movements and inundations are the movement of terrestrial and nektonic fauna according to reducing and increasing of marine waters. When there is low tide the adjacent terrestrial fauna, including large animals, reptile and birds moves to mangrove areas for the search of food, when high tide, the marine fauna moves to mangrove area in season of suitable breeding place and food material. This has been clearly explained in the Figure 10.

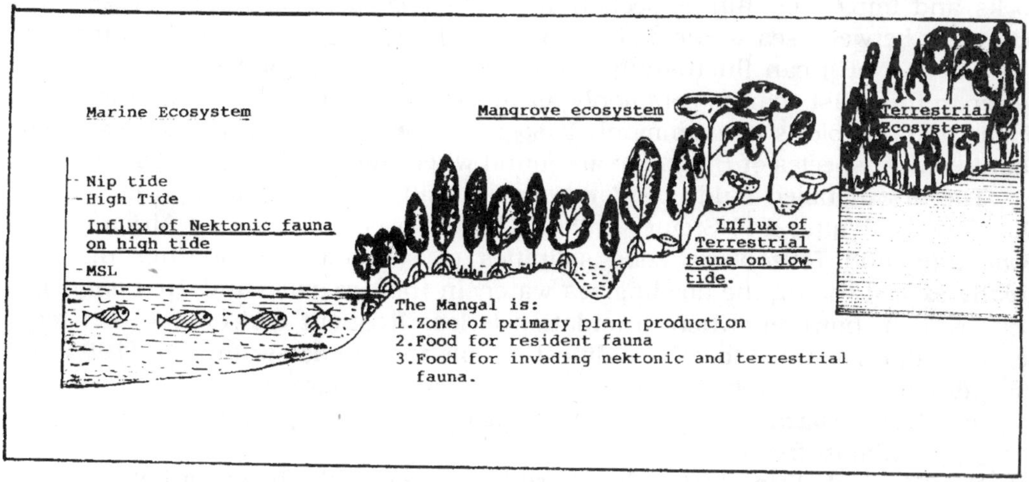

Fig. 10. Movement of Nektonic and Terrestrial Fauna

(b) Salinity

The sea is salty. The average salinity or salt content is 35 parts of salts by weight per 1000 parts of water, or 3.5 per cent. This is usually written: 35% (=part per thousand; recall that salinity of freshwater is less than 0.5%) about 27% is sodium chloride and most of the rest consists of magnesium, calcium and potassium salts. Since the salts dissociates into ions, the best way to picture the chemistry of the sea is as follows (in parts per thousand = gms per kilograms).

Table 7.

Positive ions (Cations)		Negative ions (Anions)	
Sodium	10.7	Chloride	19.3
Magnesium	1.3	Sulphate	2.7
Calcium	0.4	Bicarbonate	0.007
Potassium	0.4	Bromide	0.07

Since the proportion of the radicals remains virtually constant, total salinity may be computed by determining the chloride content (which is easier than determining total salinity). Thus 19% chlorinity approximately 35% salinity.

Salt water presence is not a physical requirement of mangroves (Bowman, 1917; Warming and Vaht, 1925; Rosevear, 1947; Egler, 1948; Daiber, 1960). Mangroves are facultative halophytes that occupy tidal areas where freshwater plants, which are intolerant to salt can not live (Wert, 1956).

Variations in soil and water salinity may not be quite so important as previously thought. It has shown that many mangroves selectively ultra-filter sodium chloride out of the water at the roots. The efficiency of the mechanism probably varies from species to species and this may be reflected in the belt they occupy in relation to salinity and flooding.

Troll and Dragendrof (1931) pointed out that water salinity could vary greatly depending upon tide height, amount of water entering from rivers and rocks and impact of rain, especially in monsoonal areas. For this reason the boundary between sea water (2.4 – 3.3% NaCl) and normal brackish water (say 1.0 – 2.4% NaCl) can fluctuate and the mean position is probably often much closer to the coast than is commonly assumed species of *Sonneratia*, other than *Sonneratia caseolaris* are commonly associated with waters of normal sea water salinity. Most species of *Brugueira* are found where the salinity is rather less.

The first detailed study of mangroves salinities and distribution of species was done by Walter and Steiner (1936). They showed that in East Africa in the *Sonneratia* and *Rhizophora* zone salinities of soil and groundwater did not greatly exceed that of the flooding sea water. In the *Avicennia* and *Avicennia* and *Ceriops* area, however, where inundation is less frequent and evaporation may severe. Salinities arose to values almost double than of sea water. Por (1983) on the basis of water salinity he classified two types of mangroves –

(i) **The estuarine** : Soft bottom mangal – salinity ranging from sea water to almost freshwater.

(ii) **The Euhaline** : Metahaline hard bottom mangal. The water bathing this type of mangal have oceanic salinities and frequency even higher metahaline areas (probably upto about 4.7%). No doubt the most important factor determining the difference between the two types of mangal in the presence or absence of freshwater influx.

The salinity depends on evaporation, water logged nature of soil, frequency of tidal inundation, and the possibility of freshwater influx. Occurrence of many mangrove species appears to be.

The sea water brought in by the tides contains on the coverage about 3.5% soluble salts and has an osmotic potential of about 24 bars. However, salinity remains more or less constant only in the limited lower range of the mangrove habitats. Even there salinity is constant only in a specific zone. As a great part of the mangroves are located on deltas or estuaries, the plants are periodically exposed either to sea or to freshwater flushes. Plants growing under such conditions have to be adapted to the fluctuations in both osmotic and ionic environments.

Segregation between constantly and periodically saline habitats also results in plant segregation. Optimal growth of many mangrove species is in salty water (Stern and Voigt, 1954). Moreover, genera such as *Sonneratia* disappear in habitats which are frequently eluted with freshwater. However, *Rhizophora mangle* plants grown on 25% sea water produced more leaves and roots than plants grown in undiluted sea water (Egler, 1951; Pannier, 1959).

(c) Sedimentation and Coastal Dynamics

In biological, ecological and geological research, there are differing notions as regards the functions of mangroves (Woodroffc, 1992). Whereas biologists view mangrove ecosystem primarily as sources of organic matter, geologists look upon coasts which are fringed with mangroves as sinks in which sediment accumulates overlong period of time. This apparent discrepancy shows the complexity of the exchange processes which take place between mangrove areas and continental shelf regions.

It would seems likely that mangrove trees catch sediment in their prop roots, thus functioning as land builders. This view was frequently held until the 1980's based on the research carried out by Davis (1938). In numerous cases there has been proof of an enormous sedimentation rate in mangrove areas, ranging between 1 mm/a to 8 mm/a (Bird and Barson, 1977). According to Woodroffc (1992), the view has been adapted, in the meantime, that mangrove forests are a result, and not the course of sedimentation in protected coastal areas, and that they only accelerate the sedimentation process. Mangroves are not looked upon as an effective form of coastal protection.

Semeniuk (1981) described the erosion of a coastal segment of North Western Australia which was covered with mangroves. He even differentiated three types of erosion —

— **Cliff erosion** : High, steep coasts form since the mangrove root only hold the upper most layers of mud. The steep coasts shift as a result of landslides in the coarse of which mangrove trees fall down and die.

— **Sheet erosion** : Continual erosion on flate coastal segments which is accelerated by burrowing invertebrates and periodic drying up and cracking.

— **Creek erosion** : Erosion as a result of the depending, widening and lengthening of tidal creeks. This process is accelerated by the tidal asymmetries which are characteristic of mangrove areas. The relatively strong odd current leads to a fast export of sediment.

Thus mangrove trees must be viewed as opportunistic forms of vegetation which follow changes in the coastal morphology. High sedimentation rates have probably been measured frequently because mangroves are primarily to be formed in places where a large degree of sedimentation is to be expected anyway, namely in estuaries, lagoons and protected bays. The effectiveness of the afforestation of mangroves for the purpose of coastal protection (such as in Cuba and Bangladesh) must therefore be investigated further in the future. Presumably mangrove do not protect coastal areas from constant erosion, but they probably protect them from the direct force of storms and trunamis (Siddiqui, 1994).

(d) Mangrove Soil

Mangroves colonise a variety of substrate that include –

(i) Silty and clayey muds

(ii) Calcareous muds

(iii) Quartzose muds

(iv) Calcareous sands or mixture of these. They are also known to occasionally colonise coarse coral reef rubble as well as cracks and hollows on rocky platforms. For transcripted and deposited by tidal and river currents.

The soils of mangroves may be sandy, muddy or peaty. Typically under well developed mangroves. They are muddy and have high salt content, sometimes more than 6.6% to less than 10% and nearly always remain saturated with water. They have very low oxygen content sometime less than 1% in the soil air.

Pore space varies between 44.6 to 63.7%. At the same time there is abundant H_2S due to almost n anaerobic condition in lower layers of soil and due to its presence, ferric compounds are reduced to various hydrated ferrous sulphides which give these soil their characteristic black colour. Organic content varies between 5% to 90% or more, the latter in peaty soil where pH is also low towards acidic soil.

Under mangrove cover the substrate undergoes physical, chemical and biological alternation. The resulting soil is distinctive of mangrove environments and is the product of several interesting factors and processes; these are —

(i) The type of parent substrate that was deposited,

(ii) The activity of plants and animals; the material; the contribute and the structures they impart to the substrate and

(iii) The effect of groundwater that is both reducing and acidic.

Well developed mangrove soils vary from permanently water-saturated to merely moist. The soils commonly contain abundant networks of ramifying cable roots, meshworks of fine nutrients (feeding) roots, pneumatophores, buried follen branches, and trunks and other organic matter in varying stages of decomposition. These are numerous open burrows, filled burrows and tunnels made by craks, other crustaceans, worms, molluscs and fish. As a result of animal activity the soil is churned up; it is also perforated by burrows and by the hollows remaining after the decay of plant roots. Depending on the number and types of those animals in the resident community that contribute their calcerous fats after death (eg. mangrove encrusting oysters, soil-burrowing molluscs). There are different amounts and types of shell material scattered in the soil.

Mangrove soils have two zones according to their availability of oxygen.

(i) Aerobic zone

(ii) Anaerobic zone

Upper levels of the soil contain more oxygen than the lower it is at upper levels that nutritive roots of mangrove are found. At deeper levels which are permanently moist and soil becomes odorous with hydrogen sulphide and may be saturated with iron sulphide which is fixed by bacteria under reducing conditions.

Hesse (1961) reported that *R. racemosa* swamps in Sierra Leone were comprised of fibrous mud, whereas, *A. germinans* swamps soils were non-fibrous. Also *Rhizophora* swamps had higher pH values, C/N ratios and contents of oxidizable sulphur, nitrogen, phosphorus and carbon. Giglioli and Thornton (1965) described the early phases of swamp evolution in the Zambia, West Africa, where *Rhizophora racemosa* pioneers in virgin alluvia composed of soft, silty soil. Proliferation of fibrous roots at the soil surface produces a 'felt like' layer which entraps sediment and increases the rate of deposition of both alluvium and leaf litter. Rosevear (1947) suggested that the fibrous mat formed by *R. racemosa* prevents further establishments of that species and conditions the soil for colonization by *Avicennia*, which requires a more consolidated and elevated substratum (Jordan, 1964). As the soil surface becomes elevated, the rhizophoretum dies and *A. germinans* replaces *R. racemosa* (Giglioli and Thronton, 1965). The many pneumatopheres of *Avicennia* further accelerate depositing and the forest floor becomes even more elevated.

Schuster (1952) reported deposition of sediments on the mangrove forest floor at every sprint tide in Java, the processes of land elevation and soil formation being accelerated by growth of beach thistles (*Acanthus* sp.) which produced large quantities of organic matter.

Soil derived by sedimentation from river water is often poor in calcium and potassium and, in mangrove swamps, tidal water is the main source of salt. In early stage of mangal development, the clay particles absorb calcium and potassium salts from sea water and a fine grained soil, rich in minerals, results. As evolution of the swamp continues, shelled animals invade and grow on the trees and substrata. The organic content of the soil increases and in the moist environment, decay processes are rapid and calcareous particles are dissolved (Abel, 1926). Wharton (1883) reported rapid corrosion of calcium carbonate ($CaCO_3$) in mangrove swamps at Aldabra. Fairbridge and Teichert (1947) concluded that pools on reef flats at low Isles (Marshall and Orr, 1931) were caused by solution of $CaCO_3$ by mangrove swamps acids. Revelle and Fairbridge (1957) suggested a carbonic acid produced by decomposition of organic matter. They also suggested that tannic acid from mangrove bark and 'humic acids' aid in decomposition.

Very little is known about the factors that form and condition mangrove mud, which may lie in an unconsolidated state to a depth of 1.5 m. Schuster (1952) discussed breakdown and modification of the substratum by bacteria, fungi actinomyetes. He mentioned the occurrence of the bacteria *Clostridium* sp. in mangrove swamps and speculated that these organisms are important in nitrogen fixation, which has been discussed later.

Most of the organic debris on and within mangrove soils is authochonous, because of the saline water, relatively high pH of surface soil water (often as high as 7.8) and anaerobic conditions at low tide, plant detritus is only partially broken down by bacteria, fungi and algae. This causes formation of peat, which is composed mainly of plant remains.

Analysis of mangrove peat has been reported by several workers. Davis (1940) gave detailed accounts of soil profile in the swamps of southern Florida and classified them according their general composition i.e. homogenous or heterogeneous or layered. He also classified them on the basis of the type of vegetation which covered the soil and the probables of types of vegetation materials. Davis reported various types of soil profile, some of which, indicated progressive soil accretion, while others did not. Giglioli and Thronton (1965) gave soil profiles from the Gabia river basin in West Africa. The profiles indicated typical alluvial soils in the process of silting.

However, *Avicennia marina* and *Arthroenernum australasicum* occurred only where high salinity of the soil water endured for long periods of time or where there were wide variations in salinity. Giglioli and King (1966) pointed out that *Avicennia germinans* grew in old soils of high salinity and that this high salt content was a function of time. *Avicennia* was apparently able to exclude *Rhizophora racemosa* because it was better adapted to high concentration of salt. *Avicennia* unlike *Rhizophora*, absorbs large quantities of salt through its roots and excretes them through the leaves (Scholander *et al.*, 1962). As shown above, the fibrous nature of the substratum also appears to be important in colonization, and it is most likely that combinations of factors, including soil salinity, regulate species distribution.

Table 8. Percentage of grain size distribution (mm diameter) in mangrove swamp surface sediments

(i) El. Salvador (Vet 1953)						
Size	1.00 mm	1.0-05	0.5–04	0.4–03	0.3–01	0.1–0.06
Percentage	0.0–5.2%	0.0–4.2	0.0–8.0	0.0–6.0	0.0–18	0.0–18.0
(ii) Java (Schvster 1952)						
Size	2mm	2.0–01	0.1–0.05	0.05–0.01	0.01–above	
Percentage	0%	2	5	30	63	
(iii) Jamaica (Chapman 1944)						
	Coarse sand	Fine sand	Clay	Salt		
	39.9	26.5	5.1	15.7		
(iv) Brazil (Friese, 1937)						
0.2 mm	0.2–0.06	0.06–0.03	0.03–0.006	0.006–0.003		
4.7–7.0	7.2–11.8	6.7–10%	6.6–10.4	22.1–24.6		
(v) India (Navalkar, 1941)						
	2.0	0.2	0.02	0.00		
	3.9–4.0	38.2–38.6	29.5–33.1	23.9–43.0		
(vi) Florida (Setroll 1963)						
	Median grain size from 10 stations varied between 0.006 to 0.700 mm					

Table 9. Grain size distribution in surface sediments of swamps dominated by *Avicennia alba* (Navalkar, 1941) and *Avicennia marina* (Clarke and Hannon, 1967)

Species	Percentage of Dry Soil.			
	Coarse sand 2.0 mm dia. particle size	Fine sand 2.0 mm dia.	Silt 0.02 mm	Clay 0.02 mm
A. alba	4.0	38.4	31.3	4.1
A. marina	75.2	3.8	1.7	4.8

(Courtesy from G.E. Walsh)

Clarke and Hannon (1969) concluded that tidal action, as modified by microtopography was the major factor which affected soil salinity over long period of time, and that succession was mainly allogenic rather than autogenic. The concept was shown to be true by Thom (1967), who demonstrated that although biotic and geomorphic processes are effective in short–term changes on actively acreting shores, physiographic processes of sedimentation and subsidence are more important over a long period of time. Physiographic changes influence salinity of the soil, degree of water saturation, soil type and drainage and therefore greatly influence the species presence.

Singh and Garge (1983) has been studied the soil of Andaman in relation to grain size, pH, salinity, pore space, total organic content, nitrogen etc. At each site, soil was analysed from the proximal zone, which is generally occupied by the *Rhizophora* species, thus it may be called as *Rhizophora* soil, middle zone, which is behind the mixed mangrove species occur. The zone may be dominated by *Avicennia, Bruguiera* or *Ceriops* species. Third distal zone which is away from the daily tidal waters, a zone which is transitive between hinterland and mangrove forest. This may be occupied by the variety of species which differ from site to site. In Table 10 grain size distribution of three types of habitats of Andaman is given.

Table 10. Grain size distribution of 3 types of habitats occur in Andaman Islands

Habitats	The Percentage			
	Coarse sand	Fine sand	Silt	Clay
Rocky substrates	12.1 – 26.8	29 – 40	27.1–51.0	35–40.3
Sandy substrate	5.6 – 6.9	67.0 –83.4	12.3 –16.4	29.5–38.4
Peaty substrate	——	10.3–15.6	23.4–30.9	49.3–73.1

(Singh & Garge, 1993)

(e) Water Logged Mangrove Soil

The ecology of water logged soil is very interesting and significant. When a soil is flooded, the rate of oxygen diffusion is greatly reduced. Depending on the depth of flooding and the amount of time the soil remains flooded. The oxygen supply to the soil can be almost completely shut off. Under these conditions, bacterial populations in the soil are drastically altered. In normal, well aerated soils aerobic bacteria are dominant. These species require oxygen for respiration. When starved of oxygen, they are reduced in population and other types of bacteria which do not depend on oxygen become dominant. Facultative anaerobes, while not requiring oxygen, do need an alternative oxidant source (e.g. nitrate ion, Iron III), mangrove for respiration. These chemical entities act as electron acceptors for various types of facultative anaerobes and hence, as the soil becomes increasingly anaerobic, are responsible for drastic and very significant chemical changes. Another groups of anaerobes that become dominant when the soil becomes anaerobic are species which are responsible for reactions such as the reduction of sulphate to sulphide and the production of methane utilizing carbon and "excess" hydrogen. A comprehensive discussion of soil microbiology.

Fig. 11a is a schematic summary of the some of the more important reactions occurring in flooded soils. The redox potentials (Eh) referred in this diagram is a convenient measure of the extent to which the soil is in a reducing or anaerobic state. Such measurements are easily made in field or laboratory studies by simple insertion of an inert platinum electrode, which 'senses' the redox state of soil. These measurements will be discussed in more detail later. Note that a large positive Eh indicate a well–oxidized or aerated state, while a large negative Eh indicates a highly reduced state. Various bacterially mediated redox reactions can be associated with intermediate Eh volumes, as shown in Figure. Referring to this diagram, it can be seen that as the oxygen supply becomes limited, it is rapidly consumed by bacterial respiration. This process has been shown to take place, over Eh range of +350 to +380 mv. When all the oxygen is consumed, and only when it is consumed, Mn^{4+} and NO_3^- are converted to Mn^{2+} and N_2 (gaseous), respectively. In turn, when Mn^{4+} and NO_3^- are completely consumed, than Fe^{3+} is reduced to Fe^{2+} and so on, until the soil eventually reaches a highly anaerobic state where the further reduction of dioxide to methane (marsh gas) occurs. The rate at which all these processes occur depends on such factors as depth and time of flooding (i.e. to what extent oxygen is limited) and amount of organic carbon present in the soil.

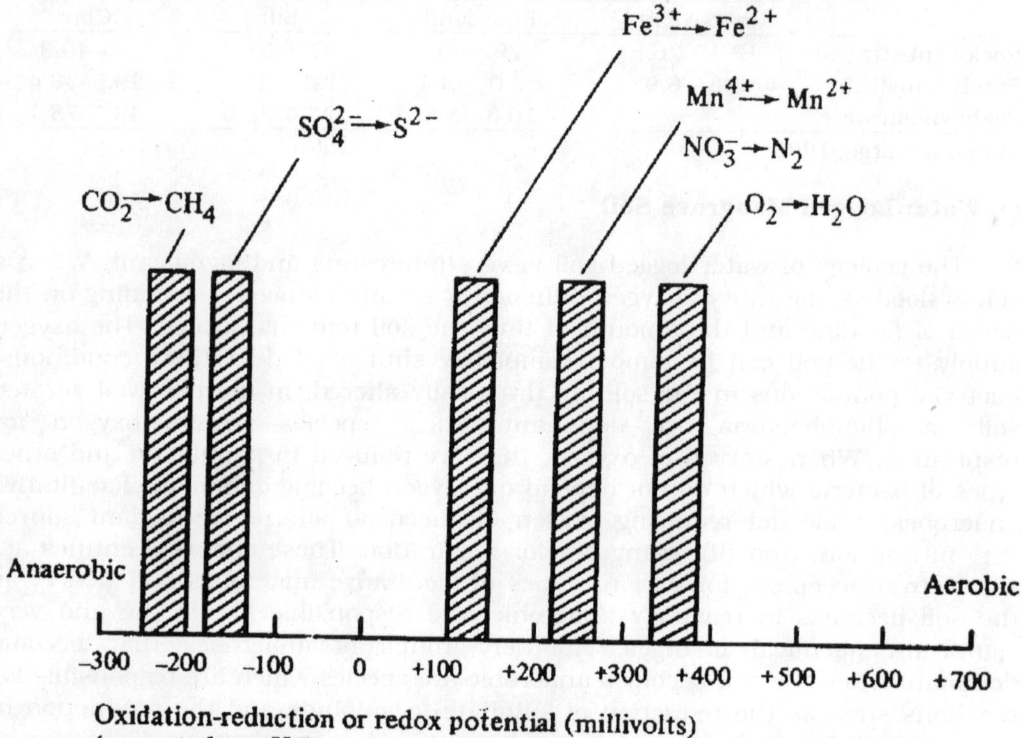

A schematic diagram of some important soil redox reactions which occur at various reducing conditions within the soil (after Delaune et al., 1976).

Fig. 11a: A schematic diagram of some important soil redox reactions.

Bacteria (except chemotrophs) require carbon compounds as energy sources and here the type of organic carbon is also critical. For example, if a sample of well-aerated soil is flooded and glucose is added to the soil, then the various reactions proceed at a very rapid rate and the soil could become highly reducing in a matter of days, or even hours. If other, more refractory carbon compounds (such as ground straw) are used instead, then the reactions will be less rapid.

The changes in the redox state of NO_3^-, Fe and Mn can have profound effects on plant growth, iron in the divalent state, for example, is at any given pH, more soluble and thus becomes more available for plant uptake. In very iron–rich soils, the iron uptake can be extreme and result in iron toxicity. Similarly, manganese becomes more available to the plants. Another very important but less direct effect is that inorganic phosphorus, which may have been previously 'trapped' or sequestered in the insoluble Fe^{III} and Mn^{IV} oxides hydroides is at increasing in phosphate availability is important, as plants can only utilize phosphorus in the inorganic phosphate forms eg. $PO_4{}^3$, $HPO_4{}^2$ and H_2PO_4.

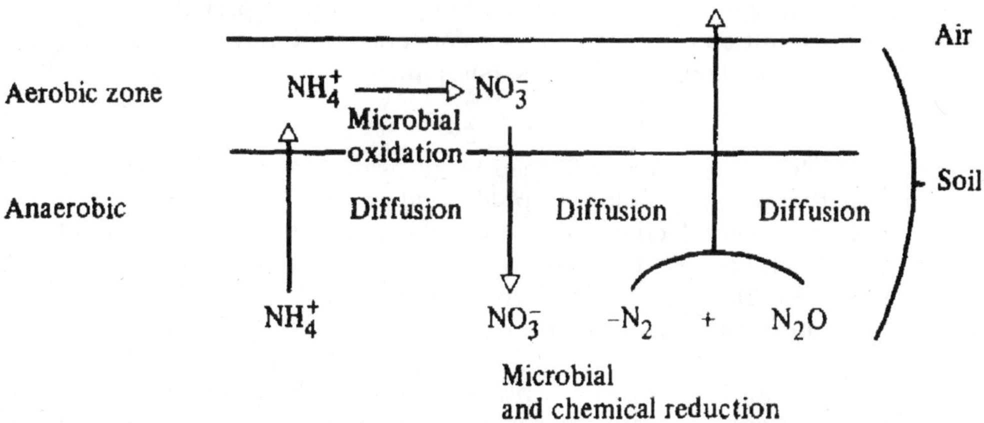

Fig. 11b. Dentrification process in anaerobic soil.

Nitrogen is an essential and often limiting nutrient for plant and can only be utilized in inorganic forms ($NO_3{}^-$, $NH_4{}^+$). Severe losses of nitrogen can occur in flooded anaerobic soils via denitrification. As shown in Fig. 11b, many flooded soils have a thin (1–5 mm) aerobic layer overlying the more highly reduced soil. Ammonium in the anaerobic zone diffuses into the aerobic layer where nitrifying bacteria can convert ammonium to nitrate. The nitrate may diffuse back into the anaerobic zone where it can be reduced to gaseous nitrogen and nitrous oxide (N_2O) and lost from the soil. This reaction scheme has been shown to be responsible for losing a significant fraction of nitrogen fertilizer amendments from rice paddy fields. However, in natural intertidal systems where inorganic nitrogen content is much lower, it is probable that less severe losses occur.

Soil pH is also an important parameter controlling the chemical status and mobility of many important elements. For example, at any given redox potential, many metals exhibit greater solubilities at a low pH. As discussed above, this can have an effect on phosphate binding in the soil. If a soil has a pH greater than about 7, it is possible that ammonia volatilization with a resulting loss of nitrogen, can occur. In some instances, however, flooding may have the effect of increasing that pH of acid soils and decreasing the pH of alkaline soils so that most flooded soils convert to about pH 7.0. The flooded soils are buffered by iron, manganese and aluminium oxides and organic acids from root exudates, are also thought to play an important role in the buffering capacities of these soils. Also, soils rich in 'labile' materials (i.e. reactive organic matter) exhibit increased microbiological activity and thus intensified reduction. This in turn may contribute to an increased buffering capacity near pH 7.0.

In saline flooded soil, the chemistry of sulphur is of great importance. For example, sulphate in sea-water is reduced to sulphide in highly anaerobic soils. Sulphide ion is highly toxic to plants and many intertidal mangrove and marsh plants have developed strategies to overcome sulphide toxicity. Most metals have highly insoluble sulphides; hence the mobility of metals in highly anaerobic soils is limited if sulphide is in excess. Over long time-periods and given the appropriate conditions in terms of iron concentration and redox potentials, etc., iron pyrite is formed according to the (simplified) equation :

$$Fe^{2+} + S^{2+} + S^{o} \longrightarrow FeS_2$$

where, S^{o} is elemental sulphur formed by re-oxidation of some of the sulphide in the soil. If the soil containing sulphide and pyrite numerals becomes aerated again e.g. by sea-level changes or man-induced effects, formation of jarosite $KFe_3^{III}(SO_4)_2(OH)_6$ occurs. This hydrolyses to form sulphuric acid, thus creating very acidic soils. Such soils, known as acid sulphate soils, are almost useless for agricultural purposes, as has been found in areas of South-East Asia and Africa.

Overall, it can be seen that a plant growing in a saline, flooded soil is in a completely different chemical environment compared with that of a well-aerated terrestrial soil. Not only do the plants have to contend with high salinity, but also with the strongly reducing conditions present. The presence of sulphide and the higher solubility of many metals can give rise to toxicity effects if the metal or sulphide is in excess. Nutrient availability can be decreased markedly by denitrification or volatilization losses of inorganic nitrogen. Alternatively, phosphate mobility may be greater. Decomposition processes tend to be slow in anaerobic soils and this could lead to decreased recycling of key nutrients. Over long time-periods, nitrogen and phosphorus may be gradually 'locked up' in organic compounds which are not decomposed or mineralized as rapidly as in well-aerated soils. Soil salinity is also a very important factor. Various halophytes have quite different salinity tolerances; and changes in salinity, either in the long term (sea level changes, sediment accretion) or short term (damming of rivers etc.), can have drastic effects on species distribution and growth.

Therefore, in order to gain a complete picture of the effects of flooded soils in plant productivity, relatively long-term studies of nutrient status, redox potential, pH, soil mineralogy and salinity changes must be made. Longer term predictions of marsh viability require information on sea-level changes, sedimentation rates, topography and hydrological regime.

Chapter – 3

MANGROVES ADAPTATIONS

Some mangrove species are occasionally found in a freshwater environment or in other inland non–tidal setting, but this is unusual and such occurrences are probably always the result of coastal uplift, or of a bay or tidal creek being cut-off from the sea (Beared, 1967; Steenis, 1963; Chapman, 1974).

Mangroves than are typically coastal plants and are subject to periodic tidal inundation. They exhibit a number of morphological, anatomical and physiological characteristics which enable them to flourish under these condition. These characteristics are —

3.1 MORPHOLOGICAL ADAPTATION

(i) Presence of various types of aerial and underground roots for prope exchange of gases and support of plant and buttresses.
(ii) Leathery shining glauceous leaves with water storage tissue.
(iii) Germination viviparous type.

3.2 ANATOMICAL ADAPTATION

(iv) Epidermis like walled and thick cuticle.
(v) Shows weakly coating.
(vi) Presence of water stomata.
(vii) Presence of trichome.
(viii) Palisade tissue very well developed.
(ix) Mucilage cells occur in some cases.

3.3 PHYSIOLOGICAL ADAPTATION

(x) High osmotic and DPD of cell sap.
(xi) Presence of ultrafiltration mechanism for ions.
(xii) Presence of salt secreting glands.
(xiii) Mangroves are C_3 plant.
(xiv) Stomata open only in morning.
(xv) Mechanism for osmotic withdrawl of water.
(xvi) Uptake and accumulation of N_2 by various ways.
(xvii) Seeds are recalcitrant.

3.4 ECOLOGICAL ADAPTATION

(xviii) Prefer to grow in saline habitat.
(xix) Wide ecological tolerance.
(xx) Periodic tidal inundation.

Warming (1883) stated that mangroves have adapted to their environment through –

(i) Mechanical fixation in loose soil,
(ii) Respiratory roots and aerating devices,
(iii) Viviparity,
(iv) Specialized means of dispersal and
(v) Development of xerophytic structure in relation to soil salinity.

Walter (1931) and Walter and Steiner (1934) concluded from studies in East Africa, that zonation was related to the capacity of mangroves to compete and survive in saline soil. Thus, they distinguished zones of *Rhizophora, Avicennia* and *Sonneratia* and stated that *Rhizophora* and *Avicennia* bore great fluctuations in soil salinity, whereas *Sonneratia* required a constant chloride content.

Macnae (1968) reviewed adaptation of mangroves with regards to growth in ill–consolidated mud, specializations of stems and leaves, relationships between root and shoot system, and vivipary.

Robyns (1971) considered mangrove to be the only truly viviparous plants. He defined vivipary as the process in which the seed remains attached in the front to the mother tree germinate into a protruding embryo with a long hypocotyle and finally falls from the tree.

Genkel (1962) speculated that mangroves evolved in the ancient tropical forest from xerotic plants in which the seeds had no dormant period and loss of fruit from the tree was delayed by chloride in the soil. The roots of mangroves are not deep and tap roots are not present.

Zieman (1972) found that height of *R. mangle* was related to root growth. In *Rhizophora*, the primary roots of hypocotyle function for any short period of time and roots functions are assumed by secondary root which extends from the main trunk.

The anatomy of aerial root has been described in details by Warming (1883), Schenck (1889), Karsten (1891), Leiban (1914), Bowman (1917), Mullan (1931) and Gill and Tomlinson (1971) gave a detailed account of root growth and anatomy.

Troll and Dragendroff (1931) gave an extensive account of the cable root system of *Sonneratia*, and similar roots systems are present in some species of *Avicennia, Lumnitzera racemosa, X. moluccensis*. For an extensive study of anatomy of respiratory root of mangroves, see Ernould (1921).

Many mangroves, such as *Avicennia marina* and *Aegiceras corniculatum* can tolerate a moderately high salinity in their sap; the salt level being controlled by salt–excretory glands on the leaf epidermis. Other species, such as *Rhizophora stylosa* retain a low internal salinity by means of salt excluding mechanisms in tree roots.

Many mangroves species are viviparous, that is, the seed germinates while still attached to the parent. *Aegiceras corniculatum, Avicennia marina* and all species of Rhizophoraceae show this character. In these cases the seedlings take root around the parent plants and are not readily dispersed by tidal currents. On the other hand some mangrove species such as *Xylocarpus australasicus* have a floating fruit stage and so can be widely distributed.

Most of the research on physiology of mangroves has stressed halophytic adaptations (Stern and Voight, 1959; Genkel and Shakoine, 1946; Conner, 1969 and Stocker, 1924, 1925).

Macmillan (1971), Bharucha and Navalkar (1942), Bole and Bharucha (1954), Chapman (1968), Shonner and Waisel (1973), Walker (1971), Laetsch (1974), Joshi *et al.* (1974), Chen *et al.* (1974), Johnson *et al.* (1971), Karmakar and Joshi (1973), Patil and Joshi (1970), Karmakar and Joshi (1968) and many others they have worked on various aspects of mangrove physiology, which we have given in this chapter.

A. Root System in Mangroves

The roots of mangroves have interesting modifications. Those that anchor the plants often occur in mud that is poorly consolidated, yet they are able to cope with this environment through the development of an extensive radially – disposed horizontal cable root system which provides a broad base upon which to grow. The surrounding soil is very deficient in oxygen, and all mangrove species have additional roots, or root adaptations, which appear to facilitate the transfer of gases into and out of the plant system. Some species have air–filled roots which arise from the cable roots and project above the soil; these are called pneumatophores. In mangrove, these are following types of roots (Fig. 12) for various purposes : –

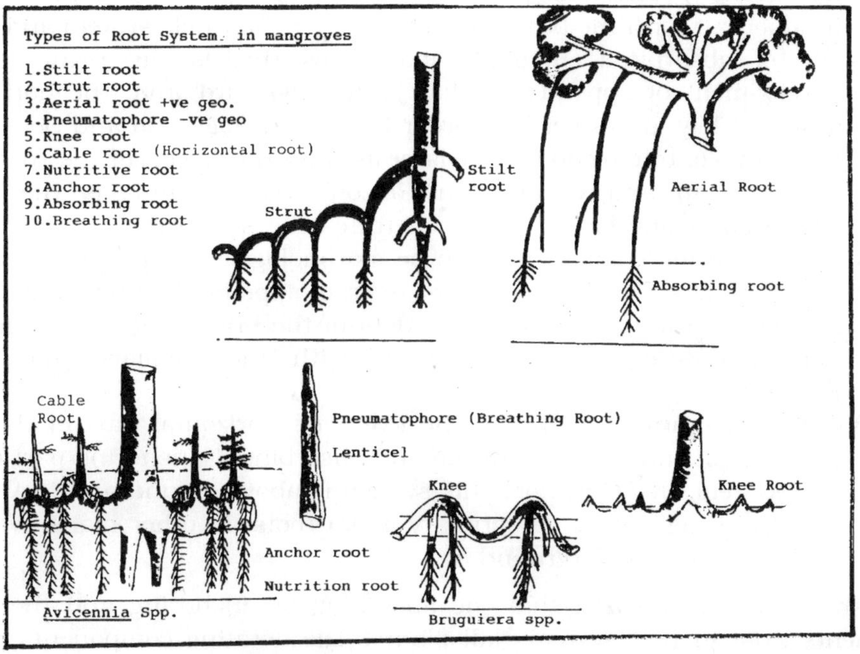

Fig. 12 : Various types of root system in mangroves

(i) Strut and Stilt roots.
(ii) Prop roots.
(iii) Cylindrical flattened pneumatophore (breathing roots).
(iv) Horizontal roots (Cable root).
(v) Absorbing roots.
(vi) Knee roots.
(vii) Anchoring roots.

All these root systems are in mangroves only in favour to cope with harsh environment and able to fulfill the needs of the plants.

If one examines the root system of the more specialized mangroves, three basic structural components can be recognized, even though these components may have different morphological origin in different species. Each component has a different function. It is the existence of these three compounds that is the most distinctive features of mangroves, rather than just the aerial portion. The functional correlations between the various parts of the root system need emphasis.

(i) Aerating Component : The erect part of the root system, which in all specialized mangroves projects above the root system, relates to gas exchange. In mangroves, this part is produced differently in different root types : The columns of the looping root system of *Rhizophora*; the pneumatophores of *Avicennia*, *Laguncularia* and *Sonneratia* ; the root knees of *Bruguiera* and *Xylocarpus mekongenesis* and the plank extension of the roots of *Xylocarpus granatum*.

(ii) Absorbing / anchoring Component : This component has a dual function : It form the basis of anchorage for the whole system and is the site of absorption. In most mangroves this is the result of the development of appendages that grow downward from the horizontal system. They are direct first–order laterals in *Avicennia* and *Sonneratia* for example, but second order laterals from the region of the root knee in *Bruguiera* and *Ceriops*. In *Rhizophora* each unit is the direct extension of one loop, which penetrates the substrate. The absorbing and anchoring components can be morphologically segregated because the more massive downward–directed root appendages provide the main anchorage; the finner ultimate orders of the branching system are the immediate sites of absorption as with the capillary rootlets of *Rhizophora*.

(iii) Cable Component : This is the extended horizontal part of the root system that unifies the aerating and absorbing / anchoring part of the root system. In *Rhizophora* this system is above ground and is a direct result of extension of the primary aerial roots. In other mangroves the cable system is underground.

It should be emphasized that the absorbing or anchoring and the cable components exist in root systems fall plants. The aerating components is the frequent additional component in swamp or mangrove forest species.

Types of Root System

Root System – Root systems, in general, are not deep due to poor aeration of the substratum but extensive and profusely branched. Production of aerial roots are very extensive forming intricate tangle making movement in it very difficult. They have been categorised as :

(a) *Strut and Stilt roots* – *Rhizophora* species produce lateral roots from the hypocotyl and bases of the stems which form arch and reach after coming in contact with substratum. Some of which produce innumerable fibrous absorbing or nutritive roots and others form another arch, and the process continues. They show secondary growth and may become up to 19 cm or more in girth and give support to the trees. They have lot of lenticles in younger stages. In Alexandra island near main sea (Andaman group of island) in an area of 375 sq.m. there were 14 trees of *Rhizophora mucronata* and 5 trees of *R. apiculata* with a total of 253 strut roots measuring approximately 408 m and in all, producing 928 arches and there were about 201 prop roots from upper branches. In the area, the strut roots have been noted to be produced in abundance only nearest to water and are much less in land side (Fig. 12).

(b) *Prop Roots: Rhizophora* produces lot of relatively thin unbranched positively geotropic roots arising from branches at various heights even at 8 to 10 metres. They appear to show branching after their tips are damaged by molluscs or some other agencies. After reaching the ground they also produce innumerable fibrous roots for absorption of water and nutrients (Fig. 12).

(c) *Cylindrical or flattened pneumatophores* : Cylindrical thin (upto 1 cm in dia. or more), negatively geotrophic roots upto 60 cm high are produced by various species of *Avicennia*. They usually do not develop in extra tidal region. In *A. nitida*, Chapman (1940) described two types of pneumatophores : (i) with smooth surface and (ii) with rough surface, the former with fewer sunken lenticels, and become firm and compact on drying but, latter are with more lenticels projecting markedly and are dry wrinkled. The former is associated with loose sandy soil and latter with compact sand, firm mud, peat or due to persistent immersion in water. The structure of lenticel also is characteristic of species. *Laguncularia racemosa* also develops pneumatophores.

Sonneratia species also produce pneumatophores which are conical and thick at the base due to secondary growth and produce exfoliating barks : *Xylocarpus granatum*, in some substratum, produces flattened blade like pneumatophores upto 30 cm high and about 5 cm wide and less than 1 cm thick (Fig. 12). They facilitate exchange of gases between atmosphere and underground roots.

(d) *Horizontal roots*: In several species like *Bruguiera* spp., *Ceriops tagal*, *Lumnitzera* spp., etc. come above the surface of the soil due to bending and again after some height curve towards soil forming knee-like structure (Fig. 12). It may grow at an angle to produce a small column and usually branches from the point of curving towards the ground.

Each branch travelling towards the ground becomes apogeotropical and produces fine branches for adsorption of water and minerals. Knee-roots also have lot of lenticels and experimentally it has been proved that they function for exchange of gases and their air space have air with oxygen content similar to external atmosphere.

(e) *Laterally compressed plank–like flattened roots*: *Bruguiera* spp. in some cases produce buttressed roots near the trunk. In *Heritiera*, some times lateral roots become laterally flattened and blade like.

In *Avicennia* there are five types of roots (Fig. 12), viz. (i) Primary adventitious roots – arising from the hypocotyl which degenerate; (ii) Horizontal or cable roots – apogeotropic usually 10–30 cm below soil; (iii) pneumatophores – aerial, negatively geotropic arising from cable roots; (iv) Anchoring roots – Rarely branched and positively geotropic arising from cable roots and are positively geotropic; (v) Fibrous roots – Highly branched absorbing roots usually, arising from the base of pneuamatophores.

Function of the Roots

In general concept that above ground component of the root system in mangroves ventilates the buried portion for respiratory purposes continues to be supported by experimental methods. That the composite root system is permeable to the mass flow of gases, with atmospheric exchange occurring through lenticels in the aerial vents or "chimneys" is evident from the large internal gas space system that develops largely by the enlargement of intercellular spaces supplemented by cell breakdown.

B. Structure of Leaves

Accounts of mangrove leaves and their structure have been given by Farming (1883), Areschoug (1902), Mullan (1931), Walter and Steiner (1936), Chapman (1944) and Metcalf and Chalk (1950). Stace (1968, 1966) has given a detailed account of epidermis and cuticle in the Rhizophoraceae and Combretaceae. Areshoug pointed out that mangroves do not possess any features that are exclusively associated with them. This is the tree of the leaves which possess contain so–called xeromorphic characters –

(i) Thick cuticle
(ii) Sunken or Chambered Stomata
(iii) Water Storage Tissue
(iv) Palisade – Like mesophyll
(v) Bundles terminating in trachieds.
(vi) Mucus cells.

Schimper (1891) argued that these features served as a protection against water loss because of the difficulty of water absorption from highly saline water : in other words, the plants suffered from physiological drought. Areschoug (1902) indeed suggested that only the thick cuticle, the sunken or chambered stoma and the mesophyll were related to water conservation.

Structurally six different types of leaves can be recognized among mangroves that have been studied.

(1) Leaves dorsiventral with hypodermal water storage :

(i) *Rhizophora mucronata, R. apiculata, R. mangle*

(ii) *Ceriops tagal*

(iii) *Xycocarpus granatum* etc.

(2) Leaves dorsiventral with hypodermal and internal water storage :

(i) *Aegieras corniculatum*

(ii) *Bruguiera gymnorrhiza*

(3) Leaves isolateral with no stomata on the upper surface none spongy mesophyll :

(i) *Excoecoria agallocha*

(4) Leaves isolateral with no stomata on the upper surface with spongy mesophyll. e.g. *Kandelia candel.*

(5) Leaves isolateral with stomata on both surfaces and spongy mesophyll acting as water storage tissue throughout leaf life :

(i) *Sonneratia*

(iii) *Laguncularia*

(6) Leaves dorsiventral with no special water storage tissue; the entire mesophyll probably acting an such :

(i) *Derris.*

The water storage tissue is a characteristic features of nearly all mangroves and at least 4 types can be recognized.

(a) Exclusively hypodermal. eg. *Rhizophora.*

(b) Derived from hypodermis plus subsequently modified assimilatory tissue. eg. *Xylocarpus granatum.*

(c) Exclusively deep seated eg. (i) *Lumnitzera*, (ii) *Lagucularia*, (iii) *Conocarpus.*

(d) Leaves without specific water storage time eg. *Derris.*

Salt glands, which will also excrete water, are formed on the surface of leaves of *Avicennia* spp., *Lumnitzera* spp., *Aegieras corniculatum* and *Acanthus ellicifolius.* The cuticle of the original glandular cells possess fine punctures through the self solution in excreted.

Leaves are usually leathery, in many cases quite shining. In *Avicennia nitida* leaves of two types have been noted (Golley *et al.*, 1962). One smaller thicker with more chlorophyll 'a' content (0.31 to 0.33 g/m^2 leaf area) and known as sun–leaves in upper exposed portion of the trees and other bigger, thinner and with less chlorophyll 'a' content (0.24 to 0.29 g/m^2) known as shade leaves. Also the sun leaves photosynthesize at many times higher rate than shade leaves. The authors have failed to note two types of leaves in Indian *Avicennia.* The stomata of most leaves have been found to be sunken and protected (Walter and Steiner, 1936) and mostly present on the lower surfaces.

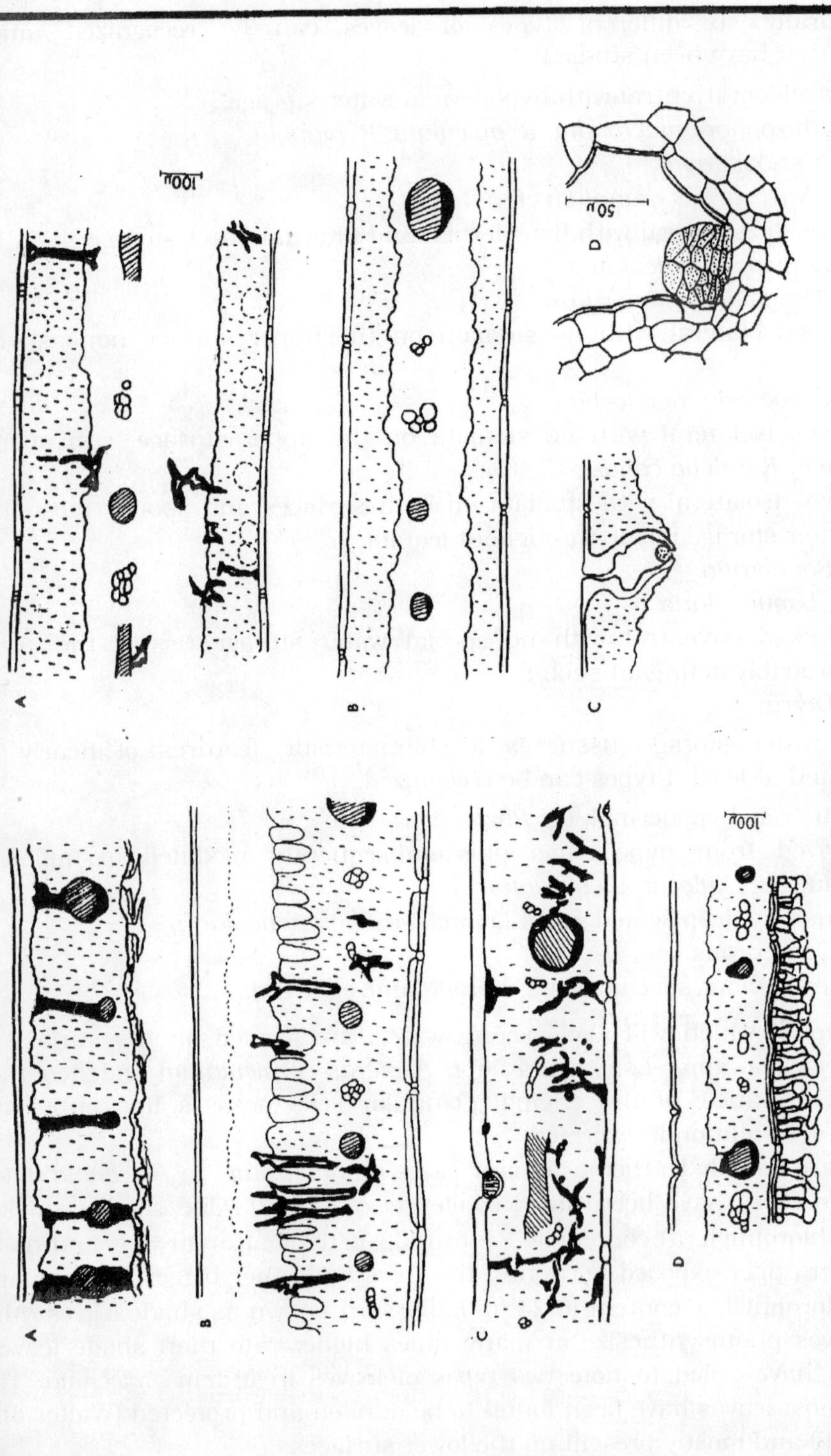

Leaf anatomy, diagrammatic isolateral leaves. (A) *Sonneratia alba*, T.S. (B–D) *Laguncularia racemosa*: (B) T.S.; (C) gland in pit on adaxial surface; (D) detail of presumed salt-secreting gland at base of pit.

Leaf anatomy, diagrammatic, dorsiventral leaves. (A) *Heritiera littoralis*, with well-developed bundle sheath extensions; (B) *Rhizophora stylosa*, adaxial hypodermis with three distinct layers; (C) *Aegialitis annulata*; (D) *Avicennia marina*. Key to symbols: Lined area, assimilating tissue; black, bundle sheath sclerenchyma or sclereids; hatched, veins; open circles, terminal tracheid clusters.

Fig. 13. Leaf Anatomy

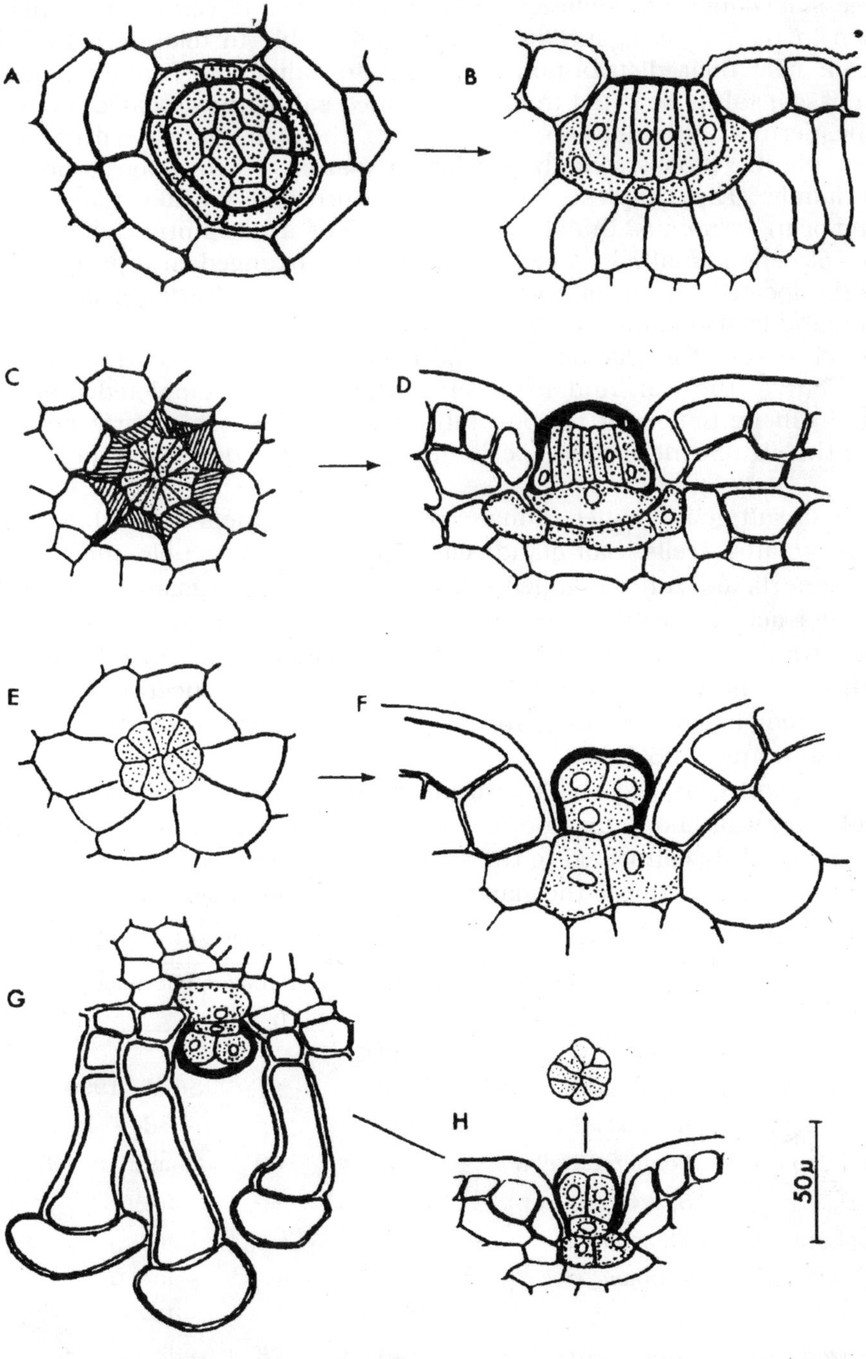

. Salt glands in mangroves. (A,B) *Aegialitis annulata*; (C,D) *Aegiceras corniculata*; (E,F) *Acanthus ilicifolius*; (G,H) *Avicennia marina*. All are from T.S. adaxial epidermis except G, which is from abaxial epidermis.

Fig. 14. Salt glands in mangroves

Salt Glands in plants

The salt glands are common in several families of plants, but this is very common in plants. Most species with salts are able to tolerate a high salinity and as a rule their distribution is limited to salines of various types. In all known cases, salt is secreted from well–defined salt glands located on or sunken into the dermis. Although the structure of the salt glands varies greatly in different plants. It is strikingly similar in plants of the same germs or even within family. The simplest type is the two–celled hairlike gland found in *Spartina* or in *Aeluropus*. In Avicenniaceae, glands are comprised of an indefinite number of cells (usually between five and nine) arranged in a group of four or more cells located on top of one stalk and two-four collecting cells (Fig. 14). A similar gland is also found in *Acanthus ilicifolious*.

In all cases, the glands have been initiated at an early stage in the development of the leaf, and their differentiation was completed much earlier than the differentiation of various other leaf tissues. This early development stresses the significant of their special function for the developing organ (Helder, 1956).

Cells of salt glands differ in many respects from the surrounding epidermal or parenclymatous cells. Salt gland cells lock a central vacuole and the number of mitochondria and other organelles is extremely high. This suggests that gland cells do not act primarily as accumulating organs, but rather as transist cells. Salts are transported outward by specific mechanisms, which consume energy originating on the mitochondrial surface. Nevertheless, concentration of ions in gland cellş may in certain cases, still be high in comparison to surrounding cells (Zeigler and Luttge, 1966).

Some gland cells are characterized by cutinized and suberized cell walls, especially the walls bordering normal parenchyma cells. In certain places, the cytoplasm is rightly connected to the strips, exactly as in caspasian strips. Thus ions can be transported into the gland cells only through symplast.

Table 10(a) : Distribution of Salt Glands in Various Species of Mangroves

Family	Species	Number of Stoma/ mm^2		References
		Upper	Lower	
1. Avicenniaceae	*Avicennia marina*	210	____	Singh, 1994
	A. officinalis	180	_____	-----do-----
2. Sonneratiaceae	*Sonneratia alba*	67	60	Chapman, 1976
	S. apetala	100	—	—do—
3. Rhizophoraceae	*Rhizophora*	100–117	—	—do—
	Ceriops	100–107	—	—do—
	Bruguiera	134–150	—	—do—
4. Lumnitzeriaceae	*Lumnitzera*	140	78	—do—
5. Acanthaceae	*Acanthus ilicifolius*	150	—	—do—
6. Eupherbiaceae	*Aegiceraus*	150	—	

In isobilateral leaves stomata occur in both surfaces. Exception are found in many species. These are normally two subsidiary cells found on either side of the pore, *Derris*, *Excoecaria* or else at the ends, eg. *Acanthus*, *Avicennia germinanus*, *A. schaueriana*. In *Laguncularia*, *Bruguiera* and *Rhizophora* there may be three–four subsidiary with per stoma.

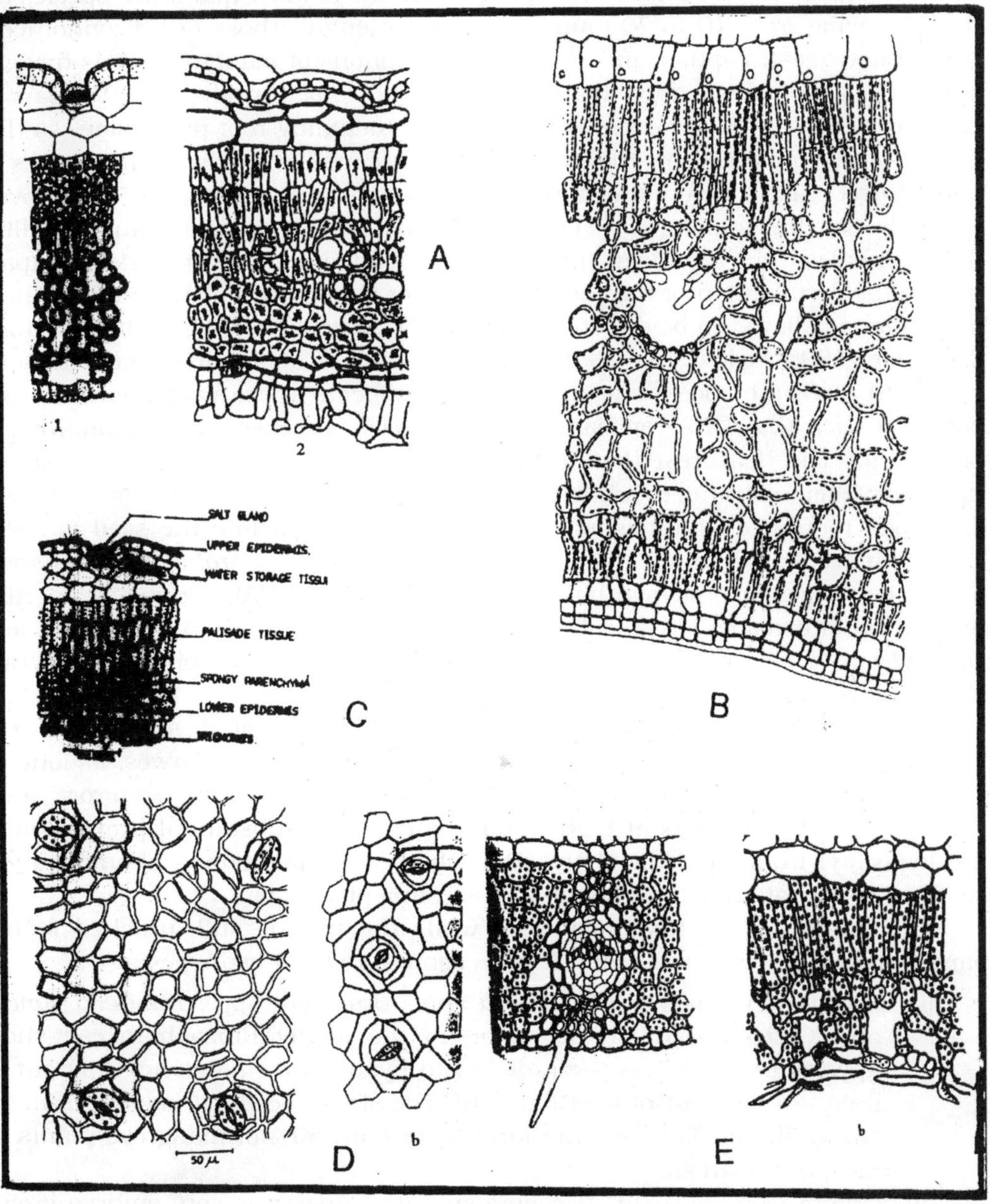

Fig. : 15. A. T.S. Leaves of *Avicennia germianans (1&2);* B. T.S. Leaves of *Candelia candel;* C. General structure of leaves; D. T.S. Leaves of *Laguncultria racemosa;* E. T.S. Leaves of *Acrostichum aureum* and *Hibiscus tiliaceous.*

In many species the stomata are sunk, but this is not so in *Bruguiera, Xylocarpus, Conocarpus, Schyphiphora, Acanthus* and *Execoecaria.*

C. Seeds and Seedlings Vivipary Germination of Seed

Compared with agricultural crop seeds. Those of most mangrove species are large, ranging from 10 to 700 mm (or more) in length, those of Rhizophoraceae being the longest. Popular and scientific descriptions of mangrove seeds have, in general, classed them as viviparous; but this is not strictly true. Vivipary is taken to imply the progression of germination, including root protrusion, so that a seedling is produced on the parent plant (an undersirable circumstances in agriculture). This is actually a very rare occurrence among the mangroves. Vivipary in the tomato, for example, leads to germination and seedling development within the moist interior of the fruit. In wheat or barley, vivipary means the germination of the seed in the ear or the mature plant. Each of these condition is believed to be directly linked to a failure of the embryo to dehydrate (Kermode, 1995) below the critical water content of about 40 percent of total weight (0.7 of water per gram dry mass). Only below these values is the water potential such that DNA replication and cell division cease, and germination is precluded. The features that characterizes embryos of viviparous seeds, therefore, their continued unabated growth, so that new seedlings with well developed roots been established on the parent plant before the seed is shed. Genetic mutants exhibiting vivipary in wheat, tomato and *Arabidopsis*, all show metabolic blocks in the synthesis of abscisic acid (ABA). This lack of ABA production is considered central to the failure to suppress DNA replication and cell division, which is linked in turn to an inability of these embryos to dehyrate in vivo (Karssen, 1995).

A pivotal step in germination and establishment of a seedling is the development and extension growth of the root, but for the mangroves, as long as the enlarging seed remains attached to the tree, the roots do not grow, even through the seed remains at high water content. The substantial protrusion of the hypocotyl from the fruit of the Rhizophoraceous mangroves should not be confined with root growth, and therefore seedling production.

In mangroves germination of seed while the fruit is still attached to the parent plant is known as vivipary. It is recognized to be of two types:

(i) Apparent vivipary – *Rhizophora* spp., *Bruguiera* spp., *Kandelia candel* and *Ceriops* spp., all belonging to the family Rhizophoraceae show apparent vivipary, as the radicle and hypocotyle emerge not only out of seed but also out of fruit (Fig. 16). The size of seedling may be from 20 cm to 55 cm. These plants form more than 50% of mangrove plants of most of the areas.

(ii) Crypto–vivipary — In *Avicennia* spp. and *Aegiceras* spp. embryo grows out of the seed but remains inside the fruit while the latter is still attached to the parent plant (Fig.16).

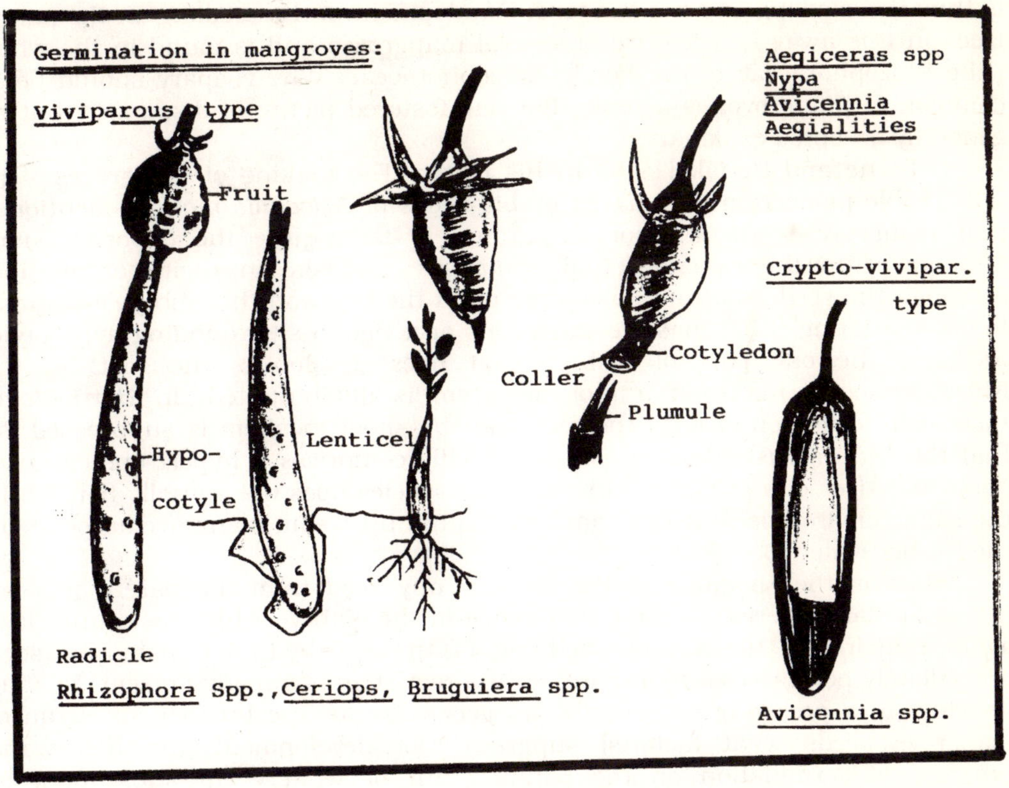

Fig. 16. Seeds and Seedlings in mangroves

Table 11. Germination Type and Seed Size in Mangroves

Genus	Apparent Vivipary (ApV)	Crypto Vivipary (CrV)	Seed size in cm
1. Rhizophora	ApV	——	20 – 70
2. Bruguiera	ApV	——	15 – 30
3. Kandelia	ApV	——	30 – 40
4. Ceriops	ApV	——	15 – 30
5. Aegiceras	——	CrV	05 – 07
6. Nypa	——	CrV	10 – 12
7. Pelliciera	——	CrV	07 – 08
8. Aegialites	——	CrV	03 – 04
9. Avicennia	——	CrV	02 – 03

The development of the embryo in seed plants is a continual process, normally interrupted by a longer or shorter period of seed dormancy, with tropical species tending to have the briefest dormancy. Thus, there is a potential continuous of development possibilities. In the extreme condition of "true vivipary" the embryo that results from normal sexual reproduction has no dormancy but grows first out of seed coat and then out of the fruit while still attached to the parent plant (Juncosa, 1982); the propagating organ (propagule) is thus the seedling and not the seed. Development of the seedling on the parent trees in this away is a feature of several mangroves and is found in the whole tribe Rhizophoraceae of the family Rhizophoraceae. Why vivipary should be so common in mangroves is a topic that has fostered plenty of discussion but no generally accepted explanation.

Osborne and Berjak (1997) in his paper "The making of mangroves : the remarkable pioneering role played by big seeds of *Avicennia marina*" mentioned that mangrove seeds are not viviparous, in fact, quite the opposite, since attachment to the parent actually imposes a stasis on their germination. Although the seeds continue to enlarge upon the tree and the embryos continue to differentiating either massive cotyledons or a vigorously extending hypocotyle, or both, the one part of embryo that does develop is the root. In the Rhizophoraceae, not even a root meristem is differentiated. In *Barringtonia racemosa*, formation of both the root and the shoot mesitem is suppressed, so that the continuously growing embryo, finally comprises a hypertrophical axis, about the sizes of a golf ball. Although this species does not actually flourish in the saline environment. It is a commonly found upstream mangrove associate of the fresher water (Berjak *et al.*, 1997).

Whereas the shedding of the familiar any seed from the parent in crop plants herdals a period or total quiescence in the embryo which losts until it is again rehydrated, the act of shedding of the still–hydrated mangrove seed immediately permits root to initiate, or, if a meristem is already present, to start growing. It is presently not possible to generalize for the broader spectrum of mangrove seeds what factor(s) suppress root development and hence the progress of germination on the parent plant or what event subsequent to shedding triggers the initiation of germination metabolism. ABA may, however, be involved in the probably best studied mangrove seed to date (that of *A. marina*). Once the maternal outer coverings which contains most of the ABA) are sloughed, the roots can start to grow at once (Farrant, 1993). Since little ABA can be detected in the tissues of the embryo itself. Its absence these may help to explain the rapid root emergence after shedding.

The Phenomenon of Seed Recalcitrance

Mangrove seeds have been classed as 'recalcitrant'. The term was coined in 1973 (Roberts, 1993) for seeds with a common property of being unable to survive any significant degree of dehydration. Probably all mangrove seeds fall into this category, and will die if dehydrated, but as long as they remain attached to the parent, their water supply is assured. Since the parent trees are usually in a wetland location, most of their seeds fall either directly into water or onto damp mud where desiccation is unlikely and where germination follows. Seeds shed onto dried mud soon lose water and viability. Seeds of *A. marina* will not survive storge for much more than two weeks, even if no water is lost (Pammenter *et al.*, 1994). In attempts to re–establish areas of mangroves that have been devastated by natural or man–made disasters (as in the Mekong Delta in Vietnam), the inability of the seeds to survive desiccation – or short–term storage periods before planting – has led to a number of disappointing results.

So the secret of the mangroves 'success in colonizing specialized intertidal habitats lies in the peculiar properties of their seeds, properties which are finely tuned to their environment. Like mammal embryos, those of the mangroves continue to grow whilst on the parent, but unlike a mammal, growth of the mangrove embryo is restricted to certain tissues whilst in other tissues, development is suppressed. With few exceptions, it is the roots that are suppressed, even if a meristem, or in certain species, multiple root meristems, are initiated. One feature that is then common to all mangroves is the speed at which the root will grow once liberated from the influence of the plant (Tomlinson, 1986) and in *Avicennia* spp., this occurs when the fruit falls into a damp environment and the seed coat splits open and falls away.

Much effort is now directed to understanding why certain cells can be dried and then revived on the addition of water without loss of genetic integrity or function, while for most living cells, dehydration leads to death. Many seeds and spores have successfully achieved a dehydration stratagem, but the mechanism remains an enigma. Mangroves have had no need for these adaptations, so their recalcitrant seeds have been a focus of research for comparison with the usual dry embryo condition of most other seeds studied. One clue has been the production of a specific class of proteins, synthesized when plant cells suffer water loss. These proteins (called dehydrins) are of relatively low molecular weight, heat stable, and with a common consensus lysine–rich sequence near the carboxyl terminus. They are produced by leaves as well as seeds, but accumulate in the cells of desiccation–tolerant embryos late in embryogenesis (hence they are called late embryogenic abundant or LEA proteins). The part they may play in desiccation tolerance is presently speculative, though it may be that their surfactant properties protect drying membranes (Close, 1996). What is relevant here is the lack of production of these lysine and glycine–rich LEAs in seeds of the mangroves (Farrant *et al.*, 1996), which may be one of a number of contributing factors accounting for their failure to survive any water loss.

Another feature, important to all seeds, is the type of storage reserves (proteins, lipids or polysaccharides) that are laid down during embryo development and then utilized at the onset of germination. In seeds that dry and can be stored, these reserves are in insoluble form and require hydrolysis to become usable as the building blocks and energy sources for growth. In *A. marina*, the reserves are mainly soluble sugars (Farrant *et al.*, 1993), which in the hydrated embryo function not only as an immediate energy source for germination, but also provide a high osmotic protectant against the sudden exposure of the embryo to saline soil and water when the seeds are shed.

Controlling germination

Although it is clear that seeds cannot normally produce roots while on the parent tree, it is also clear that they may not necessarily proceed with germination the moment they are released. There are still other factors that can control whether or not the root meristems will form (in the Rhizophoraceae) or primordia will elongate (*A. marina*). The mangrove habitat is very highly specialized in many ways, but as far as successful seed dispersal, germination and seedling establishment are concerned, an essential requirement is slowly moving tidal water and a sheltered situation. Interestingly, even off the parent plant, the embryo will not release the root primordium from growth suppression unless free water is available under aerobic conditions. *A. marina* seeds collected and held at high humidity, so that there is no water loss or gain, will not extend their roots beyond the barest protrusion, and will eventually lose viability even though the seed coats are removed (Matate *et al.*, 1997). For those seeds that fall into the water, and are carried off by the tide, extensive root growth would be highly disadvantageous, for, once cast up onto land, the exposed roots would rapidly become lethally dehydrated – probably long before the seedling could become established. So the optimal situation for root elongation is when the seeds become trapped among the littoral vegetation or upon the muddy alluvial banks around pneumatophores. The desirability of the combination of free water and a periodic aerobic exposure means that very few of the seeds produced ever establish as seedlings. In *A. marina*, those seeds that first float on the tide have a better chance for eventual seedling establishment than those shed above the high tide level, for around the multiprimordium root meristem region of the hypocotyl are many short hairs that become bristly and erect on contact with water. It is believed that air is trapped between these as the embryos float, aiding buoyancy as well as extending for several days the survival of the critically important, metabolically active root primordia.

There is an important difference between recalcitrant seeds, of which the mangroves are extreme examples, and those of desiccation – tolerant weed seeds that can remain in the soil, fully imbibed, for many years without growing. Both are metabolically active, respiring, synthesizing proteins, membranes and all classes of RNA. But, whereas seeds of the soil bank (such as those of the wild coat) are held in a state of homeostatis in which there is slow metabolic turnover but development is arrested and the cell cycle is held at G1 with no replicative amplification of DNA (Elde and Osborne, 1993), the mangrove embryo is always both metabolically and developmentally active and we know for *A. marina* that cell cycling never ceases (Berjak *et al.*, 1999).

Autoradiographic studies have shown that both protein synthesis and DNA replication in the root primordia of *A. marina* reach a low level on the day when seeds are shed, but these processes never completely stop and immediately after shedding, even if germination is still suppressed, the cells of the root meristems return to high rates of precursor incorporation into DNA (Berjak *et al.*, 1999) and cell divisions are apparent again within 24 hours (Farrant *et al.*, 1992). Whereas, the seeds of a seed bank are strictly dormant, those of *A. marina* at least and probably of all mangroves, are certainly never dormant. They are poised for immediate germination as soon as the conditions for seedling establishment success are met.

It is to this 'poised for emergence' situation that the fast and successful germination of the mangrove seed may be attributed. Most of the seeds may never germinate and may be lost either on the tide at the outward edge of the littoral belt or on mud on the landward side that is too dry and too far from the tidal zone, but those that all in the right environment are true entrepreneurs. Not only with their root tips quickly penetrate the mud, elongation of the cells behind the meristematic tips raises the whole hypocotyl and cotyledons above the mud surface, keeping the shoot region in an essentially aerobic environment.

There is a further safety ploy that protects *A. marina* during the period of seedling establishment. The root primordia are all clustered at the distal end of the hypocotyl among the hairs, but at first only four of these will elongate, another five or six remaining suppressed. But, if the seedling is dislodged from its first attachment to the mud, it has a second, and possibly even a third chance to re-establish, through root growth from these additional primordia (Farrant *et al.*, 1987). Several plant families have species that are adapted to the mangrove environment, which is saline, anoxic and unstable, but not all cope in exactly the same way. The genus *Avicennia* is the only representative of its family, and in detail is adapted in different ways from those mangroves of other families. But the description of the facets we have described for *A. marina* gives a fascinating and provoking insight into some of the diverse mechanism that plants have adopted to survive in one of the world's most challenging environments.

3.5 SALT REGULATION AND OSMOTIC ADJUSTMENT

The saline environment of marsh ecosystems creates a formidable challenge for plants living there. High salinity soils are characteristic of salt marshes. As salt water from the ocean inundates and recedes from he system in a daily cycle, sodium chloride is deposited into marsh soils where evapotranspiration amplifies soil salt concentrations (Larcher, 1980). Saline soils can be detrimental to plants in a variety of ways; therefore, plants living in saline environments have adapted mechanisms to deal with these problems.

The problem of salinity tolerance permeates many scales in the salt marsh ecosystem. The following discussion focuses on problems at both the cellular and organismal levels.

Salt accumulation in the tissues of marine halophytes is basically no different than that of other halophytes species. Several recent reports have appeared during the past several years that contain reviews of much of the literatures that is pertinent to the physiology of "coastal halophytes". Foremost among these are : 'Physiological basis of salts tolerance of plants' (Shougonov, 1964), 'Salt transport by plants in relation to salinity' (Rains, 1972), 'The biology of halophytes' (Waisel, 1972), 'Ecology of salt marshes and sand dunes' (Ranwall, 1973), 'Is any angiosperm an obligate halophyte' (Barbour, 1971) and 'Physiology of coastal halophytes' (Queen, 1974). Generally the higher salinity region of the coastal zone is a hostile environment for most plant species. Many of the problems faced by plants in this zone are related to characteristics of sea water. Of these characteristics, the most significant are –

(i) a high osmotic pressure and
(ii) an ionic mix substantially different from that of the soil solution encountered by most terrestrial plants.

Plants can not acquire water from an external medium unless the osmotic pressure within the plant exceeds that of the external medium. If the external pressure is higher than the internal pressure, water will flow from the plant to the medium. For many years investigations was of the opinion that the stress imposed on plants by high soil salinities was that of a water deficit (Bernstein and Hayward, 1958). Other worker (Statyer, 1961; Greenway, 1962) doubled the validity of this explanation They were of the opinion that adequate quantities of water are acquired under saline conditions, because plants respond to high external osmotic pressures by increasing the internal pressure to a level higher than that of external medium by accumulating additional salt from the external medium. The latter view was later confirmed (Bernstein, 1961). Thus, for those species able to develop and tolerate high internal osmotic pressures, the acquisition of water is not a factor in preventing their occurrences in a highly saline coastal habitats.

However, plant adaptation to a saline environment is not assumed by the accumulation of additional salt, while necessary for the continued acquisition of water, creates other problems. Among these are –

(i) ionic imbalances that are potentially toxic, and
(ii) internal salt concentration that are sufficiently high to be toxic to most plant species.

Challenges of Living in a Saline Environment

The high concentration of sodium chloride in salt marshes has negative consequences for plants in two primary ways: (i) it affects osmoregulation; and (ii) it has ion specific effects that can be toxic and can disrupt ionic balance (Larcher, 1980; Hagemeyer, 1997). As a result, many plants, even halophytes, respond to salinity with decreased metabolic functions.

- Osmoregulation
- Ionic Effects
- Metabolic Consequences

Osmoregulation

Water uptake and flow through a plant is driven by a water potential gradient where water flows from least to most negative water potential (measured in units of pressure, mega–Pascals). Therefore, for water to flow from the soil into plant tissues, the plant's water potential must be more negative than that of the soil. The water potential of the soil is, in part a function of the concentration of dissolved solutes in water in the soil. The presence of dissolved solutes, such as sodium chloride, reduces the soil water potential, which decreases (and even eliminates) the water potential gradient. In effect, saline soils 'hold' water osmotically making it unavailable to plants (Larcher, 1980). As water evaporates and is taken up by plants, the gradient is degraded further because solutes become even more concentrated in the soils (Hagemeyer, 1997). Some plants living in saline soils combat this problem by adjusting their water potentials through the accumulation of solutes in plant tissues.

Ionic Effects

The abundance of Na^+ and Cl^- ions within the cytoplasma of plant cells has several clearly problematic consequences (Larcher, 1980; Hagmeyer, 1987).

- Interference with energy producing membrane processes (i.e., the electron transport chains in chloroplasts and mitochondria).
- Decimation of enzyme systems in glycolysis.

- Reduction in uptake of mineral nutrients from the soil (i.e. NO_3^-, K^+ and Ca^{2+})

Metabolic Consequences

Excess sodium chloride affects several important plant functions. Energy production is impaired by high ion concentrations and as a result, photosynthesis and respiration rates have been shown to decrease in halophytic and non–halophytic plants (Pezeshkl *et al.*, 1987). Consequently, growth rates and biomass production are sensitive to salinity (Larcher, 1980). While both tolerant and non–tolerant plants respond negatively to salinity, impaired growth in salt tolerant plants is slight compared to non–tolerant species. This differential response to salinity is a clear indication that some plants have adapted mechanisms to counter the negative effects of salinity.

The plant rooted in sea water has to generate an additional hydrostatic pressure of about 25 bars (the negative water potential of water) as compared with plant rooted in freshwater. Mangrove also raise water against hydrostatic gradient to height of upto 30m. By convention, the water potential for pure water is established as zero; sea water at a concentration of 3 per cent sodium chloride (NaCl) has a negative pressure (tension) about 25 (Fig. 17).

Scholander (1968) demonstrated that the root system effectively functions as a partial to almost complete ultrafiltration system and that the water potentials of the leaf cells are always lower than that of sea water. In the living tree, under conditions of no transpiration (stomata closed), that is, no xylem flow of water to the leaves to replace water lost by transpiration, there is an additional hydrostatic pressure in the column of water in the xylem equal to –1 bar per 10 m to be overcome. This hydrostatic pressure occurs, of course, in all trees. Under conditions of transpiration (stomata open), work is done to move the continuous water column and require a pressure gradient. The pressure in the crown is decreased by transpiration. Water molecules move along a steep gradient from the leaf to the atmosphere (at a very low negative water potential for relatively dry air) via the open stomata. According to Larcher (1983), the salt resistance of vascular plants consists of various components, which may be divided into two main groups :

(i) mechanisms of salt regulation, and
(ii) salt tolerance.

Salt regulations comprises –

(i) Avoidance of intake
(ii) Elimination, and
(iii) Dilution.

While salt tolerance is defined as a property of the protoplasm to cope with high salt concentrations. Salt tolerance encompasses ion compartmentation in the protoplast and tolerance, towards toxic and osmotic effects of high ion concentrations. In mangroves, these are three mechanism of salt regulation.

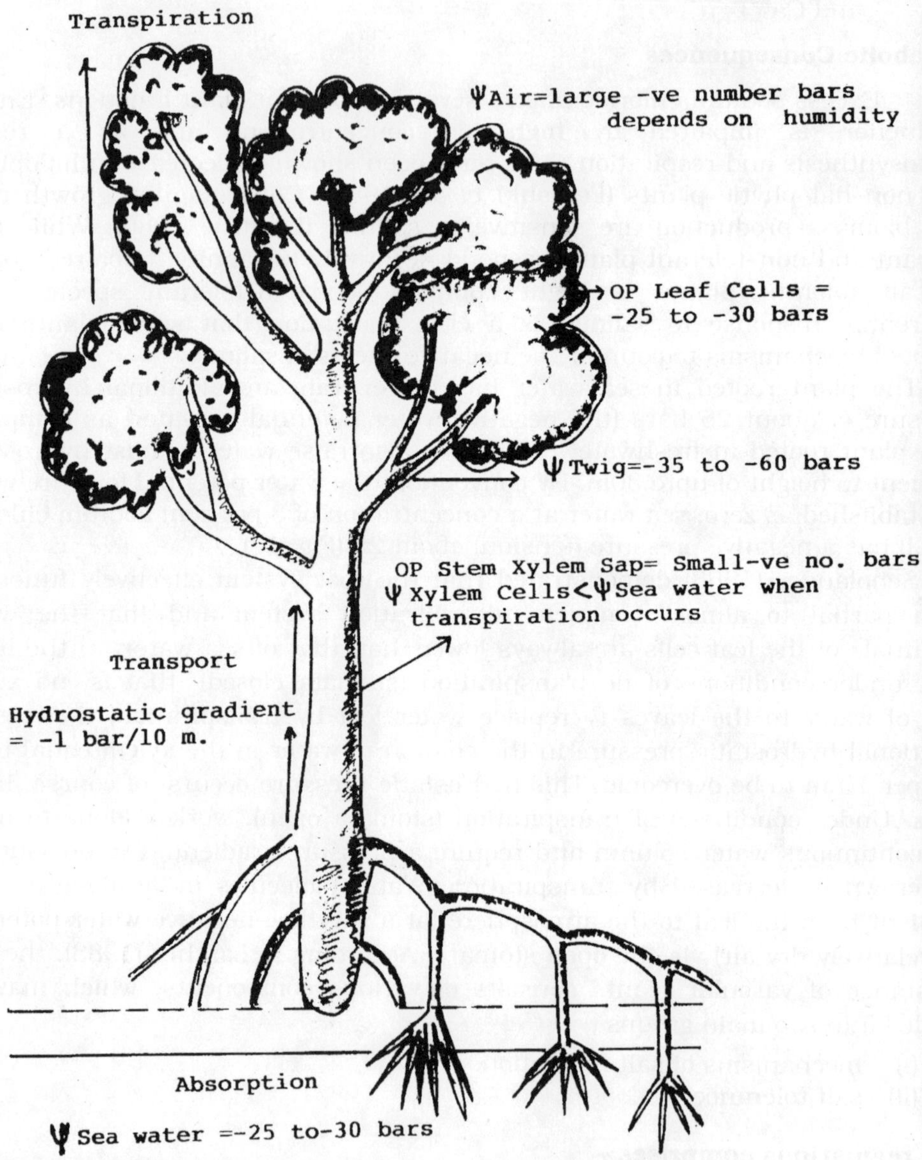

For water to move from sites of absorption to sites of transpiration a -ve hydrostatic gradient must be maintained. If ψ sea water exceeds ψ xylem cells, water should flow back into the sea.

Fig. 17. Water uptake in mangroves

(i) Salt Filtration or Salt Exclusion

In their classical work on East African mangroves Walter and Steiner (1936) postulated that the xylem gap of mangroves should be free of NaCl. Investigations of the ion content and osmotic potential of mangrove roots and leaves of different ages, as well as of transpiration rates led them to this hypothesis (Popp *et al.*, 1993). As several other workers in this field, they determined little difference in the ion content of leaves of different age and therefore, concluded that there is no permanent intake of NaCl from the xylem sap. Preliminary investigation did indeed reveal very low NaCl concentration in xylem saps of mangroves (Scholander *et al.*, 1962). A significant difference was found in xylem sap NaCl concentrations between secreting and non–secreting species, the latter being markedly lower than the former (Scholander *et al.*, 1962; Atkinson *et al.*, 1967).

Rhizophora mucronata and several other spp., *Bruguiera, Kandelia* and *Ceriops* spp. have some sort of ultra filtration mechanism in their roots enabling only selection absorption of ions (Scholander, 1968). Their xylem sap has NaCl content between 0.02 to 0.05%.

(ii) Salt excretion or Salt secretion by salt glands

Salt excretion from mangrove leaves in an order of magnitude higher than in herbaceous halophytes (Luttge, 1975; Stewart and Popp, 1987). Popp *et al.* (1993) stated that the rate of salt excretion (µmol salt m^{-2} leaf area sec^{-1}) depended on leaf age and salt concentrations in the root medium.

Avicennia and *Aegiceras* spp. and over several other species have salt glands on their leaves which secrete excess salts on surface of leaves. NaCl in the xylem sap of these plant is normally 10 times that or salt exclusion types. The excreted ions are normally Na^{+} and Cl^{-}

Table 12. Indicates the salt regulation in various species

Types of Salt regulation	Name of Species
(i) Salt excluding	*Rhizophora* spp., *Bruguiera* spp., *Kandelia* spp. *Ceriops* spp.
(ii) Salt excreting	*Avicennia* spp., *Aegiceras* spp., *Acanthus* spp., *Aegialitis* spp.
(iii) Salt accumulation	*Sonneratia* spp., *Lumnitzera* spp., *Excoecaria* spp.

Salt is only partially excluded at the root. The absorbed salt is primarily excreted metabolically via salt glands. Salt glands in *Avicennia* are scattered in individual shallow pits on tree upper leaf surface and much more densely within the abaxial indumantum, where they are not sunken but still obscurred by the palisade of 3–4 celled nonglandular hairs. Each gland consists of two to four basal vacuolated cells at the level of the epidermis, a single stalk cell with an almost completely cutinized wall, and a radiating series of at least eight terminal cells that have a thin, minutely perforated cuticle separated apically from the non–cutinized cell wall to leave a shallow cavity. The stalk and terminal cells are non–vacuolate, and the dense cytoplasm is rich in ribosomes and has a large nucleus, all features that suggest high metabolic activity.

The precise mechanism of salt secretion is not understood, but it does require energy and can be stopped by metabolic activity. Salt glands all have basal (collecting) cells, a single cutinized stalk cell, a capitate group of terminal cells.

(iii) Salt Accumulation

Sonneratia spp. and some others *Lumnitzera* and *Excocaria* spp. have been found to have excessive amount of ions in their organs and thus they absorb and accumulate ions and the leaves become quite fleshy.

Physiological adaptation to different salinity levels in mangrovephyta

Recently M. Popp (1993) has given a concluding remark about the physiological adaptation in mangroves.

(i) Avoidance of salt intake is evidently achieved by special mechanisms in the root system and not by the storage in lower plant parts. We still do not really know how mangrovephyta desalinate sea water.

(ii) Salt excretion by salt glands is very effective in controlling the salt concentration in the leaves, but is only present in a few mangrove species.

(iii) The majority of the mangrove are not salt secreting and their method of salt regulation is still something secret. Retranslocation of NaCl *via* the phloem and increasing water content / leaf area two possibilities for maintaining reasonable salt concentration within the leaves.

(iv) Sequestration of NaCl within the vacuole and storage of organic solutes in the cytoplasm in order to maintain intracellular osmotic adjustment contribute essentially to salt tolerance in mangroves.

How *Avicennia marina* copes with salt

Avicennia marina is primarily a salt excretor although it takes up only part of the salt from full sea water. The salt in the water stream that travels from root to leaves is excreted by foliar glands that are situated on the lower surface of the mature leaf. As a result, droplets of highly saline water accumulate and are raised above the lower leaf surface by a dense covering of trichomes (hairs), a mechanism which isolates the high salt solution from the leaf surface and prevents osmotic withdrawal of water from the leaf tissues.

The excreting glands are complex and multicellular, salt moves from the walls of basal collecting cells into a specialized membrane system in the body of a single adjacent, large stalk cell and from these into the terminal excretory cells from which it is eliminated.

3.6 PHOTOSYNTHESIS OF MANGROVES

The study of photosynthesis in mangroves is very limited, only few workers have worked on photosynthesis. Here the name of two persons, must be mentioned because they have worked very details in this subject, (i) Late Prof. G.V. Joshi and his students, Shivaji University, Kolhapur, India and (ii) Prof. B.F. Clough, Australian Institute of Marine Sciences.

THE WORK OF PROF. G.V. JOSHI

Mangroves constitute a group of plants which in the course of evolution has acquired complexity not only in morphological and ecological aspects but also in physiological processes. There have been numerous attempts at investigating the various aspects of their physiological adaptations and these have resulted in elucidating the intricate process of ion regulation in the mangroves. The details of these investigations have been reviewed by Walsh (1974). Even though the mechanism of salt regulation enables mangroves to have minimum salt concentration in the cellular environment, this concentrations is much higher than that found in glycophytes. .The chloroplasts in mangrove carry out a process of photosynthesis in a salt rich environment around their envelops. As the object of the present discussion describes the process of photosynthesis under saline conditions, it was thought that the mangroves would provide a material where, in the course of evolution angiospermic plants have acquired salt tolerance. Unlike marine algae investigated in the present study, the mangrove leaves are outside the sea water and do not have that high concentrations of sodium chloride around chloroplasts. With this view in mind mangroves have been investigated for the process of carbon assimilation in light and in the dark as traced by $^{14}CO_2$.

It has been rightly pointed out by Chapman (1966) that very few attempts have been made to study photosynthesis of mangroves. This is obvious from the sparse references available in the field. Succulent plants which grow under the stress of water have to keep their stomata closed during the day to economise on water. Due to this, carbon for photosynthesis is provided by massive fixation of CO_2 in dark in the form of malate and by breaking it down during the day. As mangroves also produce succulent leaves and have to economise on water, it was doubted whether the above mentioned Crassulacean Acid Metabolism (CAM) would provide carbon for photosynthesis. This was investigated by Bennet–Clark in 1933. However, it was found that the mangroves are an exception to CAM. It indicated that carbon for photosynthesis in mangroves must come from stomatal openings. This fact suggested to study the stomatal physiology of mangroves before could investigate the process of photosynthesis, investigations revealed that stomata in mangroves are open only during the early hours of the day.

The chromatographic analysis of the leaves of mangroves at Bombay by Bharucha and Rangnekar (1957) revealed massive amino acid synthesis in the plants of brackish water. In order to study the effect of saline environment on dark CO_2 fixation Joshi *et al.* (1962) exposed *Zosteria marina* Morong (marine angiosperm), *Egregia laevigata* Setchell (brown algae) and *Gigartina canaliculata* Harvey (red alga) to radioactive $^{14}CO_2$. They found that unlike *Bryophyllum* and tobacco the marine plants fix most of the carbondioxide in amino acids in contrast to organic acids of non-saline plants. They further found that cell free system of spinach leaves also mimics the pattern of marine plants when sodium chloride is present in the buffer medium. The massive amino acid synthesis of spinach cell free system can be prevented by the addition of $NADPH_2$ or NAD to the cell free preparation. It was suggested that the functioning of pyridine nucleotides are inhibited in saline mediums. The inhibition of malic dehydrogenase in salt rich environment has been subsequently reported by many investigators (Slack and Hatch, 1967; Larsen, 1967; Hason–Porath and Poljakoff–Mayber, 1969 and Osmond and Greenway, 1972). This observation along with the stimulation of transaminase recorded by Craigie (1963) in marine algae has enabled us to explain more amino and less organic acid synthesis in plants of saline environment. However, recently Winter *et al.* (1974) have shown that in *Mesembryanthemum crystallinum*, a facultative halophyte, there is more malate and not amino acid synthesis in the presence of sodium chloride. They have offered explanation for their contrary results.

The results of investigations by Joshi *et al.* (Loc. cit.) on the marine algae and on the marine angiosperm *Zosteria* were further confirmed by Webb and Burley (1965) who investigated the dark fixation of $^{14}CO_2$ by obligate and facultative salt marsh halophytes. They found that there is more amino acid synthesis from dark $^{14}CO_2$ incorporation in the halophytes which the authors attributed to various metabolic pools.

Even though, the problem of carbon assimilation in mangroves was investigated in dark, no concentrated attempts have been made to trace the path of carbon in photosynthesis in these plants. In 1965 some attempts were made in Joshi's laboratory by Chirputkar. She exposed leaves of *Rhizophora mucronata, Ceriops candolleana* and *Bruguiera gymnorhiza* to $^{14}CO_2$, and found that there was a resistance to the incorporation of $^{14}CO_2$. With great difficulty she could study the products of one hour photosynthesis. This problem was later overcome by a study of stomatal physiology which revealed that the process of photosynthesis in mangroves should be studied in the early morning. Once this break through was achieved the mangroves like *Aegiceras majus, Sonneratia acida* and *Excoecaria agallocina* were investigated for immediate and long term products of photosynthesis. Besides these, *Rhizophora mucronata, Ceriops candolleana, Bruguiera gymnorhiza, Avicennia alba* and *officinalis, Acanthus ilicifolius, Lumnitzera racemosa* and *Salvadora persica* were studied for photosynthetic efficiency. The only attempt to study photosynthetic efficiency in mangroves so far is that of Moore *et al.* (1972 and 1973). They have recorded a lower rate of photosynthesis in *Rhizophora* than in *Avicennia* or *Languncularia*.

The study of photosynthesis in mangroves must be viewed with the new concepts of photosynthesis. It is now clear that plants from arid and saline regions prefer a C_4 path i.e. a more efficient path, than the C_3 one followed by the plants not exposed to these stresses and strains. The various facets of this phenomenon are available in the symposium on "*Photosynthesis and Photorespiration*" (Ed. Hatch *et al.*, 1971). Laetsch (1974) has further elaborated the ecological adaptations of arid and saline plants. Shoemer–Ilan and Waisel (1973) have studied *Aeluropus litoralis* for the products of photosynthesis and have found that in this halophyte asparate is the major initial product and its synthesis is dependent upon the presence of NaCl in the metabolic environment. They have inferred from this that PEP–Case in this halophyte is a NaCl dependent enzyme. The NaCl dependence of C_4 path of photosynthesis is also indicated by Huber and Sankhala (1973). In the light of these researches the present study was undertaken to trace the initial as well as long term products of photosynthesis, products of dark assimilation and photosynthetic efficiency in the above mentioned mangroves which appear on the west coast of India. To study photosynthetic efficiency, the mangroves like *Rhizophora mucronata* Lamk., *Ceriops candolleana* Arn., *Bruguiera gymnorhiza* Lamk., *Sonneratia acida* Linn., *Aegiceras majus* Gaertn., *Avicennia alba* Blume, *Avicennia officinalis* Linn., *Acanthus ilicifolius* Linn., *Excoecaria agallocha* Linn., *Lumnitzera racemosa* Willd., and *Salvadora persica* Linn., were exposed to $^{14}CO_2$ under natural conditions of light and temperature from 9 to 10 A.M. for one hour. The utilization efficiency of $^{14}CO_2$ was calculated on the basis of per mg of tissue and chlorophyll.

Methodology Used

(a) Material

Healthy and fresh twigs of the mangroves were collected from Ratnagiri, Ganapatipule and Bombay. The twigs were brought to the laboratory in the shortest possible time. From the twigs, healthy leaves were removed and washed with distilled water and exposed to artificial light as a preparatory stage in the study of photosynthesis. Stomatal studies were done at various times of the day and night in the field, while photosynthetic studies were done in the laboratory between 9 and 10 a.m.

(b) Methods

(i) Stomatal studies

The method followed for stomatal studies was that of Stoddard (1965). The total number of stomata, width and length of stomatal apertures were estimated in the microscope on films obtained by nail–polish application. In this method nail–polish was applied to the middle portion of the leaf. After drying the nail–polish films were removed. The stomatal investigations were made at different times of the day/night period. The measurements of stomatal apertures were made under a calibrated microscope. Every time two impressions were taken and from each impression about 8–10 stomata were studied at different positions on the films.

(ii) Time lapse study

The leaf material weighing 2g of laminae were placed in the feeding chamber. The method followed for the study of the path of carbon in photosynthesis has been standardized in our laboratory. It consists of the following procedure. In the beginning both liquid as well as gaseous forms of $^{14}CO_2$ are supplied. However, with great difficulty a little incorporation with the gaseous $^{14}CO_2$ could be obtained. With the availability of more specific activity of $^{14}CO_2$, it has been found that the liquid phase of $^{14}CO_2$ gives more incorporation. Another advantage of the liquid phase of $^{14}CO_2$ ($NaH^{14}CO_3$) feeding is that more than one sample can be investigated under similar conditions of temperature, light and the time of the day. This enables one to compare the results of one plant material with those on another, and the experiment is done under identical conditions.

The plant material is allowed to float in Tris–HCl buffer (0.05 M concentration and pH 8.0). Feeding chambers were shaken in metabolic shaker with a constant speed. The reaction was started by adding $NaH^{14}CO_3$ (specific activity 48.3 mCi/mmole, obtained from BARC, Bombay, India). Illumination was achieved with a 360W photoflood bulb and two neon tubes placed 60 cm above the surface of the material. Illumination recorded near the surface of the material was 20,000 lux. By adding water and ice the temperature of the reaction was maintained at 20°C. The flask was continuously shaken during the period of the experiment. The reaction was terminated by adding boiling 80% alcohol.

In order to locate the initial and final products of photosynthesis, experiments were designed for 5 and 10 seconds as well as for one hour. The extract was condensed under reduced pressure and desalted. The compounds were analysed by two dimensional paper chromatography. The identity of each individual compound was confirmed by co–chromatography and autoradiography. The radioactivity in each compound was counted on 2π Proportional Counting System and low β–counting system manufactured by Electronic Corporation of India. For audiography Kodak blue X–ray films were used.

(iii) Dark fixation of $^{14}CO_2$

The procedure followed for dark fixation of $^{14}CO_2$ was similar to that employed for the light fixation. However, the experiment was carried out under complete dark conditions. For this experiment green leaves of *S. acida* were selected. The reaction was terminated in a similar manner described earlier. The distribution of radioactivity in ethanol soluble fraction in dark was analysed as per the procedure followed for light fixation.

(iv) Efficiency of carbon assimilation

In order to study the efficiency of carbon assimilation in light, eleven species of mangroves namely *R. mucronata, C. candolleana, B. gymnorhiza, S. acida, A. majus, A. alba, A. officinalis, A. ilicifolius, E. agallocha, L. racemosa* and *S. persica* were exposed to $^{14}CO_2$ for a period of one hour. This period was selected because it was observed that a steady state of photosynthetic carbon assimilation is reached by that time. The experiment was performed at 10 a.m. as the stomata of mangroves are wide open at that time. In order to obtain good physiological state of photosynthetic tissue the experiment was carried out near the mangrove vegetation and light and temperature available under natural conditions was utilized for the process of photosynthesis. Leaf tissue of equal weight was allowed to float in Tris–HCl buffer, pH 8.0 and it was supplied with equal amount of $NaH^{14}CO_3$ (specific activity 48.3 mCi/mmole). The total incorporation of $^{14}CO_2$ in the plant extract (ethanol soluble + insoluble fractions) was counted on 2π Proportional Counting System (Acid stable activity).

The vial containing $NaH^{14}CO_3$ is provided with specific activity, total activity and known amount volume of the radioactive material. In the mangroves under investigation the chlorophyll contents of the leaves were estimated. Based on these data $^{14}CO_2$ utilized by plant per mg of chlorophyll per hour was calculated by the method of Jensen and Bassham (1966) and further percentage of utilization efficiency was calculated by finding out the ratio between the number of molecules utilized to the number of molecules generated in the chamber.

The stomatal behaviour at various hours of the day and night in *R. mucronata, B. gymnorhiza, A. ilicifolius, A. majus, E. agallocha, K. rheedii, L. racemosa, S. acida* and *S. persica* is depicted. We can see that the stomata of the mangroves remain open during early hours of the day and close when the intensity of sunlight increases in the afternoon. It appears that the maximum photosynthetic carbon assimilation in the mangrove takes place from 8 to 10 a.m.

The mature and healthy leaves of *Aegiceras, Acanthus, Sonneratia, Rhizophora* and *Excoecaria* were exposed to $^{14}CO_2$ for a period of less than ten seconds to locate the initial products of carboxylation in light. The aspartate and alanine are the initial products of photosynthesis. In the case of *E. agallocha* the values are for ten seconds where activity besides aspartate, it appears in malate, citrate and alanine. It is possible that the activity from the initially formed compounds is diverted to TCA cycle. The results clearly indicate the synthesis of dicarboxylic acids in which aspartate predominates. It appears that the mangroves are C_4 type of plants and aspartate formers inspite of the absence of Kranz type of anatomy.

The distribution of radioactivity in various ethanol soluble compounds after one hour of photosynthesis in the above mentioned mangroves is studied. It is evident that the initial products have been utilized to build up amino and organic acids as well as sugars. The activity in amino acid fraction is usually more than that in organic acids. An interesting feature of sugar fraction is the presence of arabinose, rhamnose, xylose and maltose in free state. The dominance of arabinose which is not usually found in free state in other plants is of significance.

The results of one hour light $^{14}CO_2$ fixation in the leaves of eleven mangroves under natural conditions of light and temperatures have been recorded. These experiments were designed to study the efficiency of carbon assimilation in the mangroves and carried out under identical conditions. The total chlorophylls for each material has been recorded in the same table and it is clear from the table that *E. agallocha* contained the highest amount of chlorophylls while the least chlorophyll value was recorded in *S. persica*. The highest $^{14}CO_2$ incorporation per mg of tissue is recorded in *A. officinalis*. The percentage of utilization efficiency was calculated and the species were compared. It was found that *A. officinalis* again is the most efficient in the utilization of CO_2 while the second is *A. alba*, the third is *E. agallocha* and the fourth is *S. acida.* From the mangroves investigated by us *B. gymnorhiza* is the least efficient. The results justify the dominance of *Avicennia, Sonneratia* and *Excoecaria* in the mangrove vegetation of Western India. From the chlorophyll values and radioactivity, μ moles of CO_2 incorporated per mg of chlorophyll per hour was calculated and it was found that *S. persica* is the most efficient, though this species has the least chlorophyll content. In carbon fixation efficiency *B. gymnorhiza* is the least in the list. It has been found that *B. gymnorhiza* has poor K putake mechanism. It may be mentioned that *B. gymnorhiza* is the less dominant species of the mangroves.

(a) Stomatal Behaviour

In most plants, stomata tend to open during the day and close at night although they may not be open all day and closed all night. The removal of CO_2 in the dark can lead to stomatal opening thus simulating part of the light effect. However, Holdsworth (1971) has stated that stomatal opening has no obvious relationship to CO_2 uptake. The plants growing in hot climate have wide stomatal opening in the morning and partial or even complete closure sometimes after midday followed by reopening in the after noon (Meidner and Mansfield, 1968). That stomata close in the afternoon in response to temperature which are high is also the observation of Zelitch (1971). Davies and Kozlowski (1974) have reported that stomatal closure occurred at high light intensities in *Acer saccharum.* Even at 30°C the stomata remain fully open in a high light intensity if the relative humidity is in the range 50–100%. However, below 50% there is a reduction in the aperture. It is also seen that stomata open and close faster in green than in chlorotic plants with changes in humidity from 20–80%.

Fischer (1968) and Fischer and Theodore (1968) have reported the stomatal opening in isolated epidermis of *Vicia faba* with respect to KCl response. In dark and normal air the opening response was maximum at 100 mM KCl while light + CO_2 free air it was maximal at about 10 mM KCl. They further state that K is essential while Cl seems to be of a secondary importance in these processes.

Ramadas and Raghavendra (1974) have reported that exogeneous pyruvate (30 mM) enhanced the stomatal opening in the isolated epidermal strips of *Commelina benglealensis* while the other organic acids at those concentrations had excreted an inhibitory effect. Mederski *et al.* (1975) have studied the effect of leaf water deficit on net CO_2 assimilation in soybean, mung beans and maize. When stomata were allowed to regulate the gaseous diffusive resistance of the leaf CO_2 assimilation decreased with increasing leaf water deficit. When air was forced through the leaf, the rate of assimilation was not inhibited by increasing leaf water deficit.

Russell and Johnson (1975) exposed field grown soybeans at the mid–bloom stage to $^{14}CO_2$ for 1 hours beginning at either 9 a.m., 11 a.m., 1 p.m. or 3 p.m. The leaves labelled during the time periods associated with the highest net C fixation rates (11 to 12 noon and 3 to 4 p.m.) assimilated more ^{14}C, translocated more to stems and roots and exhibited a higher rate of export from the leaves.

There are very few reports on stomatal physiology of mangroves. Lewis and Naidoo (1970) have worked on stomatal structures in *A. marina* from South Africa. They found the transpiration rates high in the min–morning and low in the afternoon.

In the present investigation it has been recorded that the stomata of the mangroves remain open during early hours of the day and close when the intensity of sunlight increases in the afternoon. This again opens to a slight extent in the evening. This stomatal behaviour may be in response to temperatures which are high in the afternoon. This suggests the possibility that maximum photosynthesis assimilation in the mangroves may take place from 8 to 10 a.m.

The number of stomata in the mangroves are very much higher when compared with those in desert succulents recorded in our laboratory. For *Kalancoe* the values are 32 and 40, in *Aloe* they are 20 and in *Portulaca* they are 89 and 48 per mm^2 on the upper and lower surface of the leaf respectively. The first two are Crussulacean succulents. They have stomata wide open during the night when they assimilate CO_2 from both the sources, one from respiration and the other available in the atmosphere in order to synthesize malic acid which they use as a source of CO_2 in light when stomata are closed. This type of behaviour of carbon recycling is not seen in the mangroves.

Table 13. The number of stomata per mm^2 for the mangrove species investigated are as follows –

13(a). Mangroves having stomata on the lower side

(i) R. mucronata	166
(ii) B. gymnorhiza	247
(iii) A. ilicifolius	544
(iv) K. rheedei	223
(v) A. majus	115
(vi) E. agallocha	234

13(b). Mangroves having stomata on both the surfaces

	(UF) Upper Surface	(LF) Lower Surface
(vii) S. acida	71 U.F	78 L.F.
(viii) L. racemosa	87 U.F.	78 L.F.
(ix) S. persica	239 U.F.	236 L.F.

Lewis and Naidoo (1970) worked on stomatal structure in *A. marina* and found that transpiration rates are higher in midmornings and low in the afternoon. The results of our investigations clearly show the reasons for low transpiration in the afternoon i.e. the closure of stomata in the afternoon. It is clear that maximum CO_2 incorporation in the mangroves can take place only in the mornings. This is obvious as mangroves are tropical plants which are exposed to bright sunlight. The environmental conditions and temperatures will favour maximum photosynthesis during early hours. The water economy necessiates closure of stomata in the afternoon when temperatures are high and controls carbon economy of the plant.

(b) Initial products of carbon photoassimilation

It is clear that aspartate and alanine play an important role in carbon photoassimilation of the mangroves. No less than five mangrove species exhibit this feature and it is possible that others may have similar path of carbon photoassimilation. Investigations on these lines are in progress in our laboratory and will throw light on the problem in the near future. In the mangrove radioactivity is much more in aspartate and alanine than in the PEP + PGA fraction. This is consistent with our observation on marine algae in which several forms synthesize aspartate and alanine as immediate products of carbon assimilation.

Recently Shomer–Ilan and Waisel (1973) have exposed *Aeluropus litoralis* Parl to $^{14}CO_2$ for a period of fifteen to sixty seconds in the presence or absence of NaCl. After fifteen seconds of photosynthesis, highly labelled spots of PEP and PGA appeared on the chromatograms as well as labelled aspartate, alanine and serine. When the products of early photosynthesis in *Aeluropus* were compared with C_4 plant, *Chloris gayana*, they were found to be similar. These similarities indicate C_4 type of photosynthesis in the saline halophyte. Our results on the mangroves synthesize a C_4 acid, aspartic acid, as the immediate product of photosynthesis. Besides aspartate, alanine appears as another metabolic product of early photosynthetic carbon assimilation. Our results are more significant as they are for less than ten seconds and in most cases for five seconds. These results suggest that the mangroves are C_4 type of plants and possibly of 'aspartate formers'. Walker (1971) has reported that in some C_4 plants aspartate is apparently a more important short term fixation product than malate and the mangroves appear to be of this type of C_4 plants.

The objection to call mangroves as C_4 type of plants is the absence of 'Kranz' type of anatomy in the plants of littoral swamps. We have studied the anatomy of several mangrove species and have not found the architecture of the leaf as essential for C_4 type of photosynthesis. In C_4 plants bundle sheath cells contain large sized chloroplasts which follow C_3 type of photosynthesis and produce sugars. However, the trend of recent investigations is that all plants following C_4 type photosynthesis may not have 'Kranz' anatomy. Crookston and Moss (1970) have shown that even in C_3 plants there is a well developed bundle sheath and that too with large chloroplasts. Laetsch (1974) has also discussed the problem of C_4 photosynthesis and Kranz anatomy. He states that in the course of evolution plants of xeric and saline environment evolved structural and functional adaptation to assimilate and retain carbon in a more efficient manner and hence developed C_4 type of photosynthesis. It is possible that the mangroves resorted to C_4 type of photosynthesis without acquiring 'Kranz' anatomy. The method for efficient carbon assimilation was a necessity because the available time for photosynthesis is only a period of 2–3 hours in the morning.

Shomer–Ilan and Waisel (loc. cit) who found the products of Calvin cycle in C_4 saline *Aeluropus* with ten seconds suggest that inspite of Kranz anatomy in the grass the machinery for C_3 and C_4 type of pathways can exist in the mesophyll cells. This has also been stated by Coombs (1971). Our results on marine algae (Joshi *et al.*, 1974) also show the possible operation of both the cycles in the same cell or chloroplast. Based on these evidences it may not be wrong to conclude that the mangroves are C_4 plants without 'Kranz' anatomy. In order to have more proof to support this hypothesis investigations on photorespiration and oxygen inhibition of photosynthesis in mangroves are under progress (Joshi *et al.*, 1962).

It appears that CO_2 which enters through stomata during the early hours of the day in the mangroves is assimilated by PEP–Case as this is a faster assimilator of the gas than RuDP–Case. The immediate product is oxaloacetate which is immediately converted into asparate as the plants have possibly more amino transferase than malic dehydrogenase which is inhibited under saline conditions (Joshi *et al.*, 1962 and Osmond and Greenway, 1972). The early formed asparate is utilized for the synthesis of alanine and it provides carbon for the synthesis of sugars by Calvin cycle reactions. Our enzymatic studies, on mangroves which will be discussed in the following chapter, clearly show that these plants have more PEP–Case activity than RuDP–Case. The investigations on marine algae have shown that under saline conditions, in the prevailing concentration in the cellular environment, RuDP–Case cannot operate efficiently and the main functional enzyme is PEP–Case. It is possible that in the mangroves similar conditions exist and PEP–Case has the main role to play. Shoemer–Ilan and Waisel (1973) have also known that PEP–Case is a salt dependent enzyme in halophytes and is more operative than the RuDP–Case. The enzymatic picture so far available in the mangroves indicate that the machinery for synthesis and utilization of asparate by C_4 pathway is present in the mangroves. Our investigations of malic enzyme pyruvate–Pi–dikinase, enzyme exclusive of C_4 plants, and more amino transferases in the mangroves are in progress and the results when available should throw light on the pathway of carbon assimilation. Like Shoemer–Ilan and Waisel (*loc. cit.*) we have undertaken studies of the utilization of labelled asparate in the mangroves and the preliminary results suggest that like *Aeluropus* the mangroves can utilize asparate. This supports the hypothesis that the mangroves have the necessary machinery to utilize asparate.

Another consistent feature of the initial products of photosynthesis in mangroves is high labelling in alanine. In *Aegiceras* and *Excoecaria* this labelling is more than is asparate. It has been reported by Hatch (1971) that in 'asparate former' type of plants asparate and alanine amino transferase have much higher activity than 'malate formers' and it is consistent as asparate replaces malate as a CO_2 donor in these plants.

Chen *et al.* (1971) have studied short term products in Bermuda grass after a 5 second exposure to $^{14}CO_2$ and they have found that 84% of the activity was recorded in asparate and malate. After transfer from $^{14}CO_2$ to air, the total radioactivity is decreased in asparate with an increase in 3–PGA and alanine while it remained constant in malate. The predominant labelling in alanine was interpreted to indicate that alanine was derived from 3–PGA. Kennedy and Laetsch (1973, 1974) have reported that the large amounts of alanine are produced in the first few seconds of photosynthesis in *Portulaca oleracea.* The normal precursor–product relationship (PGA —> pyruvate —> alanine) does not appear to operate in this species since labelling in pyruvate proceeds that in PGA. They further stated that not only *P. oleracea* but other species also synthesise appreciable amounts of alanine in the first few seconds. They suggest that alanine can originate from pyruvate by transaminase reaction while the origin of pyruvate remains unknown. It appears that in *Portulaca* the path of carbon in photosynthesis is little different from that in C_4 plants. Our investigations on *P. oleracea* are also on similar lines and suggest alanine as the immediate product of photosynthesis in same plants. Johnson *et al.* (1971) have suggested that in asparate formers, alanine synthesis is in cytoplasm. However, rapid and heavy labelling in alanine in the mangroves can only occur if there is an immediate transfer of asparate from the chloroplast to cytoplasm.

In Kranz type of anatomy alanine synthesis originates from asparate and serves as a transfer pool of carbon from chlorophyll to cytoplasm and then to bundle sheath chloroplasts. As there is no definite Kranz type of anatomy in the mangroves alanine synthesis cannot play that role. Kennedy and Laetsch (1974) attribute a different role to alanine and that is acting as an immediate product. We had supplied labelled asparate to the cell free system of mangroves and it was converted into alanine. This suggests that alanine in mangroves can originate from asparate and only enzymatic study involving asparate decarboxylase can confirm it and attempts in this direction are in progress.

(d) Dark $^{14}CO_2$ Fixation

The members of the Crussulaceae, Cactaceae, Liliaceae Bromeliaceae and Compositae, amongst many other families, exhibit the peculiar metabolic behaviour of accumulating an organic acid in their tissues when kept in the dark (Bennet–Clark, 1933). According to Calvin *et al.* (1951) malate and succinate are the major products formed in the dark in *Chorella.* Saltman *et al.* (1957) again investigated *B. pinnatum* and reported that malate and asparate which are formed in the dark originate from OAA. They have further investigated non–CAM plants like tobacco and showed that the major products of dark fixation in higher plants occur in the synthesis of organic acids and amino acids.

Joshi *et al.* (1962) have studied the effect of NaCl on dark fixation of CO_2 by marine and terrestrial plants. They have shown the predominant fraction of activity in the amino acids in marine algae, *Zosteria*, a marine angiosperm and in cell free spinach system in NaCl buffer. The cell free homogenates without salt can fix CO_2 in the dark with a pattern similar to that of the intact spinach leaf. However, in the presence of NaCl the pattern is altered and greater amounts of asparate are formed. There is a significant decrease in organic acids.

In the present investigation green leaves of *S. acida* were subjected to $^{14}CO_2$ fixation in the dark and the results obtained are recorded. It is evident that there is no sugar formation in the dark. From the total incorporation, amino acid fraction has an activity double that of the organic acid one. Malate is a predominant acid synthesized in the dark while glutamine, glycine–serine, glutamate, alanine are the highly labelled amino acids in dark CO_2 fixation. It is clear that in dark no sugars are formed. In non–saline plants organic acid fraction dominates (*Bryophyllum* and tobacco) while in the saline ones amino acid fraction has that role to play.

Joshi and Bartakke (1974) have reported that in *Aloe* (CAM plant) asparate is the initial product of short term fixation in light while it is malate in the dark. Both the initial products are utilized for synthesis of other metabolites. In this plant also organic acid fraction dominates over amino acids.

Clegg and Whittngham (1970) have recorded the immediate products of dark $^{14}CO_2$ fixation as malate and asparate in potato tubers. Akagawa *et al.* (1972) who studied *Ulva pertusa* have shown that radio carbon was incorporated much more into amino acid fraction than into the organic acid fraction in dark $^{14}CO_2$ fixation. Willmer and Dittrich (1974) have reported that the first products of $^{14}CO_2$ fixation by epidermal tissue of *Tulipa gesneriana* and *Commeliana diffusa* in the light and dark are malate and asparate. According to the findings of McLaren and Thomas (1965) organic acids were the main products of dark $^{14}CO_2$ fixation by green callus tissue of *Kalanchoe crenata*, but no activity was detected in sugars or sugar phosphates and greater quantities of malate, citrate and isocitrate were found. Usuda *et al.* (1973) in maize and Kluge *et al.* (1974) in *B. daigremontianum* and *Sedum pracaltum*, found malate synthesis during dark CO_2 fixation and the activity incorporated into malate was located greatly in Carbon 4 position.

Karmarkar and Joshi (1968) exposed NaCl treated leaves of *B. pinnatum* to $^{14}CO_2$ in the dark for 20 minutes and found that the rate of CO_2 fixation was altered by NaCl and the pattern of CO_2 fixation increased with the increase in the synthesis of amino acids. Winter *et al.* (1974) have studied the time course of $^{14}CO_2$ dark fixation in leaves of the facultatively halophytic, *Mesembryanthemum crystallinum*, cultivated with the withhold 400 mM NaCl in the nutrient medium. The result of previously mentioned investigations shows that plants grown under saline conditions, incorporate ^{14}C predominately into amino acids. However, studies by Winter *et al.* (1974) on *M. crystallimum* which was grown on NaCl substratum and exposed to $^{14}CO_2$ in dark indicate more incorporation in organic acids (especially malate) than in amino acids. The authors have given explanation for their contradictory results.

Karekar (1974) has reported that amino acid fraction is dominating over organic acid fraction under dark $^{14}CO_2$ fixation in *E. tubulosa*. Similar results were obtained in the present investigation. It was found that the amino acid fraction is double that of organic acid fraction in dark $^{14}CO_2$ fixation in *S. acida*. Among the organic acid fraction, malate is predominant while in amino acid fraction glutamate, glycine–serine, glutamine, alanine are dominant. The results suggest as reported by many previous investigators that in saline environment amino acid fraction of dark fixation of $^{14}CO_2$ predominates over organic acids one. The results on mangroves support the works of Bharucha and Rangnekar (1957), Joshi *et al.* (1962), Gee *et al.*, (1965), Webb and Burley (1965) and Karmarkar and Joshi (1968), but contradict Winter *et al.* (1974). The explanation for the observations of Winter *et al.* can be found in the results of our experiments on *Kalanchoe* which suggests that in the succulent synthesis of amino acids depends on the pool capacity of the plant. In *Kalanchoe* amino acid synthesis is less while it is more in *Aloe.* It is possible that *Mesembryanthemum* does not have the capacity for amino acid accumulation.

(e) Efficiency of carbon assimilation

Eleven species of the mangroves were exposed to $^{14}CO_2$ for a period of one hour to study the efficiency of carbon assimilation. The experiment was performed from 9 to 10 a.m. near the mangrove vegetation under natural conditions of light and temperature. Equal amounts of buffer, leaf tissue and radioactive material ($NaH^{14}CO_3$ specific activity 4.3 mCi/mmole) were used so that the efficiency for carbon assimilation could be calculated under identical conditions.

The total chlorophylls, radioactivity per mg of chlorophyll and tissue and percentage of utilization efficiency have been recorded. It is clear that there are great variations in $^{14}CO_2$ fixation in the mangroves investigated. If consider total activity expressed per mg of tissue per hour, then the highest incorporation is found in *A. officinalis*, while, *A. alba* is the next best. In *B. gymnorhiza* least activity was recorded. Based on these observations the utilization efficiency was calculated and it was found that both the species of *Avicennia* namely, *A. officinalis* and *A. alba* are the most efficient carbon assimilators. *E. agallocha* and *S. acida* are third and fourth in the order of efficiency of carbon utilization.

It is evident that there are great variations in the chlorophyll contents of the mangroves and the values range from 52 to 167.61 mg per 100g fresh tissue. The lowest chlorophylls are recorded in *S. persica* and the highest values are found in *E. agallocha.* If express the radioactivity in terms of chlorophyll then *S. persica* appears to be the first and the most efficient while *A. alba* remains second on the score. However, for a plant what is more important is the carbon fixed in mg of tissue rather than mg of chlorophyll. As *Salvadora* has less chlorophyll its values are found to be more but that does not help its growth. The results expressed in terms of µmoles CO_2 fixed per mg of chlorophyll per hour do not change the pattern. *S. persica* and *A. alba* remain efficient. These results clearly indicate that both the species of *Avicennia* are more efficient than the other mangrove species. *B. gymnorhiza* showed the least incorporation among the mangroves.

Moore *et al.* (1972, 1973) have reported that *Rhizophora* plants generally show lower photosynthetic rates than either *Avicennia* or *Laguncularia.* Further, they report that in the environment in which all these three species are capable of growing. *Rhizophora* is by far the most abundant. In this instance dominance cannot be implied from potential net photosynthesis.

The results are similar to those of Moore *et al.* (*loc. cit.*) and indicate that both the species of *Avicennia* are more efficient in photosynthetic carbon fixation. At Ratnagiri and Terebhol *Rhizophora* is a common species, while *Sonneratia* shows dominance at Ratnagiri, though their photosynthetic rates are low. Dominance of *Avicennia* at Terekhol and Ratnagiri as well as throughout the Western Maharashtra shows its ability to withstand higher salinity conditions which is correlated with the efficient photosynthetic rate. It is difficult to explain the dominance of *Sonneratia* and *Rhizophora* in relation to their lower rates of photosynthesis. However, they are not as low as that of *Bruguiera* in which the photosynthetic rate is the least. This species is not found at Ratnagiri and Terekhol but it appears at Vashisti where the salinity is much diluted due to fresh waters from Koyana dam. It appears that the mangroves which have values over six usually dominate the true mangrove environments.

CONTRIBUTION OF PROF. B.E. CLOUGH

Mangroves, like other higher plants, depend on the photosynthetic reduction of carbon dioxide to form carbohydrates and other organic constituents necessary for growth and maintenance. An understanding of their photosynthetic characteristics and how these are regulated by internal and external factors is thus fundamental to any assessment of the growth and productivity of mangroves in different environments.

Unlike other tree species, mangroves commonly grow on saline, waterlogged soils. Despite being waterlogged, the high salinity and low osmotic potential of mangrove soils means that most mangroves experience moderate to severe tissue water deficits. This is so because the highest water potential that can be attained by mangroves growing in a soil solution surrounding the roots. Thus, for mangrove growing in a soil saturated with sea water, the highest predawn water potential will be about –2.5 MPa, a value that would be sufficient to cause a significant inhibition of photosynthesis and other physiological processes in nonhalophytic plants.

Photosynthetic Pathway

The nature of the pathway for CO_2 fixation in mangroves is still unclear. Joshi *et al.* (1974, 1975) reported the presence, in mangrove leaves, of phosphoenol pyruvate (PEP) carboxylase, the enzyme responsible for the carboxylation of PEP to oxaloacetate in the C_4 photosynthetic pathway. They also found that alanine and aspartate were heavily labelled after a short exposure of leaf material to carbon–14, concluding that mangroves employ the C_4 photosynthetic pathway. Consistent with this view, Joshi *et al.* (1984), using pulse chase techniques, reported that oxaloacetate was an intermediate in photosynthetic carbon fixation, as is required in the C_4 photosynthetic pathway.

On the other hand, Andrews *et al.* (1984) found that the ^{14}C and ^{13}C carbon isotope ratios in leaves of a large number of Australian mangrove species ranged from –24.6% to –32.2%. These values are typical of C_3 plants (Smith and

Epstein, 1971). Since it reflects discrimination against ^{13}C value has proved to be a most reliable character for distinguishing between C_3 and C_4 photosynthesis in higher plants. Furthermore, photosynthesis in mangroves typically saturates at moderate light intensities (Moore *et al.*, 1972, 1973; Andrews *et al.*, 1984 and Ball and Critchley, 1982), has a temperature optimum below 35°C and a CO_2–compensation point that ranges from 50 ppm to 90 ppm, depending on leaf temperature. These are all C_3 characteristics.

Thus, while there is some biochemical evidence to suggest that mangroves are C_4 plants, the carbon isotope ratio and gas exchange characteristics strongly indicate that mangroves, in common with other trees, are C_3 plants.

Photosynthetic Gas Exchange Characteristics

Light Response

Photosynthesis in mangroves shows a hyperbolic response to increasing light flux densities (Moore *et al.*, 1972, 1973; Attiwill and Clough, 1980; Andrews *et al.*, 1984 and Ball and Critchley, 1982). This response may be described approximately by the simple relationship (Thornley, 1976).

$$P = \alpha IP_m \; / \; \alpha I + P_m \qquad (1)$$

This is a rectangular hyperbola, where P is the route of gross photosynthesis, α is the quantum efficiency, given by the initial slope of the photosynthesis–light response curve at low light densities, I is the light flux density, and P_m is the maximum gross photosynthetic rate at saturating light flux densities.

In most cases the relationship between photosynthetic rate and light flux density is better described by a quadratic function of the form (Thornley, 1976).

$$P^z \theta - P(\alpha I + P_m) + \alpha IP_m = O \qquad (2)$$

a non-rectangular hyperbola, where the additional constant of θ allows greater flexibility in the shape of the light response curve. It should be noted that Eqn. 2 reduces to Eqn. 1 when $\theta = 0$. The light response curves obtained by Ball and Critchley (1982) and by Andrews *et al.* (1984) suggest that θ lies between 0.8 to 0.9.

The quantum efficiency, α, is much reduced by unfavourable environmental conditions, such as high leaf temperature, high salinity and water stress. Under optimum conditions (leaf temperature <30°C, salinity <20%, adequate nutrition), α is of the order of 0.05 µmol CO_2 per µmol quanta (Ball and Critchley, 1982 and Clough and Boto, unpublished), but may be as low as 0.02 under less favourable conditions (high leaf temperature, high salinity, nutrient deficiency) (Attiwill and Clough, 1980; Andrews *et al.*, 1984; Clough and Boto, unpublished and Bjorkman, unpublished).

Unfavourable environmental conditions also severely depress P_m. This parameter reflects the rate of supply of CO_2 and thus stomatal conductance, as well as the rates of the carboxylation reactions, either or both of which may be reduced by high leaf temperature, high salinity and high vapour pressure deficit (VPD) between the leaf and its environment (Ball and Farquhar, 1984a, 1984b; Andrews *et al.*, 1984 and Andrews and Muller, 1985). Thus the maximum light saturated photosynthetic rate may range from 5 µmol CO_2 $m^{-2}s^{-1}$ or less in very unfavourable conditions upto about 25 µmol CO_2 $m^{-2}s^{-1}$ in the most favourable conditions (Andrews *et al.*, 1984 and Clough, unpublished).

Recent work (Bjorkman, unpublished) has shown that photosynthesis may be strongly photoinhibited in mangrove leaves exposed to bright direct sunlight, particularly at high salinities, leading to a marked reduction in quantum efficiency. This can result in a significant reduction in the rate of photosynthesis in leaves exposed to direct sunlight at the top of the canopy. Other studies have shown that iron deficiency and perhaps other nutrient deficiencies that lead to leaf chlorosis, also reduce the quantum efficiency (Clough and Boto, unpublished).

The light compensation point appears to be relatively constant at about 40–50 µmol quanta $m^{-2}s^{-1}$ under a wide range of conditions (Ball and Critchley, 1982 and Andrews *et al.*, 1984).

Temperature Response

Studies in southern Florida (Moore *et al.*, 1972 and 1973) and in Australia (Andrews *et al.*, 1984 and Andrews and Muller, 1985) have shown that the rate of photosynthesis is much reduced at high leaf temperatures. In the few species examined so far the rate of photosynthesis appears to be relatively unaffected by leaf temperature over the range 17°C – 30°C, but falls sharply at temperatures much above 35°C and is close to zero at 40°C. The temperature response of photosynthesis in mangroves is thus similar to that of other C_3 plants and unlike that of C_4 plants, which generally have a higher optimum temperature for photosynthesis (Berry and Bjorkman, 1980).

As will be discussed later, high leaf temperature may influence photosynthesis indirectly through its effect on the vapour pressure deficit between the leaf and its environment. Apart from this effect, however, high leaf temperature also has an adverse effect on carboxylation reactions, with the result that the CO_2 compensation point rises with increasing leaf temperature (Andrews *et al.*, 1984). There is good evidence (see review by Berry and Bjorkman, 1980) that the decrease in carboxylation efficiency resulting from high temperature is due to stimulation of photorespiration in C_3 plants. This leads not only to a reduction in P_m, but also to a decrease in quantum yield (Ehleringer and Bjorkman, 1977).

Leaf temperatures well above the optimum for photosynthesis are common place in northern Australia on clear sunny days. Leaves exposed to bright sunshine in the tropics and held normal to the sun's rays may have leaf temperatures of 40–45°C, about 10°C higher than ambient, while exposed leaves at their normal erect orientation may have temperatures of upto 5°C above ambient (Andrews *et al.*, 1984 and Andrews and Muller, 1985). The erect inclination of exposed leaves of the Rhizophoraceae and some other mangrove species is thus clearly an important adaptation that minimises the thermal load experienced by leaves that are exposed to strong direct sunlight.

Photosynthetic Response to Vapour Pressure Deficit

It is well known that for a given stomatal conductance, the rate of water loss from a leaf increases steeply with increasing VPD between the leaf and its environment. In most cases, high rates of water loss lead to the development of leaf water deficits because the rate of water supply is less than the rate of loss, even in plants growing in saturated soils. In the leaves of most plant species this leads to partial or complete stomatal closure, associated with a concomitant decrease in photosynthesis. In addition, water deficits may have a direct effect on the carboxylation reactions (Farquhar and Sjarkey, 1982).

In practice, photosynthesis in mangrove leaves exposed to direct sunlight appears to be linearly proportional to VPD (Andrews and Muller, 1985), and there is increasing evidence that this relationship is due to co–limitation of photosynthesis by the stomata and by the rates of carboxylation reactions (Ball and Farquhar, 1984a, 1984b).

Effects of Salinity

Various studies have shown that a number of mangrove species grow best at salinities between about 4% and 15% (Connor, 1969; Clarke and Hannon, 1970; Downton, 1982; Burchett *et al.*, 1984 and Clough, 1984). However, there have been few studies of the effect of salinity on photosynthetic gas exchange in mangroves. The work of Joshi and his co–workers on the biochemistry of photosynthesis in mangroves at high salinity was mentioned earlier. Recently Ball and Farquhar (1984a, 1984b) reported that the rate of light saturated photosynthesis decreased with increasing salinity, attributing this to co–limitation of assimilation rate by stomatal conductance and photosynthetic capacity in response to differences in water status induced by the various salinity treatments. Thus, on the evidence available so far it is likely that salinity exerts its effect on photosynthesis mainly through changes in leaf water status.

Effects of Nutrient Supply

The effect of nutrient supply on photosynthesis in mangroves has been largely neglected, though nutrients such as iron, magnesium, nitrogen and phosphorus are known to be essential for photosynthesis in other plant species. A recent study carried out in Clough's laboratory has shown that low availability of iron can lead to a reduction in chlorophyll concentration and reduced quantum yield in leaves of *R. stylosa*. It is also likely that low availability of other nutrients may limit photosynthetic carbon fixation in some mangrove environments.

Photosynthesis, Water Loss and Salt Balance

As pointed out earlier, high salinities coupled with high rates of transpiration, lead to the development of appreciable leaf water deficits which at the very least may inhibit growth, and in more severe cases could be lethal. Furthermore, as already noted, high salinities coupled with high rates of transpiration lead to increased salt uptake and transport to the leaves, placing a greater burden on osmoregulatory processes in leaves and other organs. There is increasing evidence that minimisation of water loss and the achievement of a high water use efficiency are key elements in the ability of mangroves to exploit the intertidal environment. Andrews and Muller (1985), for example, obtained a value of 0.64 for the water use efficiency of *R. stylosa*, which is very high compared with the usual range for C_3 plants of 0.2–0.4, and approaching the range of 0.7–0.8 for C_4 plants.

The high water use efficiency of mangroves is achieved through close coupling between photosynthetic metabolism and stomatal conductance. This is reflected in an almost linear relationship between photosynthetic rate and stomatal conductance which, in turn, results in a more or less constant intercellular CO_2 partial pressure over a wide range of environmental conditions (Ball and Critchley, 1982; Ball and Farquhar, 1984a, 1984b; Andrews *et al.*, 1984 and Andrews and Muller, 1985). This close coordination between photosynthetic metabolism and stomatal conductance does not necessarily mean that photosynthesis is limited by stomatal conductance; on the contrary; there is increasing evidence that many environmental factors exert a direct effect on photosynthetic metabolism, which in turn regulates the stomatal conductance to maintain the intercellular CO_2 partial pressure close to the region of transition between linear and non–linear dependency of photosynthesis rate on intercellular CO_2 partial pressure (Ball and Farquhar, 1984a, 1984b); it has been shown that this leads to optimal water use efficiency (see review of Farquhar and Sharkey, 1982). Irrespective of whether environmental factors act directly on stomatal conductance or on photosynthetic metabolism, however, stomatal conductance is the operational link between photosynthesis, water loss and salt loading in mangroves. Morphological adaptations that assist in minimising water loss and optimising water use efficiency include thick cuticles, waxy coatings, and sunken or otherwise hindered stomata, which in all but a few genera (e.g. *Sonneratia, Lumnitzera, Osbornia, Laguncularia*) are found only on the abaxial surface of the leaf (Macnae, 1968; and Saenger, 1982).

From the foregoing discussion it is clear that salinity and irradiance are two key environmental factors regulating photosynthesis in mangroves. It should be noted here that salinity is thought to influence photosynthesis chiefly through its effect on leaf water status. Irradiance is important not only because of the direct dependency of photosynthetic rate on quantum flux density, but also because exposure of leaves to high irradiance leads to photoinhibition, particularly at high salinities. Furthermore, exposure to high irradiance increases leaf temperature, thereby increasing the VPD between the leaf and its environment and hence the rate of water loss. These considerations lead to the conclusion that high leaf temperatures and large vapour pressure deficits that inevitably accompany exposure of the leaf to bright sunlight in the tropics depress photosynthesis, either directly through an effect on electron transport or carboxylation, or indirectly through a reduction in leaf water status. In this sense, high salinities and environmental conditions that predispose the leaf to high rates of water loss have a synergic effect on photosynthetic production by mangroves.

Canopy Considerations

The degree of exposure and hence the extent to which a leaf experiences environmentally induced stress varies according to its position in the canopy. Leaves at the top or near the outer edge of the canopy are exposed to higher irradiance and greater thermal stress than those in more shaded situations within the canopy.

The interception of light by a closed mangrove canopy is influenced significantly by the architecture, or spatial arrangement of the canopy. As a first approximation, the vertical attenuation of light within the canopy can be described by Beer's Law.

$$I = I_O [1 - \exp(-kL)]$$

where, I is the photon flux density at the bottom (or some other, height) of the canopy, I_O is the photon flux density incident on the top of the canopy, k is an average light extinction coefficient for the canopy and L is the leaf area index.

The extinction coefficient, k, varies with leaf angle, being about 0.3–0.4 for erect leaves and closer to 0.8–0.9 for horizontal leaves. Since the orientation of the leaves of many mangroves, notably members of the Rhizophoraceae, changes progressively from erect to horizontal with increasing depth in the canopy, the extinction coefficient also increases with depth in the canopy. Furthermore, in some mangroves, again notably the Rhizophoraceae, the leaves are clustered in rosettes, often with large gaps between individual rosettes on the same branch. This heterogeneity in the spatial arrangement of leaves may lead to appreciable error when using a simple Beer's Law relationship to interpret light profiles within mangrove canopies.

While the steep inclination of exposed leaves of many mangrove species, especially members of the Rhizophoraceae, undoubtedly minimises photoinhibition and other radiation–related stress, nevertheless exposure of leaves near the top of the canopy to high irradiate could reduce photosynthesis below its potential maximum rate by as much as 20–30%. By contrast, photosynthesis in shaded leaves lower in the canopy is limited more by light than by water and thermal stress. The low light compensation point and relatively high quantum efficiency of shaded leaves often results in rates of photosynthesis comparable to those of leaves in more exposed situations (Clough, unpublished data).

These considerations lead to the conclusion that the formation of a uniform and well–developed closed canopy in mangrove forests is important for maximal photosynthetic production. In the case of isolated trees or seedling colonizing exposed mud flats, photosynthetic production per unit leaf area is likely to be substantially lower than that of trees forming part of a closed canopy forest. This is an important consideration in selecting the spacing between seedlings when replanting cleared areas of mangroves, where it is desirable to choose a spacing between seedlings that would allow the formation of a closed canopy within the first few years.

It is widely considered that mangroves grow best and are most productive in moist equatorial climates where the rainfall is distributed evenly throughout the year. This can be attributed partly to the lower salinities that usually accompany high rainfall, especially in estuaries with large catchments. More importantly, however, the persistent cloudiness that usually accompanies a high, evenly distributed rainfall pattern allows the maintenance of a favourable plant water balance because the leaves do not experience the severe thermal stress and high evaporative losses that may result from exposure to strong direct sunlight. In most cases, photosynthesis is unlikely to be much reduced directly by the lower light flux densities because of the higher quantum yield and because of the relatively low light flux densities required for light–saturation.

Salinity and the degree of cloudiness are likely to have a synergistic effect on photosynthetic carbon fixation, as well as, on other physiological processes associated with growth that are known to be sensitive to water stress or thermal stress. Unfortunately, very little is known about the effect of environmental stress on physiological processes other than photosynthesis in mangroves. These aspects need to be investigated for a full understanding of the role of environment in regulating mangrove growth and productivity.

On the basis of dominance of aspartate as initial photosynthetic product and trapping of phenyl hydrazine derivatives of OAA in the process and presence of pyruvate–pidikinase, Joshi *et al.* (1980), suggest the presence of Cu mechanism in several Indian mangroves.

However, Andrews and Clough (1980) studied the photosynthesis of many Australian mangroves and found out Isotopic discrimination of C_{13} by some 40 mangrove species, which proves that their photosynthesis is via C_3 pathways.

Adaptations of plants to saline conditions

The very good paper on 'adaptation of plants to saline conditions' was appeared in 1993 by K.H. Batanouny in "towards the rational use of high salinity tolerant plants". In this book, we are mentioning his contribution in few pages so that readers can understand the mechanism very clearly.

Adaptation and Mechanism

Salines are mainly characterized by high NaCl content and high osmotic potential. Other characteristic may also affect many salines, e.g. high concentrations of sulfates, carbonates, and bicarbonates, and high soil pH, as well as poor soil structure and disturbed aeration. Under the arid conditions which prevail in the Middle East, the aridity aggravates the problem created by salinity. These factors and characteristics which prevail in the salines of the arid zones have an apparently adverse effect on the nutrition, growth, and reproduction of most plants. Halophytes in arid regions, though having some characteristics in common with those in humid regions, have particular characteristics which enable them to evade and/or resist and tolerate salinity. Various adaptive mechanisms have been selected in halophytes during the course of their evolution. The ability of a plant to survive, grow and reproduce, or even survive without net growth, means that such a plant is endowed with particular adaptations enabling it to resist and/or tolerate salinity. Adaptations should appear and be genetically fixed, in addition to various acclimatizations, as a result of particular pressure imposed by different environmental conditions. These plants spend their lives between *Scylla* and *Charybdis*. The harmony between the various mechanisms and adaptations is precarious. Exclusion of particular ions and reduction of their absolute or relative concentrations in the cells and tissues of the plant are undertaken by a variety of mechanisms. On the other hand, the osmotic adjustment and the accumulation of salts is another mechanism. Under arid conditions, the plants should be endowed with mechanisms enabling them to keep their water balance positive, at least for some periods. The equilibrium between these various mechanisms, which occur in reality as indicated by the life of the plant, is not an easy or a simple one. However, the classification of mechanisms is necessary in order to facilitate our understanding of them and to evaluate the role of each of them.

Mechanism and adaptations have been classified by different investigators (e.g. Arnold, 1955; Waisel, 1972; Flowers, 1975; Breckle, 1986 and others). The main categories of these mechanisms and adaptations are shown in Table 13c.

Table 13c. Adaptive mechanisms of plants to cope with salinity stress

A.	**Avoidance mechanisms**	
	1.	Growth only during favourable low salinity seasons; such plants have short-time life cycles.
	2	Growth only on favourable sites with low salinity.
	3.	Limitations of root growth and root absorption activity to distinct soil horizons
B.	**Evasion mechanism**	
	1.	Selective against Na^+ and Cl^-
	2.	Diversion of the salts out of the assimilating tissue.
	3.	Compartmentation of salt within plant organs, tissues and cells.
	4.	Retranslocation of salts to roots and secretion by roots.
	5.	Leaching of salt out of the shoot.
	6.	Shedding of salt-loaded tissues and organs.
	7.	Secretion of salts by salt glands or bladders.
	8.	Production of adventitious roots and building mounds.
C.	**Tolerance mechanisms**	
	1.	Increasing of salt resistance of tissues, cell and organelles.
	2.	Production of protectants.
	3.	Osmotic adjustment.
	4.	Ion-specific adaptation of cytoplasm.
	5.	Increase of halo-succulence; leaf and stem succulence.

The halophytes growing in each habitat exhibit adaptations and acclimatization which may differ in nature and/or degree and efficacy. There is some sort of plasticity in the responses of plants to the different conditions, according to the severity of any particular condition. Each species has its own array of adaptive mechanisms. Though halophytes have attracted the attention of scientists for a long time, our knowledge of their adaptation is limited. In this study, there is an attempt to summarize the available knowledge of some of the adaptations and characteristics enabling halophytes to live under the adverse conditions of the salines of the arid zone of the Middle East.

1. Salt Secretion

Salt Glands

Salt secretion is one of the best known adaptive mechanisms which enable halophytes to regulate the salt content of their leaves and to cope with the salinity of their environment. This function is performed by special plant structures known since the last decades of the last century. These glands were discovered and described for various plant species by many investigators in the nineteenth century (De Bary, 1877; Volkens, 1884; Marloth, 1887). These studies were followed by Schtscherback (1910), Ruhland (1915), De Fraine (1916) and recently by Arisz *et al.* (1955), Zeigler and Luttge (1966), Luttge (1971), Batanouny (1973), Batanouny *et al.* (1977, 1990) and many others.

The salt glands have been discovered in 11 families including 17 genera of the dicotyledons and in one family of the monocotyledons, i.e. the *Gramineae* (Liphschitz and Waisel, 1982). These last authors state that the *Gramineae* include 16 genera, all of the *Chloroideae* subfamily, with active secreting salt glands and 17 genera of the *Panicoideae* subfamily with non–functioning or slightly functioning glands.

The dicotyledonous families which have species with salt glands are :

Acanthaceae	*Acanthus*	(Walter & Streiner, 1936)
Avicenniaceae	*Avicennia*	(Biebel and Kinzel, 1965)
Combretaceae	*Loguncularia*	(Biebel and Kinzel, 1965)
Convolvulaceae	*Cressa*	(Volkens, 1884)
Frankeniaceae	*Frankenia*	(Volkens, 1884)
Myrsinaceae	*Aegialitis*	(Scholander *et al.*, 1967)
Plumbaginaceae	*Aegialitis*	(Atkinson *et al.*, 1967)
	Limonium	(Arisz *et al.*, 1955)
	Statice	(Volkens, 1884)
	Limoniastrum	(Volkens, 1884)
Primulaceae	*Glaux*	(Rozema, 1975)
Rhizophoraceae	*Rhizophora*	(Atkinson *et al.*, 1967)
	Ceriops	(Walter & Steiner, 1936)
	Bruguiera	(Walter & Steiner, 1936)
Sonneratiaceae	*Sonneratia*	(Walter & Steiner, 1936)
Tamaricaceae	*Tamarix*	(Marloth, 1887)
	Reaumuria	(Volkens, 1884)

Structure of the glands

The structure of salt glands varies greatly in different plant species but it is usually similar in plants within one family. The simplest type of glands is found in the *Gramineae*. Glands are present on both sides of the leaves in longitudinal rows parallel to the veins. Each gland consists of two cells, a basal cell and a cap cell. The basal cell is the collecting cell whereas the upper cell is the excreting one. The cells contain dense cytoplasm and a prominent nucleus, but lack a central vacuole. The gland is characterized by cutinized and suberized cell walls (Fig. 14).

Multicellular glands comprising 16 cells are found in the *Plumbaginaceae*. The structure of the chalk (salt) glands of *Limoniastrum monopelatum* was incorrectly described by Volkens (1884) as consisting of eight cells surrounded by four subsidiary 'Nebenzellen'. Batanouny (1973) found that the gland is formed of 12 cells surrounded by four subsidiary 'Nebenzellen'. The twelve secretory cells are arranged in groups of four cells, separated by thin walls at right angles to each other and perpendicular to the surface. The four subsidiary cells surround the secretary cells. A pit is situated on the top of the gland and is surrounded by a transversal opening on the leaf surface. The twelve glandular cells have extremely thin walls, except, to a slight extent, the walls adjacent to the base of the gland. The glandular cells lack central vacuoles and have a granular cytoplasm and large nuclei. The outer walls of the subsidiary cells are strongly cutinized. Also, the upper sides of the gland and the neighbouring epidermal cells are covered by a thick cuticle. A rigid structure enclosed by cutinized walls is formed below the leaf epidermis in which the gland is firmly inserted (Batanouny and Abu Sitta, 1977).

Ecological factors affecting salt secretion

The rate of salt secretion is affected by numerous factors. These include : the salt concentration of the growth medium, ions which are present, light, temperature, oxygen pressure, the presence of metabolic inhibitors and different osmotica (Arisz *et al.*, 1955; Scholander *et al.*, 1962; Pollak and Waisel, 1970; Pollak, 1975; Batanouny *et al.*, 1990). In some salt secreting plants, e.g. *Avicennia marina*, it has been found that there was no correlation between transpiration and secretion rates (Waisel *et al.*, 1986). The latter authors state that such a correlation would have been expected if the salts brought in by the transpiration stream were the direct and sole source of secretion. The lack of such a correlation indicates that the salts which reach the leaves *via* the transpiration stream are sequestered in the cells prior to secretion. Additional evidence for this conclusion given by these authors is shown by the high secretion rates measured at night, inspite of the negligible transpiration at that time.

Though in some plants the quantitative contribution of the salt glands to the removal of excess salts is relatively small, secretion may still contribute to plant adaptation to salinity in two ways (Waisel *et al.*, 1986). These two ways are : (a) a quantitative contribution when the salt status of the leaves reaches a threshold level and secretion of even small quantities of salt will determine the survival of the plant, (b) there may be a qualitative contribution if the ionic composition of the leaves or the ratios between the essential and toxic ions become limiting. Such a change in Na^+/K^+ ratio may determine the activity of several key metabolic processes.

Kylin & Gee (1970) have shown that the Na^+/K^+ ratio has a clear effect on the optimum activity of ATPase of *Avicennia nitida* leaves. The selective removal of Na^+ and /or Cl^- may decrease the local ratios of Na^+/K^+, Cl^-/No_3^- or $Cl^-/H_3PO_4^-$ and enable the leaves to metabolize normally inspite of high total mineral content.

Selectivity of secretion mechanism for Na^+ and Cl^- against nutritive ions prevents the domination of Na^+ and Cl^- over these ions. Such selectivity has been shown for most species of salt-secreting plants (Waisel, 1961; Atkinson *et al.*, 1967; Rains and Epstein, 1967; Rozema *et al.*, 1981).

Shedding of salt-saturated tissues and organs

This is considered one of the mechanisms for removing considerable quantities of salts from the bodies of halophytes. In the case of some *Juncus* species as *J. maritimus* and *J. arabicus*, leaves are shed after being loaded with undesirable ions. The death and shedding of old leaves help the plant to get rid of considerable amounts of salts which were adsorbed during the life span of these leaves. It is interesting to know that the death of the leaves starts from the tip and proceeds downwards. The amount of salt content in the leaves is usually higher at the tip than the bases of the leaves.

In the case of articulate leafless chenopods, some branches are shed. In addition to this partial death for the sake of survival, the fleshy cortex is removed from the stems of old branches. This occurs as a result of the formation of phellogen deep in the cortex and the cork layers produced push the fleshy cortex, which is finally shed. The tissues of this succulent cortex with fleshy cells accumulate considerable amounts of salt during the life of the plant. The older the plant, the more concentrated will be the cortical tissues. Shedding of the cortex means discarding the undesired salts from the plant. Removal of the cortex also means the removal of a photosynthesizing tissue with stomata on the epidermis. Therefore the shedding of the cortex is also a mechanism by which the plant reduces its transpiring surface. This is a useful mechanism in halophytes growing under arid conditions. Shedding of the cortex is observed in *Arthrocnemum* spp., *Halocnemum* spp. and *Salicornia* spp. In halophytes with succulent leaves, e.g. *Salsola* and *Suaeda* spp., leaves are shed after being loaded with high salt contents. According to Chapman (1968), succulents such as *Allenrolfea, Halocnemum* or *Salicornia* discard portions of their fleshy cortex together with their leaves, thus releasing large quantities of salts from plants and enabling their survival. The old leaves of *Suaeda vermiculata* become black after being loaded with a high salt content and are shed, releasing considerable amounts of salts.

Osmotic adjustment

Plants growing under saline conditions must maintain a high concentration of osmotically active substance in order to compete successfully with the water-retaining capacity of the surrounding medium (Waisel, 19720. The importance of osmotic adaptation in terms of ion uptake is that in most halophytes the free energy of water is lowered by a massive increase in the ion content of the plant, which can reach very high values. Flowers (1975) gives examples that 20–40% of the dry weight in some species in the genera *Suaeda, Salicornia* and *Atriplex* may be composed of inorganic cells. The osmotic potentials of halophytes are significantly lower than those of glycophytes. More negative osmotic potentials are obtained in halophytes than in glycophytes.

The increase in negativity of the osmotic potential of cell sap associated with osmotic adjustment of plants to salinity is partly accomplished by the accumulation of ions; under certain conditions it can be achieved by the accumulation of large amounts of organic acids (Bernstein, 1961), However, the major components of the salt content are Na^+ and Cl^- ions. In an extensive survey of the literature, Arnold (1955) deducted that chlorides accounted for between 35 and 95% (average 65%) of the osmotic potential of halophytes, between 15 and 65% (average 35%) of marsh glycophytes, and between 0 and 65% (average 17%) of other glycophytes (cf. Flowers, 1975).

It is interesting to note that the high contribution of inorganic ions to the OP of the halophytes constituted another differential characteristic between halophytes and glycophytes (Onal, 1965, 1966). In some species of halophytes, e.g. *Atriplex* spp. organic ions may appreciably increase the cellular osmotic potential (Strogonov, 1964). Osmotic adaptation of plants, such as *Artemisia salina* and *Anabasis salsa*, results mostly from the accumulation of oxalates (Strognov, 1964). Oxalates play an important role as counteranions, especially in chenopods found in the semi–arid and arid zones. Oxalate levels increase with increasing Na^+ concentration in the medium (Williams, 1960). Na^+/K^+ ratios are generally high in halophytes in contrast to glycophytes, although there is a relative enrichment of K^+ in the plant, in comparison to the medium (Flowers, 1975). The enrichment of K^+ in the tissues relative to the medium is indicative of the ability of the halophytes to absorb essential elements in the presence of vast excesses of other elements.

It has been suggested that the distribution of ions within the cells is discontinuous, i.e. that the 'soluble phase' of the cytoplasm is relatively low in ions, while the vacuole contains the majority of those ions adsorbed (Flower, 1975). Waisel (1989) states that compartmentation of Na^+ and Cl^- inside cell vacuoles was claimed to be the major mechanism which enables normal metabolic activity in high–salt cells. This was based on the observation that many extracted plant enzymes lost their activity upon subjection, *in vitro*, to NaCl. However, this hypothesis could not be generalized, since the cytoplasm of various cells of salt–affected plants contains large quantities of Na^+ and Cl^-, also, the activity of various enzymes, previously thought to be salt sensitive, can be altered so as to become salt tolerant, or even salt stimulated (Waisel, 1989).

Osmotic adaptation differs not only between species but also between organs and even between cells. The osmotic adaptation is not a static phenomenon, periodic changes occur in the OP of plants.

Succulence

Exposure of halophytes to salinity results in numerous structural changes. These changes, as given by numerous investigators, include : thicker leaves, larger cells, especially those of the spongy parenchyma, smaller intercellular spaces, higher elasticity of the cell walls, fully developed water storing tissues, smaller relative surface area (surface / volume ratio), low chlorophyll content, and smaller and fewer stomata per surface area (cf. Stocker, 1933).

It has been believed that succulence is one of the major factors in salt tolerance. Increased succulence has a diluting effect on the toxic ion content of the cells, thus enabling plants to withstand the effects of larger quantities of salts. Waisel (1972) states that this system has its disadvantages due to the consequent lowering of nutritional ionic content.

Two types of succulent halophytes have been distinguished on the basis of their water relationships: succulent mesomorphs and succulent xeromorphs (Weissenbock, 1969). In the first type – typical of hydrohalophytes – all leaf cells, including epidermal cells, are succulent and the number of stomata per unit area is low (Biebel & Kinzel, 1985; Weissenbock, 1969).

The xeromorphs, which usually grow in habitats with high water stress, are characterized by succulent and large mesophyll cells, and enlarged cortex, small epidermal cells, and abundant stomata. Succulent xeromorphs are characterized by succulent and large mesophyll cells, and enlarged cortex, small epidermal cells, and abundant stomata. Succulent xeromorphs are characterized by a highly developed and a highly lignified vascular system. The cell sap of succulent xeromorphs usually contains more organic acids than the sap of succulent mesomorphs, which contains more inorganic ions. The well-known work of Van Eijk (1939) showed that sodium chloride is the most efficient salt in promoting succulence, and chloride rather than sodium plays the major role in its induction. Jennings (1968) states that halophytes differ from glycophytes with regard to their response to sodium, as a factor bringing succulence, only in degree, not in kind. It is to be noted that the effect of sodium parallels the effect of increased light and aridity. Jennings (1968) proposes that these three factors exert the same effect on plant cells, namely to bring about an increased ATP synthesis. Succulence is caused by the shift in ATP metabolism induced by sodium transport.

The ability of sodium to bring about increased succulence is two-fold. First : increased succulence has a diluting effect on the ion content of the cells which might otherwise rise to toxic levels. Second : sodium acts as a specific stimulant of growth which can be considered as tending to reduce the turgor pressure component of the water potential of the cell. The virtue of this growth response will be to increase the water potential of the leaf cells without the plant having necessarily to absorb more ions. The action of sodium can thus be envisaged as eliciting a homeostatic response within the plant to this otherwise toxic ion such that its toxic action tends to be nullified (Jennings, 1968).

A decrease in succulence, together with an increase in xeromorphic characteristics, were observed in plants exposed to sodium sulphate. Chapman (1974) states that there would be a definite relationship between the concentration of Cl^- SO_4^{2-} and the degree of succulence in typical halophytic succulents, the Cl/SO_4 ratio being strongly in favour of the Cl^- ion. Strogonov (1964) confirmed that the relative proportions of the Cl^- and SO_4 ions are important, the former promoting the development of halosucculent characters and the latter of haloxeric features.

Proline and other organic metabolites

Accumulation of compatible solutes under salt stress was thought to be a basic process of adaptation, especially for plants that do not accumulate inorganic ions for osmotic adjustment. Though some amino acids accumulate in plants to some extent, proline accumulates consistently in numerous plant species and under a wide range of environmental conditions. The metabolic effects of wilting which lead to proline accumulation have not been observed with any other amino acid (Stewart, 1981). Stewart and Lee (1974) highlighted the importance of proline in halophytes. These authors found that in most halophytes proline accounts for more than 30% of the amino acid pool (the only exception in eleven species tested being *Plantago maritima*). In a study by Batanouny *et al.* (1985) it was found that the diurnal value of proline in the twigs of the halophytic grass *Sporobolus virginicus* ranges from 20 μmole/g at 5 ppm (Fig. 18). It is interesting to note that the water conditions in the photosynthesizing (transpiring) organs are reflected in the proline content and its accumulation.

Batanouny *et al.* (1985) state that, though proline accumulates in the tissues of different plants with various ecological amplitudes as a response to water stress, yet the amount of proline could not be taken as a diagnostic feature for plants of different ecological groups. Negative relations between salts tolerance and proline accumulation have been reported (Mendelssohn, 1987).

Photosynthetic Pathways

It has been found that both C_3 and C_4 photosynthetic pathways occur in halophytes. While many succulent chenopods are C_4 plants, e.g. *Suaeda* spp., *Seidlitzea rosmarinus*, there are other succulent C_3 chenopods, e.g. *Arthrocnemum* sp., *Halocnemum* sp. and *Halopeplis* sp. Table 13d gives a list of C_4 and C_3 halophytes growing in different salines of the Middle East (cf. Frey *et al.*, 1983, 1985). Halophytic grasses which belong to the *Panicoideae* and the *Chloroideae* are salt resistant C_4 plants (cf. Liphschitz & Waisel, 1986). The salt tolerance of some of these species is achieved by salt excretion.

The presence of photosynthetic pathways in halophytic plants may be evidenced that the photosynthetic pathway is not correlated with the tolerance of or resistance to salinity. However, Leatch (1974) states that the primary adaptation of C_4 plants was probably to the saline environment and that the evaluation of structural and functional adaptations to such an environment were rather a prerequisite for efficient assimilation and growth processes.

The photosynthetic spectrum studied by Frey *et al.* (1985) showed a characteristic zonation of C_3 and C_4 species along a transact in littoral salt marshes in different places in SW Asia, in which moisture and salinity are the determining factors. The formation in which the C_4 pathway dominates are subject to stress with regard to both soil moisture and salinity. These are mainly xerohalophytes. On the other hand, the halophytic formations in which the C_3 pathway is prevalent only have to withstand salt stress. These are hygrohalophytes and mesohalophytes represented in the study of Frey *et al.* (1985) by the *Avicennia* and *Halopeplis* zone. Such plants accomplish their salt tolerance by means of different adaptations. It appears that both types of stress (salt stress and water stress) together are necessary if there is to be a selection advantage for the C_4 pathway (Frey & Kurschner, 1983).

Table 13d. A list of C_4 & C_3 halophytes growing in different salines in the Middle East

C_4 halophytes	
	Aleuropus lagopoides (L.) Thin ex Thw.
	A. littoralis (Gouan) Parl.
	Atriplex farinosa Forssk
	Halopyrum macronatum (L.) Stapf.
	Seidlitzia rosmarinus
	Sporobolus arabicus Boiss.
	S. spicatus (Vahl) Kunth.
	Suaeda fruticosa Forssk ex Gmel.
	S. monoica Forssk.
	S. pruinosa Lange
	S. vermiculata Forssk. ex Gmel.
C_3 halophytes	
	Arthrocnemum macrostachyum (Moris) Moris & Deplonte
	Avicennia marina (Forssk.) Vierth.
	Cressa cretica L.
	Frankenia pulverulenta L.
	Halocnemum strobiolaceum (Pall.) M.B.
	Halopeplis perfoliata (Forssk.) Bge ex Schw.
	Limoneum axillare (Forssk.) Ketze
	L. pruinosum (L.) Ketze
	Nitraria retusa (Forssk.) Aschers.
	Zygophyllum album L.

Though it has been recognized that some halophytes show a requirement for Na^+ as a micronutrient ion, yet it has been shown that the requirement for this micronutrient appears to be restricted to C_4 plants (Brownell and Crossland, 1972).

In most respects the effects of salinity on photosynthesis are mediated *via* water relations of the leaves and closely associated processes. It is therefore not surprising that the induction to CAM in succulent annuals such as *Mesembryanthemum crystallinum* in response to NaCl can also be brought about by any treatment which reduces water uptake, such as growth in nutrient solutions at low temperature or low O_2 partial pressure (Winter, 1979). Although these plants are most commonly found in arid saline habitats which are seasonally arid, the change in CO_2 fixation is a response to water stress, not to salinity per se (Osmond *et al.*, 1982).

Chapter – 4

FLORISTIC BIOLOGY AND BIODIVERSITY

Mangroves belong to many different families which may not be closely related in a phylogenetic sense but may have in common a number of special characteristics such as pneumatophores and a viviparous seeding habits.

The number and distribution of mangrove species are interesting biogeographical problems. In very early accounts of mangrove only *Rhizophora* species were considered as the mangroves. Firstly, Schimper (1891) described systematically about this vegetation and he listed 26 mangroves species belong to 10 families. Afterwards Davis (1940) and Lugo and Snedaker (1974) covered various aspects of mangroves in detail. However, Chapman (1976) described 11 mangroves families comprising of 55 species found exclusively in these swamps which he named as "true mangrove species". However, two families viz. Acanthaceae with three species of *Acanthus* and Sterculiaceae with two species of *Heritiera* were not included by him.

Recently, Saenger *et al.* (1983) created the working group on "mangrove ecosystems" to collate existing information on global status on mangroves. The experts of this committee had a consensus on 60 mangrove species as the "exclusive mangrove species" belonging to 21 genera and distributed over 16 families which are confined only to mangrove habitats and elsewhere too. Saenger *et al.* (1983) have described the distribution of mangrove forests in 6 geographical zones viz. –

(i) Asia
(ii) Oceania
(iii) West Coast of the Americas
(iv) East Coast of the Americas
(v) East Coast of the Africa and the Middle East
(vi) West Coast of Africa

These are two main centres of mangrove diversity. They have very dissimilar floristic inventories in both size and composition.

(i) *Eastern Group* (essentially the eastern hemisphere) — East Africa, India, South East Asia, Australia and Western Pacific. The total number of true mangrove species in this area is 40.

4.1 GENERALIZED DISTRIBUTION OF MANGROVES

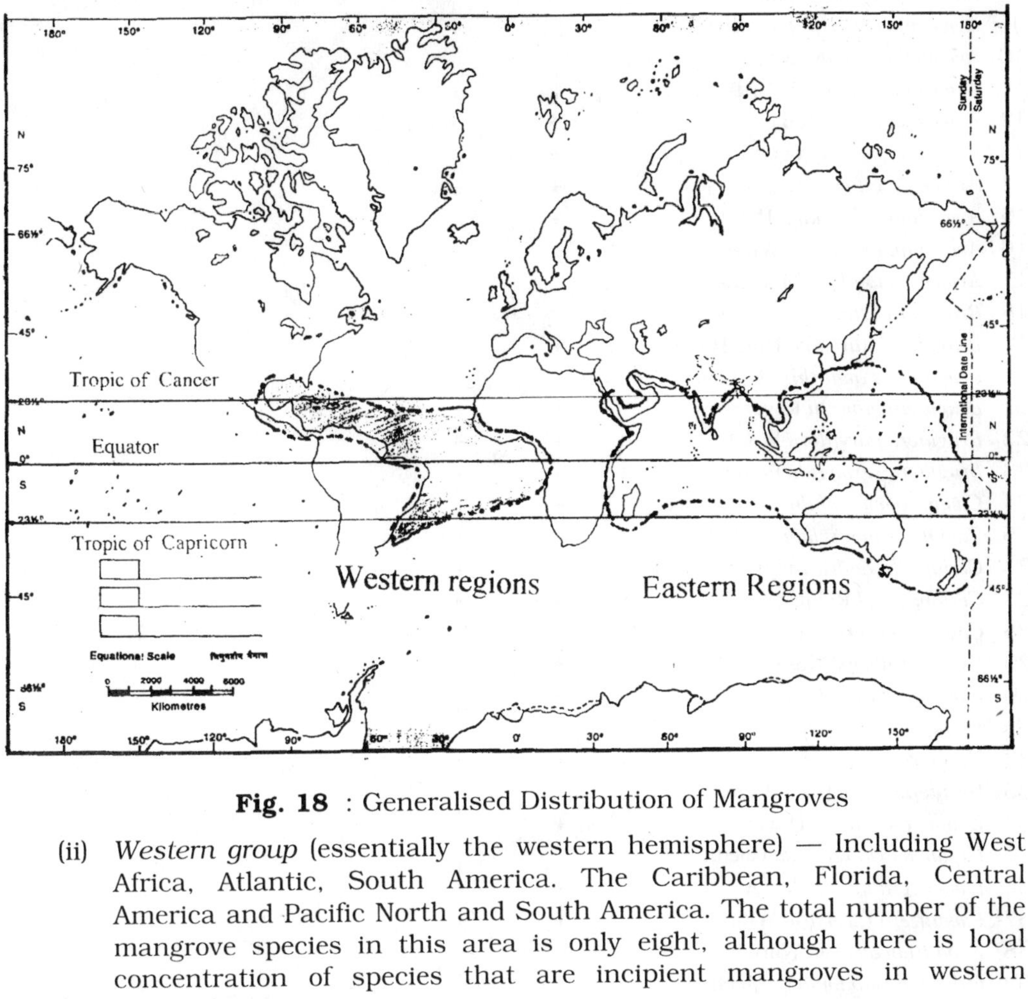

Fig. 18 : Generalised Distribution of Mangroves

(ii) *Western group* (essentially the western hemisphere) — Including West Africa, Atlantic, South America. The Caribbean, Florida, Central America and Pacific North and South America. The total number of the mangrove species in this area is only eight, although there is local concentration of species that are incipient mangroves in western Coloumbia.

Table 14. Geographical zones used to describe distribution of mangrove forests

S.No.	Name of the mangrove species	Asia	Oceania	WC Amer.	EC Amer.	WC Afri.	EC Afri. Me
		1	2	3	4	5	6
(A) Exclusive mangrove species							
1	*Acanthus ebracteatus* Vahl.	+	+				
2	*Acanthus ilicifolius* L.	+	+				
3	*Acanthus volubilis* Wall.	+					
4	*Aegialities annulata* R.Br.		+				
5	*Aegialities rotundifolia* Roxb.	+					
6	*Aegiceras corniculatum* L.Blanco	+	+				
7	*Avicennia alba* Blume	+	+				

		1	2	3	4	5	6
8	Avicennia bicolor Standl			+			
9	*Avicennia eucalyptifolia* Zipp.		+				
10	*Avicennia germinans* L.			+	+	+	
11	*Avicennia intermedia* Griff.						
12	*Avicennia lanata* Ridley	+					
13	*Avicennia marina* Vierh.	+					
14	*Avicennia officinalis* L.	+	+				+
15	*Avicennia rumphiana* Hall.f.	+	+				
16	*Avicennia tomentosa* Willd.		+				
17	*Avicennia tonduzii* Moldenke				+		
18	*Bruguiera cylindrica* L. Blume			+			
19	*Bruguiera exaristata* Ding Hou	+	+				
20	*Bruguiera gymnorhiza* L. Lam.		+				
21	*Bruguiera hainesii* Rogers.	+	+				
22	*Bruguiera parviflora* W&A	+	+				
23	*Bruguiera sexangula* Poiret	+	+				
24	*Camptostemon philippinensis* Bec.	+					
25	*Camptostemon schultzii* Mast.	+	+				
26	*Ceriops decandra* DingHou	+	+				
27	*Ceriops tagal* Robinson	+	+				+
28	*Conocarpus erectus* L.				+	+	
29	*Cynometra iripa* Kostel	+	+				
30	*Cynometra ramiflora* L.	+					
31	*Excoecaria agallocha* L.	+	+				
32	*Heritiera littoralis* Aiton ex Dry.	+	+				
33	*Heritiera fomes* Buch–Ham.	+					
34	*Kandelia candel* L.Druce	+					
35	*Laguncularia rcemosa* Gaertn.			+	+	+	
36	*Lumnitzera littorea* Voigt.	+	+				
37	*Lumnitzera racemosa* Willd.	+	+				+
38	*Nypa fruticans* van Wurmb.	+	+			+	
39	*Osbornia octodonta* F. Muell.	+	+				
40	*Pelliciera rhizophoracea* P&T.			+			
41	*Phoenix paludosa* Roxb.	+					
42	*Rhizophora apiculata* Blume.	+	+				
43	*Rhizophora harrisonii* Leechman.			+	+	+	
44	*Rhizophora lamarckii* Montrouz.		+				
45	*Rhizophora mangel* L.		+	+	+	+	
46	*Rhizophora mucronata* Lam.	+	+				+
47	*Rhizophora racemosa* G. Meyer.				+	+	
48	*Rhizophora selala* Tomlinson		+				
49	*Rhizophora stylosa* Griff.	+	+				
50	*Scyphiphora hydrophyllacea* Gaer.	+	+				
51	*Sonneratia alba* J. Smith	+	+				+
52	*Sonneratia apetala* Buch–Ham.	+					
53	*Sonneratia caseolaris* L.Engl.	+	+				

		1	2	3	4	5	6
54	*Sonneratia griffithii* Kurtz.	+					
55	*Sonneratia ovata* Backer	+	+				
56	*Xylocarpus australasicus* Ridley	+	+				
57	*Xylocarpus gangeticus* Parkinson	+					
58	*Xylocarpus granatum* Koeing	+	+				
59	*Xylocarpus moluccensis* Roem.	+	+				
60	*Xylocarpus parvifolius* Ridley.	+					
	Total	**44**	**38**	**7**	**7**	**7**	**9**
B. Some important non–exclusive species							
1	*Acrostichum aureum* L.	+	+	+	+	+	+
2	*Acrostichum danaefolium* L&F			+	+		+
3	*Acrostichum speciosum* Willd.	+	+				+
4	*Barringtonia racemosa* Roxb.	+	+				
5	*Brownlowia argentata* Kuntz.	+	+				
6	*Brownlowia tersa* (L) Koestm.	+	+				
7	*Cerbera floribunda* K. Schum.	+	+				
8	*Cerbera manghas* L.	+	+				
9	*Clerodendron ineme* (L.) Gaertn.	+	+				
10	*Cynomerta manni* Oliver					+	
11	*Dimorphandra oleifera* T&H				+		
12	*Dolichandrone spathacea* K. Schu.	+	+				
13	*Hibiscus hamabo* Sieb & Zucc.	+					
14	*Hibiscus tillaceus* L.	+	+	+	+	+	+
15	*Mauritia flexosa* Linn.			+	+		
16	*Meytenus emarginata* DingHou		+				
17	*Myristica hollrungii* Warb.		+				
18	*Oncosperma filamentosa* Bl.	+					
19	*Pemphis acidula* Forster	+	+				+
20	*Pterocarpus officinalis* Jacq.				+		
21	*Thespesia acutiloba* E&M.						+
22	*Thespesia populnea* Soland	+	+		+	+	+
23	*Thespesia populneoides* Kostel		+				
	Total	**14**	**14**	**4**	**7**	**4**	**7**

Table 15. Distribution of Exclusive mangrove species in 6 Geographical areas

S.No.	Regions	Number of species	% Occurrence
1	Asia	44(46) Singh & Garge (1993)	76.6
2	Oceania	38	63.3
3	West coast of Americas	07	11.6
4	East coast of Americas	07	11.6
5	West coast of Africa	07	11.6
6	East coast of African Middle East	09	15.0

The mangrove ranges differ globally quite clearly regarding numbers and diversity of species. Based on the number of woody plant species which may stand as an indicator of the total number of species occurring in the tidal forests, an eastern range encompassing the coasts and islands of the Indian ocean and the West Pacific with a large diversity of species is to be clearly distinguished from the western mangrove province with a much poorer diversity which includes the coasts of America and West Africa.

Biodiversity of mangrove species can be studied in following regions –

(i) Mangrove biodiversity of South and South East Asia.
(ii) Mangrove biodiversity of Australasia.
(iii) Mangrove biodiversity of Americas.
(iv) Mangrove biodiversity of Africa (West)
(v) Mangrove biodiversity of East Africa and Middle East.

(i) Mangroves Biodiversity of South and South East Asia

The country included in this regions are India, Indonesia, Japan, Pakistan, Malaysia, Myanmar, Philippines, Singapore, Srilanka, Thailand, Vietnam, Bangladesh, Cambodia, Hong Kong and Brunei Darussalem (Table 16).

In this zone, the richest countries are Indonesia, Malaysia, Taiwan and India. The total number of species in this zones are 51. The mangrove distribution and number is related to influx of fresh water. The total mangrove area in this region is some 75,173 sq. km., which represents 42% or the total area of mangroves in the world. The total length of coastline in the region in considerable, due to the very large number of islands in south–east asia. Over most of the regions coastlines are relatively sheltered and typed Tne mangrove biodiversity of this region is given in Table 16.

Table 16. Mangrove species list for South and South–East Asia.

Name of the mangrove species	1	2	3	4	5	6	7	8	9	10	11	12	13	14	15	16	17
Acanthus ebracteatus		•		•			•	•		•			•	•		•	•
Acanthus ilicifolius	•	•		•	•	•	•	•		•			•	•	•	•	•
Acrostichum aureum	•	•		•				•	•	•	•		•	•	•	•	•
Acrostichum speciosum		•		•				•		•				•		•	
Aegialities annulata								•			•					•	
Aegialities rotundifolia	•						•										
Aegiceras corniculatum	•	•		•	•	•	•	•		•	•	•	•	•	•	•	•
Aegiceras floridum								•					•				
Avicennia alba	•	•				•	•	•	•	•	•		•	•		•	•
Avicennia marina	•	•		•	•	•	•	•		•	•	•	•	•	•	•	•
Avicennia officinalis	•	•				•	•	•		•	•		•	•	•	•	•
Avicennia rumphiana								•		•			•				
Bruguiera cylindrica		•		•		•	•	•		•	•		•	•	•	•	•
Bruguiera exaristata		•	•	•	•	•	•	•									
Bruguiera gymnorhiza								•	•	•	•		•	•	•	•	•
Bruguiera hainesii								•		•						•	
Bruguiera parviflora		•				•	•	•									
Bruguiera sexangula	•	•		•				•		•	•		•	•	•	•	•
Camptostemon philippinense								•					•				
Camptostemon schultzii								•									
Ceriops decandra	•					•	•	•		•	•		•		•	•	•
Ceriops tagal	•	•		•		•	•	•		•	•	•	•	•	•	•	•
Cynometra iripa																•	

	1	2	3	4	5	6	7	8	9	10	11	12	13	14	15	16	17
Dolichandrone spathacea		•						•		•			•	•	•	•	•
Excoecaria agallocha	•	•		•	•	•	•	•	•	•	•		•	•	•	•	•
Excoecarcia indica	•							•		•	•			•			
Heritiera fomes	•						•				•					•	
Heritiera globosa		•						•									
Hieritiera littoralis	Ex •	•		•			•	•	•	•			•	•	•	•	•
Kandelia candel	•	•		•	•	•	•	•	•	•	•			•		•	•
Lumnitzera littorea		•		•			•	•		•				•		•	•
Lumnitzera racemosa		•		•		•	•	•	•	•			•	•	•	•	•
Lumnitzera x rosea													•				
Nypa fruticans	•	•	•	•			•	•	•	•	•			•	•	•	•
Osbornia octodonta								•		•			•				
Pemphis acidula								•	•				•	•	•	•	
Rhizophora apiculata	•	•	•			•	•	•		•	•		•	•	•	•	•
Rhizophora mucronata	•	•	•	•		•	•	•		•	•		•	•	•	•	•
Rhizophora stylosa				•				•	•				•	•			•
Rhizophora lamarckii							•	•		•							
Scyphiphora hydrophyllacea		•		•				•		•			•	•	•	•	•
Sonneratia alba		•		•		•	•	•	•	•			•	•	•	•	•
Sonneratia apetala	•					•	•				•				•		
Sonneratia caseolaris	•	•	•	•		•	•	•		•	•		•	•	•	•	•
Sonneratia griffithii							•			•	•					•	
Sonneratia lanceolata								•									
Sonneratia ovata		•		•				•		•				•		•	•
Sonneratia gulngai		•						•		•							
Sonneratia X urama								•		•							
Xylocarpus granatum	•	•		•		•	•	•		•			•	•	•	•	•
Xylocarpus mekongensis								•						•		•	
Total	**21**	**29**	**5**	**23**	**6**	**20**	**28**	**45**	**11**	**36**	**23**	**4**	**30**	**31**	**23**	**35**	**29**

1. Bangladesh, 2. Brunai Darrusalem, 3. Cambodia, 4. China & Taiwan, 5. Hongkong, 6. India-west, 7. India-east, 8. Indonesia, 9. Japan, 10. Malaysia, 11. Myanmar, 12. Pakistan, 13. Philippines, 14. Singapore, 15. Srilanka, 16. Thailand, 17. Vietnam.

Courtsey from : World Mangrove Atlas

(ii) Mangrove Biodiversity of Australia

The countries included are – Samoa, Australia, Fiji, Guam, New Calendania, New Zealand, Papua New Gunea, Soloman Island, Tonga, Varnatu and Tuvalu. The total number of mangrove biodiveristy in this regions are 47. The richest regions are Australia and Papua New Guinea. The total mangrove area in this regions is some 18,789 sq. km. This represent some 10% of the global total area. Since 90% of the region's mangroves are restricted to Australia and Papua New Guinea. The mangrove biodiversity of this region is given in Table 17.

Table 17. Mangrove species list for Australia.

Name of the mangrove species	1	2	3	4	5	6	7	8	9	10	11	12	13	14	15	16	17
Acanthus ebracteatus				•								•					
Acanthus ilicifolius			•	•								•					
Acrostichum aureum												•					
Acrostichum speciosum			•	•	•							•					
Aegialities annulata		•	•	•	•							•					
Aegiceras corniculatum		•	•	•	•							•	•				
Avicennia alba						•		•			•	•	•				
Avicennia marina		•	•	•	•				•	•		•	•			•	
Avicennia officinalis												•					
Avicennia rumphiana												•					
Bruguiera cylindrica				•								•					
Bruguiera exaristata		•	•	•								•					
Bruguiera gymnorhiza	•		•	•	•	•	•	•	•		•	•	•	•		•	•
Bruguiera hainesii												•					
Bruguiera parviflora			•	•								•	•			•	
Bruguiera sexangula			•	•					•			•	•			•	
Camptostemon schultzii			•	•								•					
Ceriops australis			•	•								•					
Ceriops decandra			•	•								•					
Ceriops tagal		•	•	•		•			•		•	•	•			•	
Cynometra iripa				•								•					
Diospyros ferrea				•								•					
Dolichandrone spathacea				•								•					
Excoecaria agallocha			•	•	•	•	•	•	•		•	•	•	•		•	
Hieritiera littoralis				•		•	•	•	•		•	•	•	•		•	
Lumnitzera littorea			•	•		•	•	•	•		•	•	•	•	•	•	
Lumnitzera racemosa			•	•	•							•					
Lumnitzera x rosea				•								•					
Nypa fruticans			•	•		•		•			•	•	•				
Osbornia octodonta		•	•	•								•	•				
Pemphis acidula			•	•								•					
Rhizophora apiculata			•	•		•		•	•		•	•	•				
Rhizophora mucronata						•		•			•	•	•				
Rhizophora samoensis	•						•		•					•			•
Rhizophora stylosa		•	•		•	•	•	•	•			•		•	•	•	
Rhizophora lamarckii			•	•					•			•	•				
Rhizophora x selala							•		•								
Scyphiphora hydrophyllacea			•	•		•			•		•	•	•				
Sonneratia alba			•	•		•			•		•	•	•				
Sonneratia caseolaris				•								•	•		•		
Sonneratia avata			•	•								•					
Sonneratia gulngai												•					
Sonneratia X urama				•								•	•		•		
Xylocarpus granatum			•	•		•	•		•		•	•	•	•	•		
Xylocarpus mekongensis	•		•	•		•	•	•	•		•	•	•	•	•	•	
Total	**3**	**7**	**29**	**37**	**8**	**14**	**9**	**10**	**16**	**1**	**43**	**27**	**8**	**8**	**6**	**13**	**2**

1. American Sanoa, 2. Australia, SW, 3. Australia, NW, 4. Australia NE, 5. Australia SE, 6. F.S. of Micronesia, 7. Fiji, 8. Guam, 9. New Caledonia, 10. New Zealand, 11. Palau, 12. Papua New Gunie, 13. Soloman Islands, 14. Tonga, 15. Tuvalu, 16. Vanuatu, 17. Western Samoa.

Courtsey from : World Mangrove Atlas

(iii) Mangrove biodiversity of Americas

Mangroves found in this regions are – Brazil, Coloumbia, Costa Rica, Cuba, Ecuador, Salvador, French Guyana, Guatemala, Guyana, Jamaica, Mexico, Nicaragua, Panama, Peru, Peurto Rico, Surinam, Trinidad and Tobago, USA, Venezuela and Bermuda. One of the most notable features of the American mangroves is their low diversity when compared to those of the Indo–Pacific region. They are only 13 native species of mangrove formed in this region. The mangroves cover in this region are 49096 sq. km. which represent some 27% of the total area of mangroves in the world. This figure comes close to the estimate of 28.6% by Lacorda (1993). The list of the mangrove biodiversity of this region is given in Table 18.

(iv) Mangrove biodiversity of Africas (West)

In this region, the countries included are Angola, Benin and Togo, Cameroon, Congo, Gambia, Ghana, Guinea, Liberia, Nigeria, Senegal, Sierra Leone, Zaire. The total area of mangrove in the region is some 27995 sq.km. representing some 16% of the total global area. Mangrove forests occur on all coasts from Mauritania to Angola, but the best developed are those with the heaviest rainfall, namely the coast and river estuaries of Gambia, Guinea, Gabon and Nigeria (Table 19).

Table 19. Mangrove species list for West Africa

Name of the mangrove species	1	2	3	4	5	6	7	8	9	10	11	12	13	14	15	16	17	18
Acrostichum aureum	•	•	•		•		•	•	•	•		•		•	•	•		•
Avicennia germinans	•	•	•	•	•	•	•	•	•	•	•	•	•	•	•	•	•	•
Conocarpus erectus	•	•	•		•		•	•	•	•	•	•	•	•	•	•	•	•
Laguncularia racemosa	•	•	•		•		•	•	•	•	•			•		•	•	•
Nypa fruitcans																		
Rhizophora harrisonii	•	•	•				•	•	•	•	•	•		•	•	•	•	
Rhizophora mangle	•		•				•	•		•	•			•		•	•	•
Rhizophora racemosa	•	•	•	•		•	•	•	•	•	•	•	•	•		•	•	•

1. Angola, 2. Benin and Togo, 3. Cameroon, 4. Congo, 5. Cote d'Ivorie, 6. Equatoria Guinea, 7. Gabon, 8. Gambia, 9. Ghana, 10. Guinea, 11. Guinea-Bissau, 12. Liberia, 13. Mauriania, 14. Nigeria, 15. S.T. and Principle, 16. Senegal, 17. Sierra Leone, 18. Zaire.

Courtsey: World Mangrove Atlas.

(v) Mangroves of East Africa and the Middle East

This region includes the Behrin, Egypt, Iran, Kenya, Madagascar, Maldives, Oman, Quatar, Saudi Arabia, Somalia, Sudan, UAE and Yeman.

The total area of mangrove in this region is some 10,024 sq. km. of 6% of the total global area. There are also relatively few species, which could be related to the fresh environmental conditions as much as to other historical and biogeographic factors. The total number of species in this region is 14. But most widely distributed is *Avicennia marina* (Table 20).

Table 18. Mangrove species list for the Americas

Name of the mangrove species	1	2	3	4	5	6	7	8	9	10	11	12	13	14	15	16	17	18	19	20	21	22	23	24	25	26	27	28	29
*Acrostichum aureum**			•																		•	•							
Avicennia bicolor																					•	•							
Avicennia germinaus	•	•	•	•	•	•	•	•	•	•	•	•	•	•	•	•	•	•	•	•	•	•	•	•	•	•	•	•	•
Avicennia schaueriana					•																								•
Conocarpus erectus	•	•	•	•	•	•	•		•	•	•		•	•	•	•	•	•	•		•	•	•	•		•	•	•	•
Laguncularia racemosa		•	•		•	•	•	•	•	•	•	•	•	•	•	•	•	•	•		•	•	•	•	•	•	•	•	•
Mora fruticans							•															•							
Nypa fruticans																						I*							
Pelliciera rhizophorae							•	•		•											•	•							
Rhizophora harrisonii					•		•	•		•			•							•	•	•	•	•		•			•
Rhizophora mangle		•	•	•	•	•	•	•	•	•	•	•	•	•	•	•	•	•	•	•	•	•	•	•	•	•	•	•	•
Rhizophora racemosa					•		•	•			•	•									•	•				•			•
Tabebua palustria							•	•						•								•							

1. Aruba, 2. Bahamas, 3. Belize, 4. Bermuda, 5. Brazil, 6. Cayman Islands, 7. Colombia, 8. Costa Rica, 9. Cuba, 10. Ecuador, 11. El Salvador, 12. French Guiana, 13. Guatemala, 14. Guyana, 15. Hispaniola, 16. Honduras, 17. Jamaica, 18. Lesser Antilles, 19. Mexico, 20. Netherlands Antilles, 21. Nicaragua, 22. Panama, 23. Peru, 24. Puerto Rico, 25. Surinam, 26. Trinidad & Tobago, 27. Turks and Caicos, 28. United States of America, 29. Venezuela.

* *Acrostichum danaeifolium* exists in most places that *Acrostichum aureum* occurs but it is rarely distinguished.

I*Introduced

Table 20. Mangrove species list for the East Africa and Middle East

Name of the mangrove species	1	2	3	4	5	6	7	8	9	10	11	12	13	14	15	16	17	18	19	20	21	22
Acrostichum aureum									•	•		•				•		•		•		
Avicennia marina	•			•	•	•	•	•	•	•		•	•	•	•	•	•	•	•	•	•	•
Bruguiera cylindrica										•												
Bruguiera gymmorrhiza								•	•	•	•	•				•	•	•	•	•		Ex
Ceriops tagal						•		•	•	•		•				•	•	•		•		•
Excoecaria agallocha										•												
Heritiera littoralia								•	•			•								•		
Lumnitzera racemosa		•						•	•	•		•				•	•	•		•		
Pemphis acidula		•								•		•				•						
Rhizophora mucronata					•	•	•	•	•	•	•	•			•	•	•	•	•	•		•
Rhizophora racemosa															I*							
Sonneratia alba								•	•			•				•	•			•		
Sonneratia caseolaris										•												
Xylocarpus granatum								•	•			•				•				•		

1. Baharain, 2. British Indian Ocean Terr., 3. Comoros, 4. Djbouti, 5. Egypt, 6. Eritrea, 7. Iran, 8. Kenya, 9. Madagskar, 10. Maldives, 11. Mauritius, 12. Mozambique, 13. Oman, 14. Qatar, 15. Saudi Arabia, 16. Seychelles, 17. Somalia, 18. South Africa, 19. Sudan, 20. tanzania, 20. United Arab Emirates, 22. Yemen.

Ex – Extinct in that country

I* – Introduced.

MANGROVE BIODIVERSITY OF INDIA

The first zone, Asia possesses 46 exclusive mangrove species and 14 non-exclusive mangrove species. In this zone, 28 exclusive mangrove species have been recorded from India (Blasco, 1977).

Mangrove forests are extensive and important ecosystem in the inter-tidal zone of India. They are very productive systems and they export large amounts of detritus food chain system that help support other along-shore and offshore ecosystems. Biodiversity and community pattern of mangrove ecosystems are closely related to diversity of coastal bio-climatic, geological and sociological conditions, frequency of inundation, circulation pattern and water quality of coastal area.

The magnitude and rate of change in diversity and community pattern may vary in space and time both within and between localities. However, in order to understand the dynamics of various mangrove distributional patterns of India, it is necessary to study the key factor which induce and change the environmental set-up of the ecological system. There are many factors which may control or influence the biodiversity and community pattern of mangroves. But there are some factors such as geo-morphological characteristic and geophysical energies such as river discharge, tidal amplitude and precipitation and together these factors may be important in determining the structure and composition of mangrove forest. The ecological status of the Mangroves of India is extremely complex due to the diversity of coastal geomorphology and climatic conditions.

Although, a good amounts of the work on the Indian Mangroves have been done by many workers (Navalkar, 1940; Bharucha and Navalkar, 1942; Blasco, 1975, 1977; Joshi, 1975; Bhosale, 1985; Rao *et al.*, 1972; Singh, 1989; Singh and Garge, 1993; Untawale, 1984, 1985 and 1987; etc.) in different climatic and geographical regions. In present description, an attempt has been made to evaluate a complete picture of mangrove forests of India in the light of biodiversity, community pattern and status under different climatic and phyto-geographical situations.

Mangrove Biodiversity

Mangrove are one of the dominant features of coastal ecosystems of the tropics. While ecologically they have been considered an important component of coastal watersheds, they are continually exploited for aqua-culture, agriculture, charcoal, urban development and other economic activities. Here, attention has been focused on biodiversity of Indian mangroves their ecological and economic status.

On the basis phyto-geographical position and regional situation, the mangroves of India has been grouped into 5 categories:

(i) Mangroves of Sunderbans and Ganga Deltas (SBG)
(ii) Mangroves of Andaman and Nicobar Archipelago (AN).
(iii) Mangroves of Deltas of rivers Mahanadi, Godavari, Krishna and Cauvery (MKGC).
(iv) Mangroves of Bombay, Ratnagiri, Goa, Lakshdweep, Karnatak and Kerala (BGK).

(v) Mangroves of Gulfs of Kutch and Cambay and Gujrat (GCK).

Mangroves forest of India is not uniformly distributed throughout the coast, maximum species are concentrated on east coast than west coast and to the Islands of Andaman and Nicobar. Maximum coastal area of peninsular India is devoid of mangroves due to over exploitation for urbanization, coastal development and marine pollution, only 8% of the total coastline is under mangroves. Sidhu (1963) estimated that the total mangrove area in the Indian territory is about 7,00,000 ha and regional distribution were as follows :

Table 21. Comparative data of mangrove areas of India

Major Zones	Waheed Khan (1959) sq. m.	Sidhu (1963) Ha	Blasco (1977) Ha	Blasco & Aizpuru (1997) Km^2
West Coast				
Gujarat	147.0	52,616	20,000	200
Maharashtra	96.0	62,208	20,000	200
Gulf of Kutch				
East Coast				
The Gangetic delta	465.0	4,18,888	2,00,000	2000
Orissa	1637.0	12,000	5000	50
Andhra Pradesh	74.5	18424	10,000	100
Tamil Nadu		2640	15,000	10
Andaman-Nicobar		1,15,200	1,00,000	440
Total	2419.5 (549950 ha)	6,81,976	3,56,500	3,000

(1) Mangroves of Sunderbans and Ganga deltas (SBG)

Mangrove forests of these regions have been studied by a large number of workers: [Anonymous, 1957; Mukherjee, 1975; Blasco, 1975; Chanda, 1977; Mitra and Banerjee, 1979; Naskar, 1981, 1983, 1984; Naskar and Guha Bakshi, 1982, 1983, 1984 & 1987; Chakraborty, 1993 etc.) and published many papers and books on different aspects of the mangrove forest of the Sunderbans.

The Sunderbans delta consist of low lying conglomeration of Islands about 56 in number separated by interconnected tidal rivers, creeks, and canals. The principal tributaries of the Ganga extended from the river Hoogly (West Bengal) on the west to the Meghna (Bangladesh) on the east and ingress of the sea (Bay of Bengal) exhibits a typical estuarine environment. The Sunderbans under the Indian shore, are located in the civil district of 24 Parganas of West Bengal. It falls little south to the tropic of Cancer, between the latitude 21°31'-22°30' N and longitude 88°10'-89°51' E. The climate of this region is tropical, winter is cool but never cold. Summer is warm but the day temperature rarely goes beyond 31.2°C.

(i) Mean maximum temperature °C	31.2
(ii) Mean minimum temperature °C	13.7
(iii) Relative humidity to up to	96%
(iv) Total annual rainfall is	1651.7 mm
(v) The average rainy days during the monsoon months	132

The present area of the Sunderbans is about 8.0 million hectares of which about 4.2 million ha land is under reserve forest, including the area under water. Blasco (1977) has mentioned about 2,00,000 ha mangrove area is under Sunderbans. The tidal amplitude is variable in Sunderbans delta river systems. It ranges from 3.0m to 8.7m from MSL depending upon the solar phase and position of the rivers. These flat lands were the subject of tidal inundation by saline river and sea water. The entire coastal strip of India including Sunderbans delta contains mostly saline soil. The Sunderbans saline soils are considered responsible for higher plant mortality because of high concentration of chlorides and sulphates of sodium, calcium and magnesium.

The mangrove forests of the Sunderbans cover the extensive tracts of swampy lands along with tropical sea coast, and always fringe with muddy salt water, creeks, lagoons and estuaries.

The dominant woody species are : *Aegialitis rotundifolia, Aegiceras corniculatum, Avicennia alba, A marina, A officinalis, Bruguiera gymnorhiza, B. cylindrica, B. parviflora, B. sexangula, Ceriops decandra, C. tagal, Excoecaria agallocha, Heritiera fomes, H. littoralis, Kandelia candel, Lumnitzera racemosa, Rhizophora apiculata, R. mucronata, Sonneratia apetala, S. caseolaris, Xylocarpus granatum and X. moluccensis along with two palm species – Nypa fruticans* and *Phoenix paludosa. Heritiera littoralis* and *Nypa fruticans* are found in abundant in this region. In highly exploited areas, there is dominance of *Acanthus ilicifolius* and *Avicennia officinalis* species associated at variety of soils. The natural forest is dominance of *Sonneratia apetala, Ceriops decandra* and *Avicennia marina* (Table 22).

Naskar and Guha Bakshi (1987) have recognized 5 types of community patterns in mangrove forest of Sunderbans.

(i) *Avicennia officinalis – A. alba – Sonneratia apetala* community
(ii) *Ceriops decandra – Excoecaria agallocha* community.
(iii) *Bruguiera gymnorhiza – Heritiera fomes* community
(iv) *Xylocarpus moluccensis – Rhizophora apiculata* community.
(v) *Nypa fruticans* community.

Nypa fruticans normally grows gregariously on the both side of the narrow canals of the creeks. At present, due to biotic interference, *Nypa fruticans* is now suffering threats and declining from Sunderbans. On the fresh deposited salt *Avicennia officinalis* and *Sonneratia apetala* are the fresh to come; while the next species are *Excoecaria agallocha* and *Heritiera fomes. Rhizophora mucronata* flourishes well on the saline zone also needs submerged conditions of high tides. *Ceriops* species is also flourish in similar situation.

(2) Mangroves of Andaman and Nicobar Archipelago (AN)

Mangrove forest of this region is studied by many workers (Sahni, 1959; Thothathri, 1960; Singh and Mall, 1980; Singh, 1989; Dagar *et al.*, 1991; Singh and Garge, 1933). The Andaman and Nicobar Islands consists of 410 Islands, separated by sea channels and creeks. This group of islands is situated in the Bay of Bengal between 6° to 14°N latitude and 92° to 94°E longitude. It has coastline of nearly 100 km which is 33% of the total coastal area of the Indian

Union. The topography of this terrain is mostly hilly, though some flatlands along the sea coast are seen. The rocks, in most of the cases consist of mainly non–calcareous, grey, sandstone and imbeded shells with occasional nests of poor coal, conglomerates and pale grey limestone.

The soil is generally soft and deep sandy loam varying from fine texture of alluvial flats to the gravel strewn soil of the low hills. The climate of islands is warm, humid with little temperature fluctuations.

(i)	Mean maximum temperature °C	29.9
(ii)	Mean minimum temperature °C	22.5
(iii)	Mean relative humidity in %	80.2
(iv)	Total annual Rainfall in mm	3180
(v)	The average rainy day / year	127

Parkinson (1923) has recognized 5 main types of vegetation in Andaman and Nicobar Islands –

(i) Mangrove forests
(ii) Beach forests
(iii) Evergreen forests
(iv) Moist deciduous forests
(v) Moist evergreen forests

The mangrove forests develop well in areas which are covered with brackish water during high tides. Mall *et al.* (1985) have recognized seven types of mangrove communities on the basis of species comopsition and dominance of species in diverse ecological conditions.

(i) *Rhizophora mucronata – Rhizophora apiculata* community.
(ii) *Bruguiera gymnorhiza – Ceriops tagal* community
(iii) *Rhizophora mucronata – Bruguiera gymnorhiza* community.
(iv) *Lumnitzera littorea – Avicennia marina* community
(v) *Bruguiera gymnorhiza – Avicennia officinalis* community
(vi) *Heritiera littoralis – Pongamia pinnata* community
(vii) *Acanthus ilicifolius – Acrostichum auream* comunity.

The mangrove forests of Andaman and Nicobar Islands are very rich in species diversity and and species composition. The height and girth are unimaginable in size. The dominant species are – *Rhizophora mucronata, R. apiculata,* associated with *Bruguiera gymnorhiza, B. parviflora, B. cylindrica, B. sexangula, Aegiceras corniculatum, Avicennia marina, A. officinalis, Excoecaria agallocha, Heritiera littoralis, Lumnitzera racemosa, L. littorea, Nypa fruticans, Phoenix paludosa, Rhizophora stylosa, R. lamarckii, Schyphiphora hydrophyllacea, Sonneratia alba, S. apetala, Xylocarpus granatum, X. moluccensis, X. mekogenesis.* The area under severe anthropogenic influences, are dominated by *Acanthus ilicifolius, A. ebracteatus, A. volubilis* and *Achrostichum aureum.* In Andaman, there are two types of mangrove association :

(i) Monogeneric mangrove forests – occupy by single *Rhizophora* species in fringe of creeks and cannals of many Islands. These forests, where major parts of the days are under sea water.
(ii) Mixed Mangrove forests – which occupy the flatland and lagoons of the many islands dominated by many species e.g. *Lumnitzera racemosa, L. littorea, Bruguiera gymnorhiza, Xylocarpus granatum* and other species.

The careful observation was made to find if vegetative regeneration take place. In many places trees were removed from the base and coppice was seen. Natural regeneration by germplasm was seen in case of *Bruguiera gymnorhiza, Rhizophora mucronata, R. apiculata, Avicennia marina, A officinalis, Sonneratia alba* in various Islands in undisturbed conditions.

(3) Mangroves of Deltas of Mahanadi, Krishna, Godavari and Cauveri rivers (MKGC)

Mangrove forests of these regions are studied by many workers (Rao, 1959; Rao *et al.*, 1985; Krishnamurty, 1984, 1985; Banerjee, 1985 etc.) The mangrove forests near mouths of rivers Mahanadi, Godavari, Krishna and Cauveri have not been explored adequately by any mangrove researcher. These four river deltas come under three states – Orissa, Andhra Pradesh and Tamil Nadu. The coastal region of these three states is also known as coromandal coast from Baleshwar to Kanyakumari. In recent estimates of mangrove forest of Andhra Pradesh, which are mainly located in deltas of Godavari (33,260 acres) and the Krishna (12,800 acres) estuaries. The extend of tidal flats of the deltas can be divided into three distinct regions such as –

(i) True estuarine tidal flats
(ii) Middle estuarine tidal flats.
(iii) Inner riverine tidal flats.

The vegetation of these deltas have been observed in a zonal pattern, distributed in gradual sequences from the river mouth of both side of the deltas are characterized by similar species and plant communities. In the community the dominant species are *Avicennia marina, Aegialities rotundifolia, Bruguiera cylindrica, B. parviflora, Sonneratia apetala, Ceriops tagal, C. decandra, Avicennia alba* which are able to tolerate maximum salinity conditions of the areas.

The middle part of the estuaries tidal flats, where land masses are notably influenced by innumerable meandering creeks and channels resulting in the formation of several frings and inlets, are dominated mostly by *Rhizophora apiculata, R. mucronata, Ceriops tagal, Ceriops decandra, Kandelia candel, Xylocarpus granatum, Bruguiera gymnorhiza, Aegiceras corniculatum* and other many species, which are adopted greater forces of the creeks and channels.

The most inland tidal flats of the inner estuarine regions which are subjected to greater fresh water influence and lower tidal action due to elevation, are characterized by remarkable brackish water mangroves community predominating *Nypa fruticans* and *Heritiera fomes*.

For many years, mangrove forests have been exploited in these regions. The poor appearance of the mangrove trees is due to excessive cutting and grazing. These practices have changed the nature of vegetation of these river deltas. *Avicenia* spp. was everywhere in good condition, this species was over–exploited for fuel purpose due to this another hardy species *Acanthus ilicifolious* replaced the areas. The natural regeneration is very poor and low. In some places lands became devoid of mangroves and coverage is less than 10%. These areas were gradually occupied by grasses and bushy species like *Suada* (Rao *et al.*, 1985).

The climatic condition of these regions is very hot and humid.

(i)	Mean maximum temperature in °C	31.4
(ii)	Mean minimum temperature in °C	23.7
(iii)	Mean Relative Humidity in %	85%
(iv)	Total amount of Rainfall	950 mm
(v)	The average rainy days / year	113

These delta are underlying serious hydrological, geomorphological and anthropic alterations. The vast deltaic mangroves of these regions, the perennial supply of the fresh water and admixture with sea water through tides via the coast and the mouths of the rivers have resulted in species richness. The diversity of species composition is evolved as a result of varying degree of adaptations to the salt of the sea water, brackish water mix by the species. The Mahanadi, the Godavari, the Krishna and the Cauvery systems are affected by biotic influence. The effect is more pronounced in Mahanadi within th past few years primarily due to human interference. The lack of periodical stock–taking of mangrove forestry has resulted in the disappearance of (a) various species; (b) enchroachment by man; and (c) reclamation for injudicious uses and purposes.

In west coast threatened species are *Nypa fruticans, Kandelia candel, Rhizophora* sp., *Heritiera* and *Phoenix paludosa.* These deltas are greatly affected by biotic features. Pichavaram mangrove ecosystem has been studied by Krishnamurthy (1985) in details. He stated that there was reason to believe that these mangroves had once extended from Marakhanam to the Gulf of Mannar region including Rameshwaram. However, they show only discontinuous distribution.

(4) Mangroves of Bombay, Ratnagiri (Maharashtra), Goa, Lakshdweep, Karnataka and Kerala (BGK)

Mangroves of these regions are fully explored by many workers (Navalkar, 1951; Bharucha and Navalkar, 1942; Bhosale 1974; Bhosale and Joshi, 1973; Bhosale *et al.*, 1983; Kokate, 1985; Untwale, 1984, 1985 & 1987; Untwale *et al.*, 1980; Jagtap, 1985; Rao, 1985;, Rao and Chakraborti, 1982; Rao and Nair, 1984, etc.). The total area under mangrove forest is not certain but recent census indicates that–

(i)	Bombay region	31,000 ha
(ii)	Goa	2,000 ha
(iii)	Karnataka	8,000 ha
(iv)	Lakshdweep	Nil
(v)	Kerala coast	1671 ha

The climatic condition of this region is very humid and rainfall is very high.

(i)	Mean maximum temperature °C	29.3
(ii)	Mean minimum temperature °C	21.4
(iii)	Mean relative humidity in %	89
(iv)	Total amount of rainfall in mm	813
(v)	The average rainy days / year	121

Mangrove forest of Bombay region is result of industrialization, biotic pressure, settlement of new colony. In Bombay city and its neighbourhoods like

Elephanta islands, the presence of tanneries (Mahim Creek) and discharge of the effluents has resulted in a large scale of destruction of mangrove forest of this regions. This area is mainly occupuied by stunted growth of *Avicennia marina* and *Acanthus ilicifolius* species.

The total estimated area of mangrove vegetation in Goa is about 0.3% to 0.5% of the total mangrove cover of Indian coast. When compared to other states having 3701 sq. km. are of which about 0.5% area is covered by mangroves. Total 60 species of mangroves have been identified in the world of which 45 species occur in Indian mangroves (Blasco, 1977). In Goa, 15 species belonging to 10 genera and 7 families were reported. The dominant mangroves are – *Rhizophora mucronata, Sonneratia alba, S. caseolaris, Avicennia officinalis, A. alba, Excoecaria agallocha, Kandelia rheedii, Ceriops tagal, Bruguiera gymnorhiza, B. parviflora* etc. Seven rivers occur along the Goa coast. The Mandovi and Zuari estuaries along with the Cumbarjua canal form the largest estuarine complex in Goa.

In Karnataka apart from encroachment for agriculture or aquaculture operations there is relatively little exploitation of fringing type mangroves which is not a continuous strip, but appear rathre in belts or patches. The present estimates of existing and prospective mangrove habitats are 8000 ha. In view of its physical and economic importance, attempts have been made since 1989 to increase their density by a comprehensive programme of protection, conservation, re-vegetation (afforestation) and restoration ecology. The ongoing project covers about 398 ha of mangroves plantation in Coondapur region. The sea board in Karnataka, in all is very much influenced by tidal rivers 'hole' and back water creeks. Among the principal estuaries of the seaboard are the Netravati–Gurpur estuary, the Mulky–Pavaye estuary, the Udyavara–Pagola estuary, the Swarna–Sita Kodi estuary, the Charaka–Haladi kollur estuary.

It has been noticed that the salinity of the water determines the composition of the mangrove vegetation of this regions. Grasses or sedges represent the pioneer successional stages. *Porteresia* in the down stream where salinity is more and *Cyperus* in the upstream where salinity is less. Following community has been recognized in Karnataka area (Rao, 1993).

(i) *Rhizophora–Bruguiera* community
(ii) *Avicennia – Sonneratia* community
(iii) *Sonneratia – Kandelia* community
(iv) *Excoecaria – Clerodendron* community and
(v) *Acanthus – Aegicaras* community

The chief component of the secondary formation is that of *Acrosticum aureum* often intrudes into the *Acanthus–Clerodendron* habitat. The succession of mangroves solely depends on the accretion of silt along the habitats. With increased accretion, *Rhizophora mucronta* is dominating. Under less turbulent tidal relief *Sonneratia caseolaris* and *Aegiceras corniculatum* are the common plants of the flora (Rao, 1993).

It is interesting to note that parts of Karnataka, like Kundapura, Mangalore, Karwar also once hoasted of a rich mangrove vegetation. As in Kerala, much area was reclaimed and lost irretrievably. Perhaps the least effected due to remoteness.

(5) Mangroves of Gulfs of Cambay and Kutch and Gujarat (GCK)

The mangrove forest of these regions is studied by Kulkarni and Junagadh, 1959; Blasco, 1975; Untawale, 1987; Singh, 1996. This coastal region is in the North–West of India, is the driest region in the country due to low rainfall.

(i)	Mean maximum temperature °C	32.4
(ii)	Mean minimum temperature °C	18.2
(iii)	Mean relative humidity in %	75
(iv)	Total amount of rainfall in mm	450
(v)	The average rainy days / year	80–90

In these regions mangals are very poor and very scattered along coastal regions. Mangals are the only source of firewood and also represent the main source of green fodder. Kulkarni and Junagad (1959) have stated that over 2500 camels belonging to Rabenis and estimated to be dependent on browsing in the mangrove forest of this tract. The most important woody mangrove *Avicennia* species is found in this region. The mangrove forest of these regions are unusual in that they occur in an arid climate. In Gujrat, the largest mangrove area is in the Gulf of Kutch, Bhuj, Jamnagar, Surat, Gulf of Cambay, stretching along the entire coast to the India and Pakistan specially Indus delta. The total area of the mangrove forest is nearly 158.8 km^2.

The major environmental constraint to effective solar salt production is the amount and distribution of rainfall. In general, it has been found that most successful salt operations are in climate where average annual rainfall is less than 1000 mm with at least 8 dry months. These condition in the huge sea salt industrial complex in the Gulf of Kutch, that is reason that mangrove forest of this region has been completely removed by salt maker.

Some mangroves are found in Gulf of Cambay and mouths of Narmada and Tapti but due to Asia's largest industrial estate and one of the 1990 industrial complexes. Vapi to Southern end of Gujrat, they are releasing heavy disposal of industrial waste toxic substances and garbage in mangrove areas resulting complete loss of mangrove species from these regions. The common species of mangrove found in this region are *Acanthus ilicifolius, Avicennia officinalis, A. marina, A. alba, Aegiceras corniculatum, Ceriops tagal, Bruguiera gymnorhiza* and *Rhizophora* species.

The Asia–Pacific region has the richest mangrove flora. As regard the centre of origin and place from dispersal of mangrove flora in this region are the Indo–Malayan. Form this place, mangrove species have been distributed in all directions. Perhaps this is one of the reason that eastern coast of India and northern part of Australia have greater numbers of mangrove species in comparison to western and southern parts of India and Australia respectively. The land mass was the greater barrier for migration of mangrove flora from eastern to western.

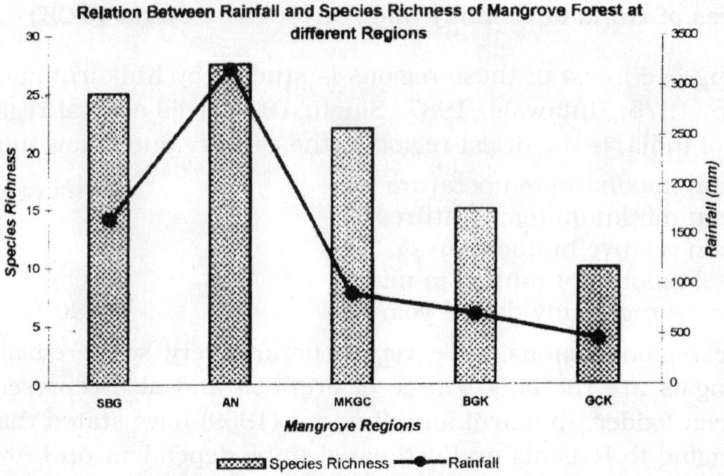

Fig. 19.

The eastern side come to have a greater (25) species composition in comparison to western side (12). These differences in species biodiversity are due to ecological, geo-morphological, phyto-geographical and hydrological conditions of mangrove forest. The biodiversity and species composition of 5 regions of mangrove forest of India have been studied in relation to climatic conditions and environmental set-up. It has been observed that (Fig. 19) the diversity of species in related to mainly the total rainfall of the phyto-geographical regions. The percentage of ground coverage and plant heights are also related to influx of fresh water (Fig. 20).

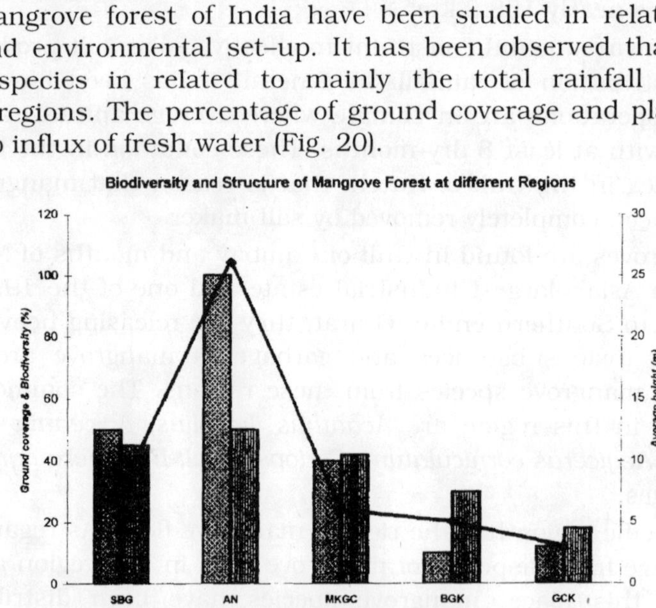

Fig. 20.

The loss of coverage and species of mangrove in any part of India is due to human activities (Fig. 21). If conditions prevailed for long duration than loss is more pronounced and severe, than only protection is the only measure for the conservation of mangrove forests of India.

Successional Pattern of Indian Mangrove

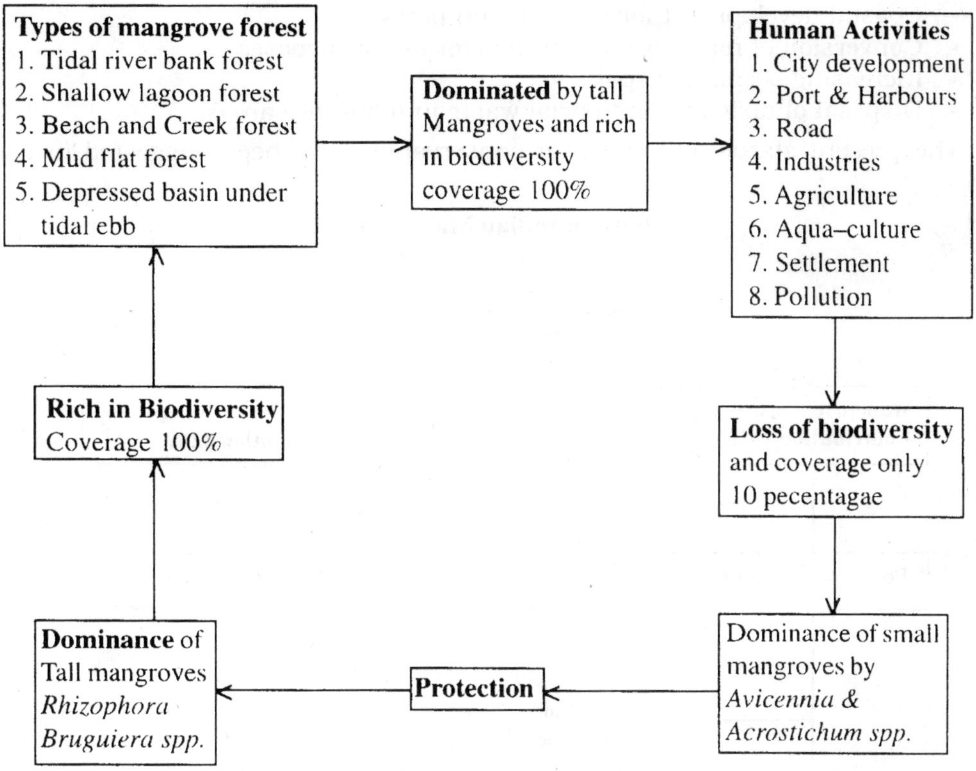

Fig. 21.

This study is based on the ecological and economic conditions of Indian mangroves. The survey of the global status of mangroves forest has found that vast area of mangroves has been destroyed or they are in process of destruction. The mangrove forests of India have under gone many drastic and in most instances irreversible changes. Throughout India, firewood and domestic fuel are decreasing in supply, while at the same time the demand is rapidly rising. To offset the deficit, mangrove forests are used directly by local people for firewood. In those states where annuaal eaxtraction in less than the annual re–growth of wood, the mangrove forests serve as sustained–yield resource and could exist in perpetuity in that state. In 1991, workshop on the status of mangroves of south–east Asian coastlines was held in Malaysia. It was concluded that this region has lost large areas of mangroves nearly 50%. And this pattern is likely to continue as greater demands are placed on forest and fisheries resources, along with land–use changes alaonga coastlines and in upland water–sheds, the result will necessarily cause a change in the ecological characteristics of tropical estuaries. Many of the species guilds and biodiversity components are sensitive to change in physical conditions. The main cause of the destruction of the mangrove ecosystems are –

- Over population near mangrove areas
- Increasing human settlement in mangrove areas
- Increasing urbanisation in mangrove areas.
- Coastal development for tourism and ports.
- Conversion of mangrove forest ares for other purposes.
- Increasing ocean transport
- Disposal of toxic and industrial waste in mangrove areas.

The present status and steps for conservation have been suggested in Fig. 22.

Status of Indian Mangroves

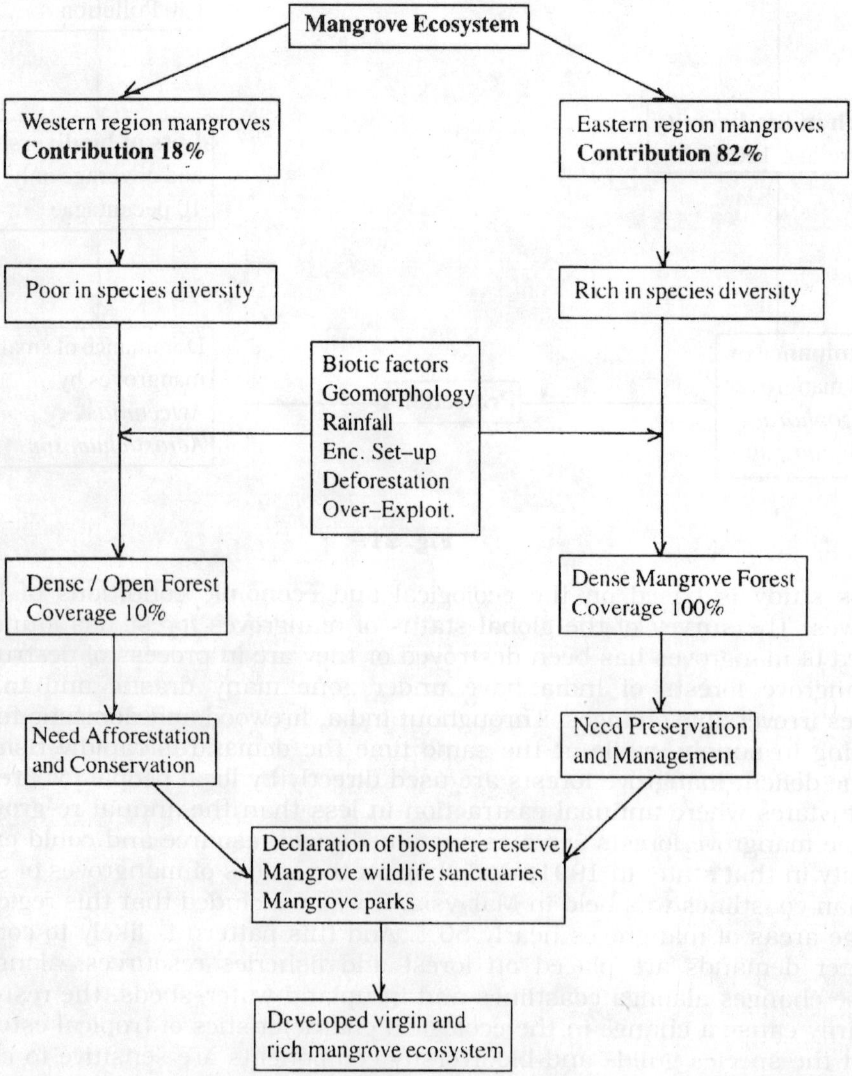

Fig. 22

Table 22. Biodiversity of Indian Mangroves at Different Regions

S.N.	Name of Species	SBG	AN	MKGC	BGK	GCK
A. Exclusive Species						
1	*Acanthus ilicifolius* vahl	p	p	p	p	p
2	*A. ebracteatus* L.	----	p	p	----	----
3	*A. volubilis* Wall	----	p	----	----	----
4	*Aegialitis rotundifolia* Roxb.	p	p	----	----	----
5	*Aegiceras corniculatum* Blanco	p	p	p	p	p
6	*Avicennia alba* Blume	p	----	----	p	p
7	*A. marina* Vierh	p	p	p	p	p
8	*A. officinalis* L.	p	p	p	p	p
9	*Bruguiera cylindrica* L. Blume	p	p	p	-----	-----
10	*B. gymnorhiza* L. Lamk	p	p	p	p	p
11	*B. parviflora* (Roxb.) W&A	p	p	p	p	-----
12	*B. sexangula* (Loir) Poviet	p	p	-----	-----	-----
13	*Ceriops tagal* (Per.) CBR	p	p	p	p	p
14	*C. decandra* (Griff.) Ding Hou	p	-----	p	p	----
15	*Cynometra ramiflora* L.	-----	-----	p	-----	----
16	*Excoecaria agallocha* L.	p	p	p	p	----
17	*Heritiera littoralis* Aiton	p	p	p	-----	----
18	*H. fomes* Buch Ham	p	-----	-----	-----	----
19	*Kandelia candel* (L.) Druce)	p	-----	p	p	----
20	*Lumnitzer littorea* (Jack) Voigt.	-----	p	-----	-----	----
21	*L. racemosa* (L) Gaertn.	p	p	p	-----	----
22	*Nypa fruticans* Van Warml.	p	p	p	----	----
23	*Phoenix paludosa* Roxb.	p	p	----	----	----
24	*Rhizophora apiculata* Blume	p	p	p	p	p
25	*R. mucronata* Lam.	p	p	p	p	p
26	*R. stylosa* Griff.	----	p	----	----	----
27	*R. lamarckii* Montrouz	----	p	p	----	----
28	*Schyphiphora hydrophyllacea* L.	----	p	p	----	----
29	*Sonneratia alba* J.Smith	----	p	----	p	----
30	*S. apetala* Buch Ham.	p	p	p	----	p
31	*S. caseolaris* L. Engl.	p	----	----	p	----
32	*Xylocarpus granatum* Koaing	p	p	p	----	----
33	*X. mekongensis* Parkinson	----	p	----	----	----
34	*X. moluccensis* (Lam) Roen	p	p	p	----	----
	Total	**25**	**28**	**23**	**15**	**10**
B. Non–Exclusive Species						
35	*Acrostichum aureum* L.	p	p	p	p	p
36	*A. speciosum* Wild.	----	p	----	----	----
37	*Afzelia bijuga* A. Grey	----	p	----	----	----
38	*Barringtonia racemosa* Roxb.	p	p	----	----	----
39	*Caesalpinia nuga* Aif.	p	p	p	----	----
40	*C. bonducella* Fleming	----	p	----	----	----

41	*Cerbera floribunda* K. Shum	p	p	p	----	----
42	*Clerodendrum inerme* L.	p	p	p	p	p
43	*Derris scandens* Benth	p	p	p	----	----
44	*Dolichandrone spathacea* Sch.	p	p	p	----	----
45	*Guettarda speciosa* Linn.	p	p	p	----	----
46	*Hibiscus tiliaceus* Slander	p	p	p	p	p
47	*Liacula spinosa* Vent	p	p	p	p	----
48	*Pandanus tectorius* slander	p	p	p	p	p
49	*Pongamia pinnata* Vent	p	p	p	p	----
50	*Thespesia acutifolia* Backer	p	p	p	p	----
51	*Susuvium portlacastrum* L.	p	p	p	p	p
52	*Sarcolobus carinatus* Wall.	p	p	p	p	----
53	*Heliotropium currassavicum* L.	p	p	p	p	----
54	*Arthrocnenum indicum* Moq.	p	p	p	----	----
55	*Artiplex stocksii* Boiss.	p	p	----	----	----
56	*Salicornia brechiata* Rox.	p	p	p	p	----
57	*Suada fruticosa* frost.	p	p	p	p	p
58	*S. maritimia* L.	----	p	p	p	----
59	*Salvedora oloides* Dam.	p	p	p	p	p
60	*Salsola foetida* Deletz.	----	----	p	p	p
61	*Porteresia coaretata* Roxb.	p	----	p	p	----
	Total	**22**	**25**	**22**	**16**	**08**

4.3 TAXONOMIC DESCRIPTION OF FAMILIES AND GENERA OF MANGROVE

(A) True Mangrove Species

Family	*Taxonimic Descriptions (Genus–Species)*
1. Acanthus	1. *Acanthus ebracteatus*
	2. *Acanthus ilicifolius*
	3. *Acanthus volubilis*
2. Avicenniaceae	4. *Avicennia marina*
	5. *A. officinalis*
	6. *A. germinans*
	7. *A. lanata*
	8. *A. alba*
	9. *A. eucalyptifolia*
	10. *A. africana*
	11. *A. bicolor*
	12. *A. schauriana*
3. Bombaceae	13. *Camptosteman schultzii*
	14. *Camptosteman philippensis*
4. Comboetaceae	15. *Lumnitzera littorea*
	16. *Lumnitzera racemosa*
	17. *Lumnitzera rosea*
	18. *Laguncularia racemosa*
	19. *Conocarpus orectus*

5. Euphorbiaceae	20. *Excoecaria agallocha*
	21. *Excoecaria dallachyana*
	22. *Excoecaria indica*
6. Leguminoceae (Fabaceae)	23. *Cynometra ramiflora*
	24. Cynometra iripa
7. Meliaceae	25. Xylocarpus moluccensis
	26. Xylocarpus granatum
	27. Xylocarpus mekongenesis
8. Myrsinaceae	28. *Aegiceras corniculatum*
	29. Aegiceras floridum
9. Myrtaceae	30. *Osbornia octodonta*
10. Palmae	31. *Nypa fruticans*
11. Pellicieraceae	32. *Pelliciera rhizophorae*
12. Plumbaginceae	33. *Aegialitis annulata*
	34. *Aegialitis rotundifolia*
13. Rhizophoraceae	35. *Rhizophora apiculata*
	36. *Rhizophora mucronata*
	37. *Rhizophora stylosa*
	38. *Rhizophora lamarckii*
	39. *Rhizophora mangle*
	40. *Rhizophora harrisonii*
	41. *Kandelia candel*
	42. *Ceriops tagal*
	43. *Ceriops decandra*
	44. *Bruguiera gymnorhiza*
	45. *Bruguiera cylindrica*
	46. *Bruguiera sexangula*
	47. *Bruguiera exaristata*
	48. *Bruguiera parviflora*
	49. *Bruguiera hainesii*
14. Rubiaceae	50. *Schyphiphora hydrophyllacea*
15. Sonneratiaceae	51. *Sonneratia alba*
	52. *Sonneratia apetala*
	53. *Sonneratia caseolaris*
	54. *Sonneratia griffithi*
	55. *Sonneratia ovata*
16. Sterculiaceae	56. *Heritiera littoralis*
	57. *Heritiera fomes*
	58. *Heritiera globosa*

(B) Mangrove Associated Species

1. Pteridaceae (Fern family)	1. *Acrostichum aureum*
	2. *Acrostichum speciosum*
2. Apocyanaceae	3. *Cerbera manghas*
	4. *Cerbera odallam*
	5. *Cerbera floribunda*
3. Bignoniaceae	6. *Dolichandrone spathacea*
4. Leguminosae (Fabaceae)	7. *Pongamia pinnata*

		8. *Coesalpinnia ruga*
		9. *Caesalpinnia bonducella*
		10. *Afzelia bijuga*
		11. *Derris scandehce*
5.	Malvaceae	12. *Hibiscus tiliaceus*
		13. *Thespesia acutifolia*
6.	Myrtaceae	14. *Barringtonia racemosa*
7.	Pandanaceae	15. *Pandanus tectorius*
8.	Palmae	16. *Liacula spinosa*
9.	Rubiaceae	17. *Guettarda speciosa*
10.	Verbenaceae	18. *Clerodendron inerme*
11.	Lythraceae	19. *Pemphis acidula*

DESCRIPTION OF SPECIES

1. FAMILY : Acanthaceae

This shruby family comprises overall three exclusive – mangrove species viz.–

(i) *Acanthus ebracteatus*
(ii) *Acanthus ilicifolius*
(iii) *Acanthus volubilis*

A genus of some 30 species in tropical Asia and Africa. It is distinguished by its usually spiny leaves. Spicate terminal inflorescences.

Genus : Acanthus L.

(i) Open flower 3.5–4 cm long, corolla blue or violet. Bracteoles persistent in fruit upto 1 cm long. Plants usually with spiny leaves — *Acanthus ilicifolius*.

(ii) Open flower 2.0 –2.5 cm; all parts smaller than previous species. Corolla always white. Bracteoles small inconsipicuous, early deciduous or absent.

(a) Plants usually spiny with thick stem. Leaves usually widest below middle. Bracteoles usually present but early deciduous — *A. ebracleatus*

(b) Plants usually slender delicate speawlong stems, leaves usually widest above the middle. Bracteoles never present — *A. volubitus*.

(1) *Acanthus ebracteatus* Vahl.

Low spiny shrub, upto height of 8m aerial root are common, generally from lower surface of receiving stems. Leaves decussate usually with pair of spines insertions of each leaf. Leaves usually widest below the middle. Flower 2.0–2.4 cm corolla white bracteoles small, early deciduous. Ripe fruit greater than 2.0 gm.

Distribution : The species described characteristic associate of mangroves with a total range from India to the western Pacific, Tropical Australia and the Philippines.

Ecology : Usually common to landward region of mangal and mostly found in association with fern *Aerostichum aureum.*

(2) *Acanthus ilicifolius* L.

A. neoguineusis Engl.

A sprawling shrub, scarcely woody upto 2m aerial roots formation has been noted. Leaves decussate, with the pair of spines, leaves glabrous, blade upto 20cm long, gradually tapering. Inflorescence as terminal forming bracteate spikes 10–20cm. Corolla zygomorphic light blue to violet, fruit usually 2–3cm long and 1cm wide; usually with 4 rugose angular seeds.

Distribution : Quite common species, noted abundantly at all mngrove region of India, Japan, Australia, Thailand, Java and New Guinea, Queensland.

Ecology : Flowering – November/December. This species associated with *A. ebracteatus*.

(3) *Acanthus volubilis* Wall

This species is much less robust in its vegetative parts than the other 2 species and is described as unarmed. The inflorescence usually has fewer flowers and is shorter. This species has been reported by Parkinson (1923) and Thothathri (1962) in Andaman (India). But the author could not get it anywhere in Andaman.

2. FAMILY : Avicenniaceae

The genus *Avicennia* L. was initially included in the family Verbenaceae, from which it differs in important features, now has been placed in separate family *Avicenniaceae*. This family is known as true mangrove family. This family is second important mangrove family which comprises a total of 9 exclusive mangrove species. Though in India only 3 species are found.

The most recent revision of the genus *Avicennia* is that by Moldenke (1969, 1967), who produces arguments for separating the genus from verbenaceae mainitaining a separate family as originally proposed by Endlicher (1838) and followed subsequently by certain other taxonomists. For a full list of the taxonomic literature relating to the genus reference should be made to Moldenk's paper.

Genus : Avicennia L.

A pantropical genus about 9 species, occupying diverse mangrove habitats, either within the normal tidal range or in back mangal and then with a high tolerance of hypersaline conditions. The genus is confirm in its gross morphology and anatomy. Bark colour is often diagnostic characters seen in the field.

Trees height is vary from 10m to 30m in normal condition. In grazing areas, the height may be 1 to 2m and known as invador or pioneer species. This species is known to very hardy and resistance species to biotic factor. In West coast, this species has been remain as mangroves.

(4) *Avicennia marina* (Forsk) Vierh.

A. intermedia Griff.

A. mindanaense Elmer.

This species is widely distributed, both longitudinally, latitudinally, with a range from East Africa and the Red sea, along tropical and sub–tropical coasts of the Indian ocean to the South China sea, throughout much of Australia into Polynesia as far as Fiji and South to the North Island of New Zealand. In India distribution of the species in all mangroves areas of East and West coast and Andaman Nicobar islands.

Flowers orange–yellow, not more than 6mm in diamter with 4 equal lobes of the corolla that are glabrous. The flowers are sweetly scanted. The leaves are rounded apically, more or less ovate, and whitish beneath. The flower heads are capital and remain so.

Trees upto 20m commonly known as "White mangrove" because of its distinct white bark. Cable or horizontal roots send pencil shaped aerial vertical roots called pneumatophenes. These may be smooth walled or rought due to presence and absence of lenticals. Leaves opposite, simple shortly petiolate. Leaf blade abtuse rounded or acute, upper surface is bright green, white silvery–white below. Fruit bean shaped, small, germination crypto viviparous.

(5) *Avicennia officinalis* L.

A. tomentose Willd.

A tree more or less similar to *A. marina* upto 15m, rather rare to some part. Bark is greying and rough, bright green inside. Leaves, dull green on upper side white lower side ivory–white tomentose base, obovate to broadly oblong. Flower yellow in clusters and arranged in corymbs.

This species has wide range from South India through Indo–Malaya to New Guinea and Eastern Australia. It is one of the more distinctive *Avicennia* species of the eastern tropics and is recognized by its large orange–yellow flowers, that are 10–15 mm in diameter and therefore larger than any other of the eastern group.

(6) *Avicennia germinans* L. Stearn

A. nitida Jacq.

A. tomentose Jacq.

A. germinans is characterized by its ovate leaves, usually with a blunt apex, the lower leaf surface and the twigs are conspicuously glaucous white in contrast to the dark green of upper leaf surface. The leaves are 5 to 8 by 2 to 4 cm with short petiole. The flowers are the largest in the genus, white and zygomorphic.

This species is widespread from Southern Florida and the Bahamas throughout the Caribbean to Mexico, Central America and the Pacific coast of the South America to Peru and the Gelapagos. Its distribution on the Pacific coast of the America seems discontinuous. It is recorded in Atlantic cost of South America to Northern Brazil.

(7) ***Avicennia lanata*** **Ridley**

A. officinalis L. var. *Spathulata* Kuntz.

This species is very similar to *A. marina* leaves cover with white woody idumentum. Flower is also very similar to *A. marina* but with very hairy outer petals and calyx. Calyx lobes recovered. Distribution is very restricted. It has been reported at Malacca and Pahang in the Malay Peninsula, Western Sarawak and Singpore.

(8) ***Avicennia alba*** **Blume**

A. marina (Fersk.) Vierh var. *alba* (Blume) Bakh.

The flower of this genus is very similar to *A. marina* type. Ovary conspicuously hairy. The leaves of *Avicennia alba* are characteristically lanceolate, pointed at the apex, with white undersurface. The bark is slightly roughened but not fissured and is somewhat brown.

The distribution of this species is very wide. From India to Indonesia, through the Malaya Archipelago to the Philippines, New Guinea, New Britain and northern Australia.

(9) ***Avicennia eucalyptifolia*** **(Zipp. ex Miq. Moldenke)**

A. officinalis var. *eucalyptifolia*

This species has been recognized by its ovate – lanceolate to lanceolate leaves, which may be 3 to 5 times as long as they are wide. The apex is narrowed to a point. Bark is smooth, greyish, light brown to greenish yellow and flaking in patches.

Distribution in Australia, Papua New Gunea, New Caledonia and New Zealand.

(10) ***Avicennia africana*** **Palisot de Beauvois**

Details of this species is not avaliable. This species has been reported from West Africa.

(11) ***Avicennia bicetor*** **Standley**

A. tonduzii

The flowers are distinctly zygomorphie and those of *A. germinans* in having the petals hairy inside. The white corolla sometimes has a yellow throat. It can be almost pure yellow in some population. The distinctive features is the inequality of the stamens. They are inserted at the same level of flower.

This species has a limited distribution and is recorded for Costa Rica and the Pacific coast or Columbia at the mouth of the Buenaventura river.

(12) ***Avicennia schaueriana*** **Stapf and Leechman**

The flowers are larger than those of *A. bicolor*, the corolla being 10 to 12 mm long and almost wide and the corolla is glabrous or at most slightly hairy. The lobes are narrow and are not reflexed. The ovary is uniformly hairy.

This species is widely distributed from the lower lesser Antilles and the Atlantic coast of Northern South America from Guyana south to Uruguay.

3. FAMILY : Bombaceae

This is a long family, distributed mostly in tropical. Plants are mostly trees with large or digitally lobed leaves. Flowers are often large with numerous stamens. Only one genus of the family are mangrove found in Littoral zones.

Genus – Camptosteman Masters

A genus with 2 species of evergreen trees in back mangal from Borneo, the moluccas and northern Australia to New Guinea and the Philippines.

Plants can be recognized by the narrow, alternate leaves, which are dull green scaly, pubescence, cupular epicalyx and closed calyx.

There are two species, which may not overlap in distribution.

C. schultzii – Leaves glabrous above, stamen numerous, ditheate.

C. philippinensis – Leaves sparsely but uniformly lapidote above. Stamen 5, polythecate.

(13) *Camptosteman schultzii* Masters

This is a medium size trees to 25–30m tall, trunk flanged at the base, bark dark, seedy unfissured, exposed root with blunt pneumatophores to 30cm tall. Leaves spirally arranged, pateolate, the petiole 2 to 3 cm long, blade 8 to 10 cm oblong or ovate to broadly lanceolate, margin entire, apex acute. Flowers in axillary 3 to 6 flowered cymes, epicalyx cupulate, calyx companulate, petals 5 imbricate, white stamen numerous ovary superior, bilocular.

Distribution in southern New Guniea, northern Australia, Indonesia and Borneo.

(14) *Camptosteman philippinense* (Vidal) BeCC

This is a medium size tree with present of lepidate scales in the upper as well as lower surface of the leaves and the 5 stamens, which are described polythecate. Ovary tri to tetralocular with axile placentation.

This species has been recorded from the Philippines, Borneo and the Celebers.

4. FAMILY : Combretaceae

A moderately large tropical woody family of some 20 genera and 500 species. Plants are mostly trees or shrubs. Leaves alternate, subopposite or opposite, sometimes ternate, petioled entire and usually simple stipules 0. Flower usually in spike or racemes. Sepals 4–5, petals 4–5, stamens 4–5 or 8–9. Ovary inferior, 1 celled. Fruit drup seed one without albumen. This family represented three genera of mangroves

(i) Lumnitzera
(ii) Laguncularia
(iii) Conocarpus

Genus – Lumnitzera Willd.

This is known as true mangrove species. Plants are medium size tree found in middle zone of mangal. The leaves are shiny, light green alternate leaves which after have notched apex bearing a small gland.

Distribution of species very wide ranging from east America to western Pacific (Fiji and Tonga), tropical Australia and Indonesia. There are three species which are found in –

(i) Flower red to purple in terminal – *L. littorea*

(ii) Flower white in axillary racemosa – *L. racemosa*

(iii) Flowers pink in axillary or terminal – *L. rosea*

(15) *Lumnitzera littorea* (Jack) Voigt

L. coccinea Wight and Arnold –

A tall tree upto 25 m but at many places it is bushy in habit. Bark dark black and tissued. It has thin knee like pneumatophore. Leaves dark green, thick, 5cm long 3 cm broad. Narrowly abovate to oblong lanceolate. Flowers in terminal racemose bright red petals. Stamens usually 7.

Distribution is very wide, found in Australia, New Zealand, Malaysia, Queensland, New Guinea, India and Indonesia.

(16) *Lumnitzera racemosa* Willd.

Comparatively medium size tree and restricted in some places. Bark grey, fissured, thin knee shaped pneumatophores. Leaves clustered towards and of branches, small, fleshy slightly broad to previous species, oval to abovate. Flowers white in axillary racemose. Stamen 10, double number of petals, ovary inferior fruit hard ovoid. Crowned by persistent calyx tubes.

Distribution as in above species in Australia, New Zealand, India, Malaysia and Indonesia.

(17) *Lumnizera rosea* (Gaud.) Persl.

Launcularia rosea Gaud.

A fern with pink flowers, which is interprated by Tomlinson *et al.* (1978) as a hybrid of *L. racemosa* x *L. littorea*.

This species has been recorded on Hinchinbrook Island, Queensland.

Genus – Laguncularia Gaertn.

A monotypic mangrove genus with a distribution restricted to Amrica and West Africa.

(18) *Laguncularia racemosa* (L.) Gaertn.

This species is typically restricted to the landword fringe of the mangrove community, but also pioneers readily into disturbed sits where it can form pure stands. Pneumatophores are only facultatively developed, in some situations they are abundantly developed, in others they may be absent. They are identified by the opposite, bluntly ovate, somewhat fleshy leaves and the conspicuous pair of glands on the distal axial parts of the petiole.

Medium size trees, trunk solitary or clustered. Bark rough, fissured grey. Phyllolaxis decussate, stem terate. Branches with lenticel. Leaves with short petiole, elliptic to oblong, margin entire. Inflorescence spicate, aggregated to form loose terminal panicles. Flowers unisexual or perfect, calyx 5, petal 5 rounded greenish yellow and hairy. Stamens 10 in 2 series 5 inserted to different levels. Fruits 1 seeded nutlet, 2 cm long endosperm absent, germination epigeal.

Genus – Conocarpus L.

A genus of 2 species : are with a very restricted distribution in East Africa and the other widely distributed in coastal communities in tropical America and West Africa.

(19) *Conocarpus erectus* L.

The species is frequently considered a "true mangrove", but it is better regarded as a mangrove associate because it lacks many important characters which is essential to call mangroves such as (pneumatophore and vivipary). Conocarpus is a back–mangal.

Plants are medium sized dioceous trees. Bark rough and fissured. Leaves spirally arranged, scattered, glabrous, silver hairy. Petiole short. Blade folded in bud ovate lanceolate. Lower leaf surface with several. Inflorescence terminal, compact. Flowers clustered in terminal globose heads. 5mm diameter. Each head with about 25 flowers. Flowers somewhat diamorphic, each with an inferior ovary. Petals absent, disc well developed and hairy. Seed small, angular germination hypogeal.

5. FAMILY : Euphorbiaceae

One of the largest plant families with over 7000 species, most commonly tropical. Plants are mostly shrubs or trees. Leaves usually alternate, rarely divided or compound. Stipules usually small. Flowers usually small or minute usually compound. Perianth often small, sometimes absolete often dissimilar in the two species. Male flowers Torus, sometimes forming an intra–staminal disk–glands or of lobes alternate with the stamens of the outer series. Stamen 1 many, free connate, anther 2 celled. Disk hypogynous. Ovary superior. Fruit usually capsule.

Genus – Excoecaria L.

A genus of 35 to 40 spcies in tropical Africa and Asia to the western pacific. The following spcies which are found in landword zone of mangroves. There are three species of *Excoecaria* only are *E. agallocha* is true mangrove species.

(20) *Excoecaria agallocha* L.

Stillingia agallocha (L.) Baill.

Plants are medium sized tree having white latex, dioecious. Bark grey, fissured, lenticels prominent. Leaves spirally arranged, terminal buds. Stipules minute, ephemeral. Leaves simple. Coriaceous, somewhat fleshy, with terate petiole. Apex rounded, slightly emarginate. Inflorescence axillary, pale green, 3 to 7 cm long initiated as Catkin like structure. Male inflorescence upto 7cm long at maturity, with a series of spirally arranged, often glandular bracts, each bracts subtending to a male flower.

Distribution of this species from east Africa, India and Ceylone to Hainan and the Ryu Kyu island through Malaysia and Papuaria, including tropical Australia and into the Pacific as far as Nive and Samoa.

(21) *Excoecaria dallachyana* (Baill) Barth

E. agallocha var. *dallachyana* Baill.

This taxon is distributed from south Queensland and New South Wales, within the species a number of varieties have been named based on leaf dimensions and the size of the reproductive parts (Pax and Haffman, 1912).

(22) *Excoecaria indica* (Willd.) Muell

This species has been reported from south East Asia, Malaysia, south and east India, and the Soloman islands, but not the Philippines. It is distinguished from *E. agallocha* by its thorny trunk.

Leaves lanceolate, globose, smooth capsular fruit, upto 3 cm in diameter.

6. FAMILY : Fabaceae (Subfamily – Caesalpinoides)

One of the largest tropical family. Plants are herbs, shrubs and trees. Leaves stipulate and usually alternate, pinnate or digitate or simple. Few genera of this family are the mangroves and some are mangrove associates, which are found in middle and distal zone. The species *Cynometra* is very prominant mangrove species found in most of the mangrove forest.

Genus – Cynometra L.

A pantropical genus about 70 species of trees or shrubs, mainly restricted to moist low land forest. The 2 species distinguished in the back mangal area, but are also recorded in land to an altitude of 400 m.

(i) *Cynometra ramflora* L. : Style straight, in line with dorsal part of full ovary, fruit without a prominant lateral beak. Ovary wall glabrous inside. Sepals not curved, distally when reflexed at authesis.

(ii) *Cynometra iripa* Kostal : Style bent, oblique to the dorsal line of the ovary, becoming a prominant lateral peak in fruit. Ovary wall hairy inside. Sepals curved distally when reflexed at anthesis.

(23) *Cynometra ramiflora* L.

Trees to 25m high with smooth thin, gray bark, the trunk flutted somewhat at the bark but not to more than 1m. Leaves distichous short somewhat zigzag developing rapidly from lateral buds. Stipulates intrapetiolar, narrow, caducous and leaving no sear. Leaves glabrous, essentially peripinnate but with short pair of opposite leaflets.

Flowers zygomorphic axillary racemes, sometimes with 2 inflorescence per node. Flowers about 15 per axis, each substituted by a short, narrow serious bract, the lower bracts often empty. Floral envelope white, but turning white.

Petals 5 free. Stamens 10, free with small author. Ovary 1, inserted slightly eccentrically. Fruits 1, one seeded, seed without endosperm.

Distribution in India, throughout south–east Asia and Malaysia to the Pacific but not Australia.

(24) *Cynometra iripa* Kostel

Trees 20–25 m high with smooth thin, gray bark, with similar morphologically, can also occur in back mangal. Sepals curved distally when reflexed at anthesis.

Distribution is restricted to Australia.

7. FAMILY : Meliaceae

Trees with alternate, ex–stipulate leaves. Flowers regular usually small and panicled, white, yellow or pink; sepals and petals 3–6; stamens twice as many as the petals, the filaments united to form a deep tube. Fruit a dehiscing capsule, or indehiscent. Seeds exalbuminous or with fleshy albumen, embryo flat; cotyledon fleshy.

Distribution : frequent in the warm regions of Asia and America, rarer in Africa; genera 37; species about 300.

These are two species of this family confined to mangrove habitats.

(i) *Xylocarpus*

(ii) *Carapa*

These genera are found in landword zone of mangroves.

Xylocarpus Konig

There are three species of this genus confined to coastal and mangrove region. All the three species of *Xylocarpus* is found in India, South East Asia, East Africa, tropical Australia, Indo–Malaya regions. Plants are medium size tree, deciduous, or evergreen, sometimes shrubs, root system. Often developed pneumatophores. Leaves alternate, without stipules, peripinnately compound, the rachis about 10–20cm long. Leaflets 2–3 pairs, each with a shat corky petiole. Inflorescence a lateral panicle. Bracts are bracteoles minute usually ephemeral. Flowers unisexual, pedicellate, tetramerous. Calyx 4 valvate. Petals 4, creamy white, valvate more or less free rounded, stamens tube white spherical. Disc well developed, ovary tetralocular with axile placentation.

(i) *X. granatum* : Trunk base often enlarged, with well developed buttress with ribbon like surface root system. Inflorescence not more than 6 cm.

(ii) *X. moluccensis* : Root system not elaborated, buttresses absent. Bark longitudinally fissured. Inflorescence often exceeding 8 cm in length.

(iii) *X. mekongensis* : Trunk surface fissured dark brown to greyish. Buttresses absent but pneumatophore upto 30 cm 3–4 cm diameter are characteristic.

(25) *Xylocarpus moluccensis* (Lamk.) Roen

Tree upto 15m, crocked brown to grey longitudinally fissured bark without any developed aerial root system and buttresses. Leaves long upto 30 cm leaflets usually 4 to 6 rarely 2, more often 6 flowers pinkish yellow, 1/4" to 1/2" diameter, fruit small in size like an orange.

This species is found in India mainly in Andaman and Nicobar island, Sunderbans. I have not found distribution of this species in any other parts of mangrove regions.

(26) *Xylocarpus granatum* Koen

Syn. – *Carapa obovata* Blume

X. obovatus (Blume) Juss.

X. bonadirensis Mallai

A moderate height buttressed tree upto 20m, bark smooth yellowish grey, peeling in thin irregular petals with ribbon like surface roofs. Leaves alternate on 20 to 28 cm long, rachis, leaflets usually 2 to 4 and 7.5–10 cm long ovovate and broad blunt apex. Flowers cream coloured, fruit woody, smooth with 15–25 cm diameter.

Generally found in middle to landword shore of the mangal with conspicuous buttresses and extensive aerial root system in ribbon or flank like structures.

Distribution of this species is very wide, India, China, south east Asia, Fizi, Guam, New Zealand, Papua New Guniea, Soloman islands, Vietnam, Bangladesh, Taiwan, Indonesia, Myanmar, Thailand.

(27) *Xylocarpus mekongensis* Pierre

Syn. – *Carapa moluccensis* Watson

X. gangeticus (Prain) Parkinson

X. australasicus Redly 1938.

This species is more or less similar to the previous one, besides, some diagnostic characters. The striking difference are the bark texture and the presence of well developed aerial roots in the form of peglike pneumatophores. Some species is used under name *X. australasicus* in some recent accounts for plants from western Australia and New Guinea. The erect peg–like pneumatophores upto 30cm, originate from the extended horizontal cable roots which grow away from the base of the trunk. The species is more consistently deciduous and other species and in Queensland and New Guinea.

Distribution of this species restricted in some part of India. I have noted in Andaman and Nicobar islands. It has been reported in Bangladesh, Indonesia, Myanmar, Singapore, Thailand, American Samoa, Fizi, New Caledonia, Palau, Papua New Guinea and Western Samoa.

8. FAMILY : Myrsinaceae

A large family of over 1000 species throughout the tropics and sub–tropics in about 30 genera distributed, but in mangroves only one genus is present. Evergreen shrubs or small trees with simple alternate and often gland dotted leaves without stipules. Flowers 5 merous, bisexual, calyx persistent and often enlarging in fruit, corolla with the lobes often twisted and overlapping to the right; stamens opposite the corolla lobes.

Genus – Aegiceras Gaertner

A genus of 2 species restricted to mangrove communities in the Asian tropics with a distribution from India and Ceylon to South China and Hong Kong through Malaysia to the Philippines, New Guinea and tropical Australia, but not known for the Pacific islands.

A small tree (12–15 x 1–2) with the habit of a mangrove. Leaves scattered, long obovate apex rounded, margins recurved, coriacous and glabrous, lateral nerves about 10 pairs, rather inconspicuous, flower white, fragrant, about 3/8 inch long on slender pedicels in umbels, petals pointed reflexed in flower. Fruit 1–2m long, cylindric, curved and pointed.

There are two species of Aegiceras confined to mangrove areas –

(i) *A. corniculatum* – Leaves large 11 by 6 cm. Inflorescence umbels.

(ii) *A. floridum* – Leaves small – 6 by 3 cm. Inflorescence racemose.

(28) *Aegiceras corniculatum*

Syn. – *A. majus* Geartn.

A. fragrans Koing.

A small tree upto 6m high, bark brown with redish tings. Leaves alternate, glabrous, obovate, apex rounded. Flower white in umbels, fragrant. Fruit curved 4 to 8 cm long. Germination crypto–viviparous type. Sepals 5, petals. Stamen 5 opposite to corolla lobes. Ovary monocarpellary with free central placentation.

This species has wide distribution from Ceylon to South China, India through the Malaya. Archipelago to Polynesia and north eastern Australia as far south as New South Wales. It seems most characteristic of the outer (seaward) mangal fringe and occurs typically as an isolated low shrubs, never forming a conspicuous part of the community.

(29) *Aegiceras florum* Roemer

Syn.– A. nigricans A. rich

The species has a much more restricted geographic distribution and is recorded from northern Borneo, Java, the Moluceas and Celebes throughout the Philippines to Indo–China.

9. FAMILY : Myrtaceae

Tree with simple opposite or alternate, exshtrulate leaves. Flowers usualy showy, white or pinkish bisexual, calyx with 4–5 segments; petals as many, imbricate, stemens very marshy. Fruit usually fleshy and crowned by the persistent calyx.

This large, essentially tropical family of 80 genera and 3000 species includes, 1 isolated aspecies associated with mangrove communities.

(30) *Osbornia octodonta* F. Mull.

Small evergreen, hermaphrodite trees or more usually low shrubs with a gray, soft spongy flaking and quite thick bark. Plants usually with several irregular slender trunk not forming a regular shape. Leaves opposite decussate on 4 angled twigs; leaf obovate usually 2 to 4 by 1.5 cm, margin entire apex rounded sometimes slightly emarginate. Petiole 1 to 3 mm. Stipules absent. Flowers axillary, petals absent, stamens yellow, numerous (upto 48). Ovary inferior with axile placentation endosperm absent.

Distribution is very limited. It has been reported Indonesia, Malaysia, Philippines, Australia, Papua New Guinea and Soloman islands.

10. FAMILY : Palmae (Arecaceae)

One of the larger, most natural and economically important families with other 200 genera and about 2600 species, typically plants with aerially unbranched stems, smooth and marked with ring like sears or covered by the persistent bases of petioles. Flowers small, in spike or panicles and enclosed in spathes, perianth 6, divided in two whorls, stamen 3 to 6. Fruit with a fibrous or scaly covering.

(31) *Nypa fruticans* (Thunb.) Warenb.

Nypa fruticans Thunberg

A monoecious palm with a thick, subterranean somewhat fleshy, dichotomously branched, rhizomatous axis upto 70 cm in diameter, with oblique raized leaf seams. Leaves 15–30 ft. long, unarmed somewhat resembling those of the coconut palm. Leaflets 4–6 feet long, rigid, glaucous beneath; petiole very strong and thick. Fruit 4–6 in long turbinate, compressed and angular, fibrous with a large white seed.

Distribution : This plant is an unusual mangrove because of its colonial habit. Distribution of this species very wide. Bangladesh, Brunei, Cambodia, China, Taiwan, India, Indonesia, Japan, Philippines, Singapore, Thailand, Vietnam, Australia, Guam, Palau, Papua New Guinea, Soloman island and Panama.

11. FAMILY : Pellicieraceae

This is a monolypic family with the genus *Pelliciera*. It is distinguished by its regularly pentamerous flowers, enclosed by a large pair of bracteoles, mangal habitat, one seeded indehiscent fruit and the large naked embryo.

(32) *Pelliciera/Planchon* and *Triana*

Pelliciera rhizophorae Triana and Planchon

Small tree, 5 to 10m high with somewhat tiered, distally arcuate branches. Trunk swollen and markedly fluted below. Bark dark, rough fissured. Leaves spirally arranged with regular a2/5 phyllotaxis, each sessile, dark glossy green, apex bluntly rounded. Flowers leaf showy, pentamerous upto 12cm wide. Sepals 5 free inbricate, white. Petals 5 free white or rose pink. Stamens 5 free. Ovary narrow cylindrical woody. Seed single without endosperm.

Distribution : This species has a limited distribution on the Pacific coast of Central and northern south America, from Buenaventura Bay in western Columbia to Costa Rica, Ecuador, Nicaragua and Panama.

12. FAMILY : Plumbaginaceae

A cosmopolitan family herbs or shrubs often maritime. Flower regular, hermaphrodite, axillary or terminal subpaniculate or racemose. Calyx 5, inferior gamosepalous tubular. Corolla hypogynous gamopetalous tubular infundibuliform. Stamens as many as te corolla lobes. Ovary free one called often 5 sulcate. Distribution of this species more or less throughout the world chiefly maritime or in saline or rocky desert. This family contain 8 genera, bout 200 species.

Genus : Aegialitis R. Brown

Plants are confined to mangrove habitats. Small shrubs with perfect flowers. Bark dark smooth fissured or flaking with age. Leves 15 cm long, somewhat fleshy, spirally arranged erect, and clustered terminally on the shoots. Flowers pentamerous, in many flowered terminal irregularly one–sided cymes. Calyx 5, petals 5, stamens 5. Ovary superior, unilocular with a single basal ovule. There are two very similar species are known –

(i) *Aegialitis annulata* R.Br.: Corolla with 9–10 lobes. Leaves dull.

(ii) *Aegialitis rotundifolia* Roxb. : Corolla with 12 lobes. Leaves shining above.

(33) *Aegialitis annulata* R. Brown

It has been reported from Indonesia, Australia and Papua New Guinea.

(34) *Aegialitis rotundifolia* Roxberg

This species I have seen in Andaman and Nicobar Islands, India, East coast of India, Myanmar and Bangladesh.

13. FAMILY : Rhizophoraceae

Littoral trees or shrubs, leaves simple, opposite, smooth, leathery with large stipules. Flowers bisexual, calyx with few or many, valvate, pointed lobes, persistent in petals as many as the sepals, often lobed or lanceolate with stamens usually twice as many. Fruit crowned by tree calyx usually germinating on the tree and seedling down along pointed radicle.

The members of this family are the true mangrove. They are remarkable for their peculiar manner of growth and seedling with sending down a large number of fantastically arched branching roots. They are firmly anchored in the soft mud, and while the securing for themselves stability and protection against storms, they protect the soil on which they grown from erosion by tidal currents. Some also send up from the roots, through the soft mud, 'wood knee' believed to facilitate the access of air to the roots. The seed germinates on the tree, sending down a long, hard pointed, pendulous radicle which dropping perpendicularly. Point foremost into the soft mud is firmly planted in site or floating upright in the water. It drift about until its hard submerged point touches the soft mud. Where it soon strikes root and establishes itself. The popular name mangrove is indiscriminately given to many kinds of trees found growing in swamps and along seashores. Such as the littoral members of this family and the species of new genera of mangroves.

There are 4 genera of this family which are known as true mangrove –

(i) *Rhizophora* – Littoral plant, seed germinating on the tree. leaves more or less elliptic, acute or tapering at both ends. Sepals and petals 4 : stamens 8. A pantropical genus of 8 species, which is very uniform in its vegetative construction. Evergreen trees growing to 30 mm tall.

(ii) *Kandelia* – Shrubs, leaves 3–5 in lay stamens many. Littoral plant. Seed germinating on tree. Leaves more or less ablong or obovate, rounded or notched at the top, sepals and petals 5–6.

(iii) *Ceriops* – Small tree, leaves 2–4 in long, stamens 10–12.

(iv) *Bruguiera* – Littoral plant seed germinating on the tree. Leaves more or less elliptic. Sepals and petals 8–14, petals lobed and emarginate. Stamens 16–28.

The distribution of species is eastern and western hemisphere.

(1) Eastern species are – *R. mucronata*
R. stylosa
R. apiculata
R. samoensis
R. lamarekii
R. selala

(2) Western species are – *R. mangle*
R. harrisonii
R. racemosa

(i) Genus : Rhizophora

(35) *Rhizophora apiculata* Bl.

Syn. R. conjugata Arn.

This species is perhaps the most distinctive in the genus, with large, dark glossy green leaves and flowers usually in pairs on a stubby axis, matching below the rosett of leaves. Flowers in two (ranly 4) on stout peduncles. Each flower sessile with a pair of massive cupular corky bracteoles. Petals glabrous. Stamens usually 12 sessile. Stigmas almost sessile.

(36) *Rhizophora mucronata* Lamk.

Usually irregularly grown and much branched with a very rough bank. Leaves 4.7 in long 2–4 in broad, elliptic acute at both ends mucronate shining, pale beneath and with black dots; petioles 1–2 m long. Flowers yellowish white, nearly 1m across or short pedicals in axillary lymes, petals hairy on the margins. Fruit dark–brown, with a radical 1–3 feet long.

Very common, fringing all tidal creeks and also along the sea shores. In some places it forms inpenetrable belts of vegetation at water's edge, many miles long and as much as a miles in depth, regularly inundated by tides. It grows gregariously or mixed with the other kinds of mangroves. Flower and fruits apparently at all times of the year. The wood is good for fuel and the bark gives a useful tanin and a dye which is used for dyeing of stern.

The flower of this species differ from those of *R. stylosa* only in the sessile stigma, inflorescence in the 2 species are very similar. the leaves of *R. mucronata* are typically much layered than those of *R. stylosa*.

(37) *Rhizophora stylosa* Grift

R. mucronata Lamk. var. *Stylosa schimp.*

A medium sized tree, ranges from southern India to the Eastern Pacific (Samoa), southward to New South Wales, and Westward to Indo–China. It is not known from South America. This species becomes a progressively more conspicuous element of mangrove vegetation in an easterly direction. In the western pacific this species seems to be particularly vigorous and will survive in extreme habitats (coral reefs, rocky coastline) though often depanperate.

Flower dull white and born on long multibranch dichotomous peduncle. Stigmas on a slender style 4–5 mm long, seedlings smooth, not exceeding 30 cm. Leaf blade narrow (to 7 cm) and short (to 12cm).

(38) *Rhizophora lamarckii* Montr.

R. conjugata var. *lamarckii*

This is interpreted as a sterile F, hybrid between *R. apiculata* and *R. stylosa* by Tomlinson and Womersley (1976). It was originally considered to be endemic to New Caledonia on abundant circumstantial. At its type locality at canala it grows not only with its putative parents but also with *R. selala* and *R. samoensis*, so that here we have the largest concentration of *Rhizophora* anywhere (5 out of a possible 8 taxa). Its wider distribution is now well established (Tomlinson, 1978; e.g., Queensland, Papua New Guinea and the Solomon Islands). It is always consistently intermediate in its morphology between its putative parents (Duke and Bunt, 1979; Tomlinson, 1978; Tomlinson and Womersley, 1976) and occurs only where the 2 grow together. The pollen appears to be sterile and seedlings are rarely produced.

Duke and Bunt (1979) note that in Queensland *R. lamarckii* is multitrunked and rambling, forming communities of gnarled and stunted individuals in higher tidal contours and lower to middle tidal reaches in areas of moderate to high rainfall. The frequently rather battered appearance of trees suggests that once established *R. lamarckii* is very tenacious and plants may be considerably old; unfortunately, precise age cannot be estimated.

(39) *Rhizophore mangle* L.

This species is distinguished by its usually rather small leaves, with the blade 10 to 12 by 5 cm, the petiole about 2 cm long, pale and extending into the midrib, which is pale green to almost white below, and particularly by the blunt apex. Stipules are 6 to 8 cm long. Older shoots are light brown. Fruits are elongated, 2 to 3 cm long, often asymmetric and tapered in the apex, which is attenuated to the persistent style 2 to 3 mm long. The fruit is characteristic olive green with a roughened surface. Peduncles in more robust shoots may be 6 cm long but are usually shorter, the pedicles about one–third this length. Seedlings are 15 to 20 cm long, blunt, and without conspicuous lenticles.

Ranges from western Africa to the Pacific coast of tropical America. In Africa its precise latitudinal limits are not certain, but it is recorded as far south as Angola and as far North as Mauritania. In the Americas it has, on the Atlantic side, a wide distribution to about latitude 25°N in Florida and to eastern Brazil, and on the Pacific side from Mexico to northern Chile, where its southern range is limited by the cold, dry climate. A better understanding of its possible co-occurrence with *R. samoensis* in Pacific South America is desired.

(40) *Rhizophora harrisonii*

There is circumstantial evidence that the species is a hybrid between *R. mangle* and *R. racemosa* but, if so, its geographic distribution is peculiar, being greater than one of its putative parents. It co-occurs with its 2 parents in West Africa (Keay, 1953) and the Atlantic coast of tropical America. However, Breteler (1977) claims that it occurs on the Pacific coast of tropical America in the absence of *R. racemosa*. Since the Panama Isthmus was last open over 3 million years ago, it seems that the situation needs re-examining, because it is unlikely that a species could exist in a vegetative condition for this length of time if *R. harrisonii* is infertile. However, Prance (1975) has since recorded *R. racemosa* on the Pacific coast of Colombia.

(ii) Genus : Kandelia (Wight and Arnold, 1834)

A monotypic genus, ranging from the Ganges Delta, Burma, through Southeast Asia to South China, the Ryukyu Islands and south Japan. In the southern part of its range it is confined to northeast Sumatra and northern Borneo, an unusually restricted southern limit for mangrove Rhizophoraceae. It is distinguished from related genera in the Rhizophoreae by the slender calyx lobes, the numerous, glabrous petals, the essentially unilocular ovaries, and the absence of either buttresses or specialized aerial roots. The hypocotyl of the seedlings is characteristically slender and tapered at each end.

(41) *Kandelia candel* (L.) Druce 1914.

K. rheedii Wight and Arnold 1834

Small understorey tree growing to 7 m. Buttresses and pneumatophores absent. Bark smooth, grayish or reddish brown. Leaves opposite, bijugate, 6 to 13 by 2.5 to 6 cm, oblong-elliptic to narrowly elliptic-lanceolate, apex obtuse, base cuneate, margin entire, somewhat reflexed. Petiole 1 to 1.5 cm long, terrate. Stipules to 2 cm long, overlapping as in *Rhizophora*. Inflorescence axillary, bifurcating with 4 or more flowers, peduncles 2 to 3 but up to 5 cm long, slender. Bracteoles cup shaped, the pair below the flower either enclosing the calyx cup or separated from it by a pedicel 2 to 3 mm long. Flowers 1.5 to 2 cm long, white; calyx lobes (5–6) linear, 15 to 20 mm, reflexed at anthesis. Petals (5–6) to 15 mm long, scarcely 1 mm wide, with a filamentous appendage between the lobes, the lobes themselves with 3 or 4 further linear appendages 5 to 6 mm long. Stamens numerous (30 to 40) inserted on the rim of the calyx cup, filaments uneven in length, 8 to 15 mm long, anthers minute. Floral disc cup shaped, enclosing the shallow nectar cup. Ovary inferior, essentially unilocular but with evidence of 3 carpels, ovules 6, anatropous, attached to basal projecting placenta. Style filiform, 1 cm long, stigma minutely 3 lobed. Fruit 1.5 to 2 cm long with persistent reflexed sepals, peduncle elongating. Seedling hypocotyl upto 40 cm long at maturity, slender, pointed apically and tapered at each end.

The architecture in the sapling stage corresponds precisely to Attims's model, with continuous growth and continuous to diffuse branching. The floral biology is not well known but seems to be the least specialized of all mangrove Rhizophoraceae, since the flowers apparently attract a diversity of small flying insects. Petals and stamens are ephemeral. In its floral biology it most resembles *C. decandra*, although its flowers are much larger. The generalized floral construction of this species may represent the ancestral condition in the tribe.

Kandelia occupies a narrow niche in the mangrove forest, since it is nowhere abundant and typically occurs in the back–mangrove communities or on the banks of tidal rivers farther inland. It still has some pioneering propensity, as recorded by Steup (in Ding Hou, 1958).

(iii) Genus : Ceriops (Arnold 1838)

Evergreen, small to moderately tall (upto 20 m high) tree with perfect flowers, typically with a distinct trunk and short basal buttresses apparently originating as short basal stilt roots. Pneumatophores sometimes developed as looped surface roots, as in *Bruguiera*. Sclereids absent from all parts. Bark smooth or slightly fissured, pale grayish or white, often with a red tinge, peeling in thick strips from the buttressed portion. Twigs jointed with swollen nodes. Leaves opposite, blade on the order of 7 to 12 by 3 to 5 cm, with a well–developed terete petiole 1 to 2 cm long; blade ovate to slightly obovate or elliptic–oblong, glabrous, without prominent veins, apex rounded or slightly emarginate, never apiculate, primordia including minute protruding hydathodes on lower surface. Cork warts lacking. Buds 1.5 to 4 cm long, characteristically spear shaped, flattened, the individual stipules thickened dorsally and including copious mucilage drying to a varnishlike consistency, both within the free space of the bud and covering young expanded portions to make them shiny. Inflorescence solitary, axillary with a short (5 mm) or moderately long (1.5 cm) axis and few (2 to 8), rarely many, flowers in compact bifurcating cymes, the flowers and axes protected by thick enveloping bracts and bracteoles. Flowers small (5 mm long) each enclosed immediately in a cup–shaped pair of partially fused bracteoles. Calyx with (5–6) pointed valvate lobes. Petals 5(–6) free, keeled, 3 mm long with apical appendages and marginal hairs, at first white but becoming brown with age. Stamens twice as many as the petals, with short filaments and at first enclosed in pairs by the petals (*C. tagal*), or with short filaments and uniformly spaced (*C. decandra*). Disc within the stamen ring well developed, lobes enclosing the base of the stamens. Ovary semi–inferior, calyx tube short; locules 3 with 2 ovules per loculus; style slender and usually with 3 obscure stigmas. Fruits conical mainly by extrusion of the upper part of the ovary, surface brown and roughened, calyx tube warty, lobes persistent. Seedling and ultimately the cotyledonary tube penetrating the fruit. Hypocotyl upto 20 cm long, 5 mm at the widest part, terete or ridged, detached from the cotyledon at the moment of release. In *C. tagal* n = 18 (Taxon 27:29, Kilwa Rd., Dar–es–Salaam, Tanzania).

There are 2 species widely distributed from East Africa and Madagascar throughout tropical Asia and Queensland to Malensia (New Caledonia and the Soloman Islands) and through Micronesia north to Hong Kong. The precise eastern limit is not known. In Melanesia *C. tagal* is recorded only for the very north of New Caledonia (Diohot, e.g. *Schmid 1594 as C. timoriensis*) and only from Winton Bay, Malekula (*Gillison RSNH 3556*), New Hebrides. The species are typical constituents of the inner mangrove, often forming pure stands on better drained sites, becoming stunted in exposed and highly saline sites, but rarely losing their simple trunk. They are easily distinguished from *Bruguiera* and *Rhizophora* by their glossy green, rounded (never pointed or apiculate) leaves, flattened buds, slender and rather short hypocotyls, and pale–barked trunk with restricted flaky–barked buttresses. The clusters of small shiny flower buds or short peduncles encased in varnishlike material are also characteristic.

Species and Varieties

Although similar vegetatively, the 2 species can be distinguished in technical details as follows :

1A. Inflorescence axis relatively long and uniformly slender (10–20 by 2 mm). Petals enclosing the stamens in pairs at anthesis and opening explosively, the apex of the petal with 3 clavate appendages. Stamens with long, slender filaments, much exceeding the blunt anthers. Hypocotyl terete or slightly ridged .... *C. tagal.*

1B. Inflorescence axis short and thick, expanded distally (10 mm or less by 3–4 mm). Petals not enclosing stamens at anthesis, the stamens in a single series not opening explosively, the apex of the slightly keeled petals with a fringe of filamentous appendages. Stamens with a short filament, equalled or exceeded by the anther, which ends in a longer or shorter appendage. Hypocotyl sharply ridged .... *C. decandra.*

Fruits and seedling characters have been used to separate the species as follows :

1A. Hypocotyl long, upto 25 cm or more, warty throughout; calyx lobes in fruit usually widely spreading or reflexed, calyx smooth. Fruits pointed apically, 15–20 mm long at maturity .... *C. tagal.*

1B. Hypocotyl not exceeding 15 cm long, warty toward apex, calyx lobes in fruit more or less erect, calyx lenticulate or warty. Fruit blunt apically, 10–15 mm long at maturity .... *C. decandra.*

These characters are not always easy to apply because they depend in part on the development of the fruit and seedling. There seem to be no constant vegetative features that distinguish the species. Wilkinson (1981) lists anatomical features of seedlings in which the 2 species differ.

(42) *Ceriops tagal* (Perr.) C.B.Robinson 1908

C. candolleana Arn.

C. timoriensis Domin

C. boiviniana Tulasne

Although several specific names have been recognized, current opinion relegates them all to one rather variable entity. The distinguishing features are –

(i) A single resting bud (the most usual condition)

(ii) A single sylleptic branch, which is commonly associated with a supernumerary bud between branch and petiole.

(iii) An inflorescence, which also may be associated with a supernumerary bud internodes on distal shoots are shortened, characteristically in relation to flowering.

The 2 species differ strikingly in their floral mechanism. Most information about pollination is available for *C. tagal*, the flower buds appear to open mainly in the evening, emitting a faint but fragrant odour, and they may remain open the following day. Anther dehiscence occurs in bud, and at anthesis the petals are closed, enveloping the stamens in pairs exactly as in *Bruguiera*. Pollination is probably mainly by night–flying insects; moths have been observed visiting the flowers, presumably for the small quantity of nectar secreted by the disc, but bees may be daytime visitors. Pollen release is explosive and is triggered by a delicate touch of the petals. This would presumably throw pollen onto a suitably positioned flower visitor. The tension that sets this mechanism is generated by the enclosed stamen pair held back by the pouched petal.

In *C. decandra*, the stamens are not enclosed at anthesis, and they retain their antesepalous or antepetalous position without displacement. Antepetalous stamens are oppressed to the subtending petal, but there is no explosive release of pollen; the short filament is not suitable for developing tension. Details of flower visitors are lacking for this species, but the floral mechanism is apparently much less specialized than that of *C. tagal*.

(43) *Ceriops decandra* (Griff)

C. roxburghiana Arn.

Ecological Distribution

Ceriops tagal is typically a plant of the inner mangroves, apparently on better drained soils; flooding with salt water is infrequent but produces high soil salinity by evaporation. Under these circumstances it will form pure stands. *Ceriops decandra* more often grows within the tidal zone mixed with other Rhizophoraceae. The 2 species differ in the major features of floral biology, as is discussed later; otherwise, they are sufficiently similar that a general account of vegetative features illustrates both. In Queensland *C. decandra* seems to have larger and darker leaves, but this may be a consequence of habitat differences.

Seedling and Sapling

Dispersal and establishment of seedlings follow the pattern usual for the Rhizophoraceae. The sapling is initially unbranched, but once branching begins it may be continuous or intermittent, depending on the vigor of the plant. Branching is by syllepsis, with the lateral axes developing in pairs within the terminal bud. This pattern is repeated on successive orders of branches but without any marked tendency to develop plagiotropy by apposition, as in *Rhizophora* and *Bruguiera*. Consequently the crown remains relatively narrow. Vegetative branching becomes less frequent on distal axes. This minimizes the congestion of axes.

The terminal buds of *Ceriops* are distinctive in their construction. The individual stipules, which are shorter than those in the other Rhizophoraceae, have a marked dorsal thickening so that the bud as a whole is flattened. There is much free space even within this flattened bud, since only a single younger pair of leaf primordia even stipules is present at any one time. In addition the fulid, presumably secreted by the stipular collectors, dries to form a semisolid plug of material that occupies the free space within the bud. The same material coats the surface of the newly extended internodes and young appendages in the manner of a varnish that flakes and falls off as the organs expand.

With the onset of flowering, which occurs in quite small saplings, the full complement of lateral appendages developed by the terminal bud is present so that each leaf may subtend the following :

(iv) Genus : Bruguiera Lamarck

Trees growing to 40 m high, evergreen, with continuous growth, hermaphroditic; short aerial roots of sapling developing into stout buttresses of adult trees; pneumatophores kneelike. Leaves opposite, bijugate; petiole, entire; blade glabrous, usually ovate to oblong–ovate, with a cuneate to acuminate base, apex usually bluntly pointed. Stipules in pairs, 1 to 4 cm long, sometimes reddish, orientation as in *Rhizophora.* Flowers either solitary or in shortly stalked, bifurcate, 2 to 6 flowered cymes. Bracteoles never enclosing base of flower. Flowers either large (2 to 4 cm long) or smaller (1 to 1.5 cm long) with either about 13 (large flowered) or about 8 (small flowered) pointed, initially valvate, fleshy calyx lobes. Petals delicate, as many as and alternating with the calyx lobes, bilobed and with marginal, interlocking hairs, the lobes commonly with appendages distally and usually with a rigid filament between them. Stamens with slender, somewhat contorted filaments, twice as many as sepals and enclosed in pairs by the pouched petals, dehiscing precociously, the pouches exploding when triggered. Ovary inferior, obscurely 2 to 4 locular, with 2 ovules in each loculus. Floral disc cup shaped, style slender with an obscurely 3–lobed stigma. Fruit 1 seeded with persistent, erect, or reflexed lobes. Hypocotyl terete or obscurely ribbed, cigar shaped, blunt apically. Fruit falling with the seedling, the cotyledonary collar not extending.

The genus is distinguished from other mangrove Rhizophoraceae in its numerous (more than 6) calyx lobes and the propagule, which consists of both seedling and fruit, unlike the other genera in which the seedling is detached by abscission of the cotyledonary node so that cotyledons and fruit remain on the tree, abscissing independently after the seedling is detached. *Bruguiera* is also consistent in its explosive mechanism of pollen release; this is triggered by a visiting pollinator, in which only *Ceriops tagal* of the other Rhizophoreae resembles it.

The genus is somewhat imperfectly segregated into 2 groups of species; the large, solitary–flowered group (e.g. *B. gymnorrhiza*) and the small, many–flowered group (e.g. *B. parviflora*). The difference in flower size is correlated with difference in leaf size and fruit size. This distinction was the basis for the recognition of 2 genera : *Bruguiera* (sensu stricto) and *Kanilia*, representing the first and second group, respectively, by Blume. Subsequently Miquel reduced *Kanilia* to a section of *Bruguiera* and applied the name *Mangium* (from Rumphius) to the other section. However, Ding Hou (1958) pointed out that the distinction between the 2 sections was obscured by the later discovery of the 2 species *B. hainesii* and *B. exaristata*, which in some features are morphologically intermediate. The first species sometimes has solitary flowers; the second sometimes has flowers in pairs and the size difference between flowers is not great. On the other hand, there is a fairly sharp distinction between species on the basis of flower orientation in relation to contrasted classes of pollinators (birds versus insects). The morphological continuum is thus transcended by a functional discontinuity, but this is insufficient justification to revive the sectional names.

The following key provides diagnostic information about the 6 species recognized in *Bruguiera* :

1A Inflorescence 1(–2) flowered; flowers recurved, 2–3 cm long, calyx lobes .... about 13; sinus between lobes of petals empty or at most occupied by short bristles not exceeding the lobes. Fruit stout, 0.5–2 cm in diameter. Large–flowered species pollinated mainly by birds 2

1B Inflorescence (1)–2–6 flowered; flowers oblique or erect, 1–1.5 (–2) cm long, calyx lobes about 8(–10); sinus between lobes at petals always occupied by a long bristle exceeding the length of the lobes. Fruit slender, 0.4–0.8 cm in diameter. Small–flowered species pollinated by insects.. 4

Large–flowered species

2A Leaves obovate, 5–9 by 3–4 cm. Flowers sometimes in 2's, 2–2.5 cm long, yellowish green. Petals 9–11 mm long, usually without a bristle in the sinus between the petal lobes .... *B. exaristata* Ding Hou

2B Leaves elliptic–oblong, 10–20 by 5–7 cm, usually longer than 10 cm. Flowers 3–4 cm long, reddish. Petals 12–15 mm long, with a bristle in the sinus between the petal lobes. 3

3A Tips of petal lobes blunt, without filamentous appendages ....*B. sexangula*

3B Tips of petal lobes acute, each extended into 3 filamentous appendages .... *B. gymnorrhiza* (L.)

Small–flowered species

4A Calyx lobes slender, short (less than 3 mm), only one–fourth to one–fifth the length of the calyx, erect or at most slightly spreading in fruit. Petals 1.5–2 mm long .... *B. parviflora*

4B. Calyx lobes stout, long (more than 3 mm), half the length of the calyx, horizontal or reflexed in fruit. Petals 4–9 mm long 5

5A. Mature flowers 10–12 mm long. Petals 3–4 mm long. Calyx tube at mouth ... 2 mm in diameter, calyx lobes 8 and completely reflexed in fruit .... *B. cylindrica* (L.)

5B Mature flowers 18–22 mm long. Petals 7–9 mm long. Calyx tube at mouth 5 mm in diameter; calyx lobes 10 and extended at right angles in fruit .... *B. hainesii* C.G. Rogers

Details descriptions are provided for one representative of each group. A complete synonymy is provided by Ding Hou (1958); see also Ding Hou (1957).

(44) ***Bruguiera gymnorrhiza*** **(L.) Lamk.**

B. rheedii Blume

B. conjugata Merr.

Trees to 30 m or more high with short buttresses. Bark rough, black, fissured in a regular chequered pattern. Shoots plagiotropic by apposition, developing terminal short shoots (Aubreville's model). Leaves elliptic–oblong, coriaceous, 8 to 22 by 5 to 10 cm, apex bluntly pointed, base cuneate, Petiole up to 4 cm long, often glaucous with a white wax, as on the young stems, leaves often reddish beneath. Stipules to 4 cm long, often reddish. Flowers solitary, upto 3.5 cm long, reflexed at anthesis, pedicel naked, upto 2.5 cm long. Calyx usually reddish with 12 to 14 lobes that remain erect in fruit. Petals to 15 mm long. 2 lobed with the deep median sinus occupied by a slender bristle, additional 3 to 4 bristles on the acute petal lobes. Stamens enclosed in pairs by the petals. Calyx cup deep, style slender, filliform with 3 to 4 stigmatic lobes, calyx tube scarcely elongating in fruit. Hypocotyl to 25 cm long, cigar shaped, blunt apically, slightly angular.

This tree is a characteristic element of the middle mangrove community throughout its range, extending into the transitional landward communities. Ding Hou (1958) describes it as the "largest and probably the longest lived of the mangrove community". It may grow on exposed shores where the mangrove belt is thin or depauperate.

It has the broadest range of the genus and indeed of all mangroves, extending from East Africa (including Madagascar) through Ceylon and the Malay Archipelago into Micronesia and Polynesia (Samoa) northward to the Ryukyu Islands and southward in tropical Australia (Western Australia and Queensland).

Bruguiera gymnorrhiza offers a good example of Aubreville's model and may be contrasted with *Rhizophora* (Attim's model), although there are some close resemblances. The difference between these 2 architectural models leads to important differences in crown shape, however, and particularly in crown plasticity. The sampling axis is orthotropic and with radial symmetry as a result of the bijugate phyllotaxis. It goes through a period during which it remains unbranched, with the length of time depending on the vigour of the sampling. This vigor is reflected by the spacing of the leaves; axes with numerous, close-set leaf scars suggest slow growth; axes with few distant leaf scars suggest rapid growth.

Branching is by syllepsis, as is usual in the family. The branches are markedly plagiotropic and sympodial by apposition. After a limited time of horizontal growth, the axis turns erect and continues its activity as an erect short shoot; producing a rosette of leaves as a consequence of the bijugate phyllotaxis. A renewal shoot arises in the axial of a leaf at the base of the erect shoot; this extends horizontally to repeat the deelopmental process. The system is proliferated when 2 instead of 1 renewal shoots develop, the 2 renewal shoots usually diverging at an angle of about 60° to form an irregular hexagonal pattern.

(45) *Bruguiera cylindrica* (L.)

Rhizophora caryophylloides Burm.

B. malabarica Arnold

Tree to about 20 m high with short buttresses, bark grayish. Leaves elliptic, 7 to 17 by 2 to 8 cm, with a bluntly pointed apex and cuneate base, petiole to 4 cm long, often reddish. Stipules to 3 cm long. Inflorescence usually 3–flowered, peduncle about 1 cm long. Flowers greenish, erect at anthesis. Calyx tube 4 to 6 mm long, 2 mm in diameter, with 8 lobes about as long as the tube. Petals shortly bilobed with 2 or 3 bristles at the apex of each lobe; each petal enclosing a pair of stamens. Calyx tube somewhat enlarged in fruit, the lobes reflexed, detached with the seedling. Hypocotyl upto 15 cm long at maturity, grooved or angled.

The plant usually grows like a small tree in the inner mangroves but is described as specifically characteristic of newly established substrates, being also somewhat susceptible to prolonged submersion. Because of its pioneering propensities, it may form pure stands, but it is also said to be very slow growing.

It is described through the Malay Archipelago as far as New Guinea and north Queensland.

This species has the same habit as *B. parviflora*, with which it may occasionally grow. The floral mechanism is also the same and involves explosive pollen discharge after triggering by small insect visitors.

(46) *Bruguiera sexangula* (L.)

B. eriopetala

B. malabarica

This species resembles *B. gymnorrhiza* closely, but differs in the somewhat smaller, thinner leaves and most clearly in the absence of long appendages on the blunt petal lobes. Wyatt–Smith (1953a) suggests that the calyx is smaller and more distinctly ribbed. The twigs and petioles lack the white waxy covering that is often characteristic of *B. gymnorrhiza*; unlike that species, the calyx is never a conspicuous scarlet, although it is occasionally reddish. The aerial roots at the base of the trunk, which appear in the sapling, may be quite well developed.

(47) *Bruguiera exaristata* Ding Hou

It seems remarkable that this distinctive species should have remained overlooked until so recently, but the reason is probably its range (northern Australia, Timor and south New Guinea) in areas that are poorly explored botanically.

It is distinguished from the 2 other species of the large, solitary–flower group in its appreciably smaller, yellowish green flowers (never reddish) with an 8(–10) lobed calyx, smaller leaves, and limited stature (not exceeding 20 m). In Queensland it is a common constituent of the landward part of mangal. The most frequent flower visitors are honey eaters.

(48) *Bruguiera parviflora*

This species is representative of the small, many–flowered group. It is distributed throughout Southeast Asia, from the Mala Peninsula to tropical Australia, the New Hebrides, and the Solomon Islands, but it is not recorded for New Caledonia. It occurs on the inner mangrove fringe and river banks and has somewhat the characteristics of a pioneering species. Good field characters are the light green stipules that form a slender tube and the delicate light green foliage with slender petioles scarcely 2mm wide.

Tree growing to 24 m with smooth, gray, obscurely lenticellate bark. Leaves 7 to 13 by 2 to 4 cm, elliptic, apex bluntly pointed, base narrowly cuneate. Stipules 4 to 6 cm long. Inflorescence 3 to 4 flowered, peduncle 2 cm long with obscure bracteoles. Flowers yellowish green, erect at anthesis, pedicles to 13 mm long. Calyx tube ridged to 9 mm long; calyx lobes usually 8, short, that is, less than one–fourth the length of the calyx tube. Petals to 2 mm long, each petal lobe with bristles, each petal enclosing 2 stamens. Style 1 to 1.5 mm long, stigmas 2 or 3. Fruit with calyx tube enlarged 2 to 3 fold, calyx lobes remaining erect. Hypocotyl smooth, to 15 cm long, truncate at the apex, scarcely 5 mm in diameter, at first green and erect, becoming brown and pendulous.

In this species the flowers remain erect and so point distally, that is, outward in the crown. They are visited by small insects including butterflies, which seem capable of the delicate triggering of the explosive pollen release and are probably effective pollinators.

The hypocotyl becomes arched a few days after falling to the ground so that its radicular end is usually in contact with the mud. The cortex at this end becomes stellately expanded and aids in fixing the hypocotyl, facilitating the penetration of the extending radicle, which becomes a taproot.

This species confirms more closely to Attim's model in contrast to the clear–cut Aubreville's model shown in *B. gymnorrhiza*. Shoots have a strong tendency to remain orthotropic, and the crown of saplings is rather narrow and conical.

(49) *Bruguiera hainesii*

B. eriopetala

This species resembles the other small–flowered *Bruguieras* but is distinguished by its larger flowers and the usually 10–lobed calyx, which stands out at right angles to the calyx tube in the fruit. It is somewhat intermediate between the 2 flower types in the genus. The inflorescences are usually 2– or 3–flowered, but solitary flowers occur. In its erect flower position it suggests that it is suited to small insects that are the putative pollinators of *B. parviflora* and *B. cylindrica*, but there are no records of lower visitors.

14. FAMILY : Rubiaceae

A very large, cosmopolitan and natural family with about 500 genera and over 6000 species represented in the tropics mainly by woody plants. Interpetiolar stipules with associated glandular structures (collectors) are a diagnostic feature; they protect terminal buds and usually persist on mature stems. The family includes several commercially important species, notably coffee.

Scyphiphora is a frequent but minor mangrove constituent in the Old World; a mangrove associate from the New World is also described.

Scyphiphora Gaertn. f.

A monotypic genus with a range from south India and Ceylon, Indo-China and Hainan, through the Malay Archipelago and Philippines to tropical Australia and New Caledonia, northward to the Solomon Islands and Palau. It is an uncommon constituent of mangroves and more exposed coastal sites including beaches.

(50) *Scyphiphora hydrophyllacea* Gartn. f.

An erect shrub, rarely exceeding 2m. Leaves opposite, decussate, branching mainly by syllepsis but discontinuous (Attim's model, Halle *et al.* 1978). Leaves glabrous, somewhat fleshy or coriaceous, oblong–ovate, 4 to 9 by 2 to 5 cm, acute at base, apex rounded to bluntly pointed or slightly emarginate, margin entire; petiole 1 to 2 cm long. Interpetiolar stipules short (to 3 mm), rounded, minutely hairy on the margin, at first enclosing younger parts; with well–developed collectors internally at the base. Flowers perfect, protandrous, usually tetramerous, in axillary regular, usually condensed cymes with 3 to 7 (upto 13) flowers, bracteoles obscure. Peduncles 2 to 15 mm long; flower sessile or shortly but obscurely pedicellate. Calyx tube glabrous, 3 to 5 mm long, free portion of calyx scarcely 2 mm long entire with 4(–5) obscure teeth. Corolla tube 3 to 4 mm long with 4(–5) at first contorted white (or slightly red), bluntly pointed petal lobes, reflexed at anthesis, throat of corolla tube occluded with dense hairs. Stamens 4(–5) inserted on the mouth of the corolla tube with a short (1mm) filament, anthers 2 mm long, medifixed and dehiscing introrsely before the flower opens. Ovary inferior, indistinctly separate from calyx tube and pedicel; 2–locular, with 2 ovules in each cell, 1 above the other, the upper ovules erect, the lower pendulous. Style slender with a club–shaped bilobed stigma, the lobes eventually spreading. Fruit green, glabrous, upto 1 cm long, crowned with persistent calyx, with 8(–10) longitudinal ridges. Outer layers of fruit fleshy, inner corky. Seeds 4 (or fewer) separated horizontally by an incomplete septum, each with a straight embryo, endosperm present, testa thin. Germination hypogeal.

The terminal buds are protected by the hoodlike stipules that secrete a copious varnishlike substance (presumably originating in the palisade of collators internally at the base of each stipule pair) that forms a plug enclosing the bud. This material also coats the exposed surfaces of younger organs, especially inflorescences. The stipules are persistent. Branches are wide–angled and sylleptic, inserted a little above the node so they characteristically extend from above the mouth of the stipules. Growth is seemingly intermittent but irregular. There are apparently no supernumerary buds so that reserve buds occur only at nodes that do not develop sylleptic branches. This restricts the reiterative capacity of the plant.

The flowers are in fairly regular compound but congested dichasia, but the order of flowering does not follow their age very precisely. Flowers in one head are often synchronous in development. The anthers dehisce introrsely before the flower opens. The large and sticky pollen grains form a mass that is retained between the club–shaped stigmas and the stamens, being prevented from falling into the corolla tube by the weft of stiff hairs at the mouth of the corolla. The petals reflex, exposing the still closed, club–shaped stigmas; pollen is further exposed by elongation of the stipule.

15. FAMILY : Sonneratiaceae

A small tropical family of 2 small genera in the Indo–Malayan region. *Sonneratia* is restricted to mangrove communities, it has essentially solitary flowers, numerous stamens, vestigial or no petals, a fruit that does not dehisce regularly, and seed without extended tails. *Duabanga* includes 2 species in lowland tropical rain forest and has a more restricted range. It is distinguished from *Sonneratia* by its several flowered inflorescence, conspicuous petals, 12 (or more) stamens, and especially the capsular fruit, dehiscing by 4 or more valves to release the numerous tailed seeds; vegetatively it lacks the conspicuous pneumatophores found in *Sonneratia*.

The family is considered close to Punicaceae, Crypteroniaceae and especially Lythraceae; similarities between seedling *Sonneratia* and species of *Lythrum* are particularly striking. The following account is based on that by Backer and van Steenis (1954), where a complete synonymy is given.

Genus : Sonneratia L.f.

A genus of 5 to 6 species, ranging from East Africa through Indo–Malaya to tropical Australia and into Micronesia and Melanesia. It is a typical constituent of mangrove communities throughout its range, often forming a seaward fringe, and it is recognized by its tall conical pneumatophores arising from horizontal roots, its opposite, simple, orbicular leaves, large flowers, and large, globose fruits with persistent calyx. It is sometimes referred to as the "firefly mangrove" because these insects congregate on the trees at night. Trees are relatively uniform in their vegetative features, the following account describes these for one representative species. Where vegetative differences are known, they are included in the key.

Tentative key to Sonneratia species

1A Calyx 4(–6) lobed, ovary and calyx not exceeding 2cm long, ovary 5–8 celled, stigma expanded, mushroomlike fruit 1.5–2 cm in diameter. Leaves narrow (less than 5 cm wide), but sometimes wider, gradually tapering toward the apex .... *S. apetala* Buth–Ham.

1B celled, stigma capitate but not expanded, fruit 3–7 cm in diameter. Leaves usually 5 cm wide or wider, abruptly narrowed to the rounded or even emarginate apex......... 2

2A Calyx flat, extended horizontally, not enclosing the ripe fruit, at most obscurely ribbed. 3

2B Calyx cup shaped, enclosing the base of the fruit, prominently ribbed 4

3A Twigs not pendulous. Leaves obovate to suborbicular, petiole scarcely developed, midrib green throughout; veins conspicuous, prominent on the upper blade surface; petals absent; filaments white .... *S. griffithii* Kurz.

3B Twigs slender, pendulous. Leaves elliptic–oblong or oval–obovate to narrowly elliptic, petiole short, midrib often red at base. Veins inconspicuous, not prominent; petals usually present, filaments red below, white distally .... *S. caseolaris* (L.)

4A Apex of fruit not depressed at base of style. Tube of the fruiting calyx smooth, lobes reflexed, petals present. Leaves ovate to oblong–ovate, with a short thick petiole .... *S. alba* J. Smith

4B Apex of fruit depressed at base of style. Tube of the fruiting calyx finely warted, lobes ascending, petals absent. Leaves broadly ovate, as broad as long, with distinct, narrow petiole .... *S. ovata* Backer

Wyatt–Smith (1953b) gives field characters for the 3 common species in Malaya.

(51) *Sonneratia alba* J.Smith 1819

S. griffithii

This is said to be the most widely distributed species, it ranges from east Africa and Madagascar to Southeast Asia, the Malay Archipelago to the Phillipines, and tropical Australia to Micronesia, the New Hebrides, and New Caledonia. According to Chai (1982) it is a pioneer species, colonizing newly formed sandy mud flats in sheltered situations.

It is distinguished by the numerous white petals that are tinged red apically and the cupular calyx under the fruit with smooth (not hairy) lobes that are usually reflexed in a characteristic manner. The leaf is obovate to oval, with the apex broadly rounded and even emarginate. The seeds are falcate and smooth.

(52) *Sonneratia apetala* Buch–Ham. 1800

This species is perhaps the most distinctive in the genus but is the least common, being restricted to southern India and Burma. It is recorded as very rare in Ceylon, with a population of only 6 trees near Muttur in the estuary of the Koddiyar River (MacNae and Fosberg, 1981). It is a small to medium–sized tree growing to 15(–20) m high with narrow leaves gradually tapering toward the apex. The flowers (1.5 to 2 cm) are consistently smaller than those of the other species, and the usually 4 calyx lobes are twice as long as the tube. The ovary is 5– to 8– celled, and the stigma is broad and mushroom shaped at anthesis. The fruit (1 to 2 by 2 cm) is proportionately smaller than in the other species, the calyx lobes remain flat and do not enclose the fruit.

The specific name is not entirely appropriate, since other species of *Sonneratia* are apetalous, whereas apetaly occurs in flowers of populations of species that may otherwise retain petals.

(53) *Sonneratia caseolaris* (L.) Engler 1897

S. acida L.f.

S. lanceolata

Trees to 15 m with continuous growth but diffuse branching (Attims's model), branches horizontal or drooping. Leaves glabrous, opposite, without stipules, shortly petiolate to almost sessile. Adult leaves broadly ovate or obovate, usually with a blunt apex, somewhat fleshy with an entire margin, 3 to 5 by 4 to 7 cm, juvenile leaves lanceolate, less fleshy, often with extended, red petioles. Apex in most leaves with a minute recurved spiculus serving apparently as a domatium rather than a hydathode. Nodes (including bracteolar nodes) with 2 lateral pairs of circular glands, 1 pair on each side of the stem. Axis angled, usually almost square in transverse section. Flowers ephemaral, opening in late evening and lasting 1 night. Flowers solitary or in few-flowered dischasia, terminal at the ends of outer twigs or in the axils of distal leaves with atleast 1 pair of subtending bracteoles. Flowers with a shallow, green calyx tube with 6 to 8 somewhat longer valvate teeth; petals slender, reddish, alternate with calyx teeth (or absent), inconspicuous, and best seen in flower bud. Stamens numerous, inserted on the inner rim of the calyx tube, filaments slender, 2 to 3 cm long, white above, reddish below, anther bilocular, medifixed. Ovary globose but flattened or even depressed above, style simple, about twice the length of the stamens, with a capitate stigma; ovary with numerous (up to 20) locules, each with numerous ovules on essentially axile placentas. Fruit a green, somewhat leathery "berry" with persistent subtending calyx; seeds numerous in the fleshy pulp of the placenta, the individual seeds irregular and angular. Germination epigeal. Root system including an extended series of cable roots giving rise to narrow, shallowly descending lateral roots and erect pneumatophores. Pneumatophores at first greenish gray with a flaky back, extending to as much as 2 m at maturity, tapering, woody via secondary thickening and with numerous narrow second-order roots developing horizontally in the substrate. The characteristic habitat of the species is river banks and tidal areas with mud banks. It ascends the mouth of the river farther than other species.

This description may serve for he other species, which are distinguished by leaf shape, flower size and fruit shape, as presented in the earlier key.

This species is common in the inner mangroves and extends inland along tidal creeks as far as the influence of salinity extends. It has a distribution from Ceylon throughout Southeast Asia, the Malay Archipelago and Phillipines, north to Hainan, east to the Solomon Islands and the New Hebdrides (but not New Caledonia) and into tropical Australia.

This species is distinguished by the large fruits in which the calyx lobes expand horizontally (not erect) in fruit and the red petals, which contrast with the greenish or yellowish white of the surface of the sepals. The leaves do not have prominent veins and are usually pointed at the apex. The seeds, which tend to float and not sink, are smaller than those of *S. alba* and *S. ovata*.

Sonneratia lanceolata has been distinguished from *S. caseolaris* by its narrower leaves, white stamens and uniformly round (not constructed) flower buds.

(54) *Sonneratia griffithii* Kurz

This species has a restricted distribution along the shores of the Andaman Sea, northward to Bengal and southward to the upper Malay Peninsula. Though described as locally common, it is little collected. It is close to *S. caseolaris*, from which it is said to differ in the prominent leaf veins and absence of petals.

(55) *Sonneratia ovata* Backer

S. alba

This species has a distribution from Thailand through the Malay Peninsula and Malay Archipelago to the Gulf of Papua in New Guinea and into

Queensland. It is recorded by Chai (1982) as an inland river–bank species but still within salt–water influence. It lacks petals and is distinguished by the finely warty calyx, the lobes of which are red on the inner side, and by the erect calyx segments in fruit.

16. FAMILY : Sterculiaceae

A tropical and subtropical family of trees and shrubs with alternate simple leaves. It includes several commercially important species, notably *Thoebroma cacao* (cocoa). One genus is represented in the Indo–Malayan mangroves.

Genus : Heritiera Aiton.

A genus of 29 species (Kostermans, 1959) mainly of rain forest trees with a distribution from eastern tropical Africa (2 species) and the remainder from India to the Pacific (Tonga and Niue), with *H. littoralis* introduced farther east in Polynesia and to Hawaii. The genus is characterized by its leaves with a scaly indumentum on the lower surface, the leaves either digitately compound or more usually simple (unifoliable by reduction). The fruit is usually a wind–dispersed samara, but in water–dispersed species the wing is reduced to a vestigial keel.

A number of species have a coastal distribution. The genus is predominantly found in inland forests, but *H. littoralis* is a characteristic constituent of the back mangal, occupying almost the entire range of the genus; two others may be described as mangrove associates, both with a limited distribution. Several of the inland species are important timber species, for example, *H. sylvatica* Vid. in the Philippines.

Key to mangrove or salt–tolerant species of Heritiera

1A. Petiole exceeding 2 cm (to 4 cm), fruit globose with a distal slightly recurved beak, plants of back–mangal fringe and lowland forests in northern Borneo .... *H. globosa* Kosterm.

1B Petiole not exceeding 2 cm (1–2 cm), fruit ovoid, extended, and flattened on dorsal side, without a distal beak; plants of back mangal, distribution not as in 1A. 2

2A Fruit knobby with a ventral ridge together with a transverse, circular ridge; plants with a limited distribution in tidal swamps in the Ganges and Irrawaddy Deltas. Pneumatophores sometimes developed .... *H. fomes* Ait.

2B Fruit smooth with a rudderlike crest, but without a transverse circular ridge; plants with a wide distribution in back mangal, from East Africa to Melanesia. Pneumatophores absent .... *H. littoralis* Buch–Ham.

(56) *Heritiera littoralis* Dryand.

H. minor Lam.

Trees upto 25m with a buttressed trunk to 60 cm in diameter, evergreen, monoecious. Bark fissured, dark or gray. Leaves spirallv arranged, simple (but regarded as a compound leaf with 1 leaflet by reduction), stiffly coriaceous, oblong or ovate–elliptic, usually 10 to 20 by 5 to 10 cm (but upto 30 by 15 to 18 cm), with a usually obtusely pointed apex and somewhat cordate at the base. Maring entire. Veins prominent below. Leaves dark green above, grayish white beneath because of indumentum of overlapping, stellate scales (similar scales on young twigs). Pctiolc 1 to 2 cm long and 1 cm wide, shortly bipulvinate. Stipules in pairs at each node, short, subulate up to 1 cm long, caducous but conspicuous in resting buds and forming bud scales.

Flowers unisexual in complex tomentose panicles in axils mainly of distal foliage leaves on current growth, often subterminal, that is, arising immediately below resting terminal bud. Flowers 3 to 4 mm in diameter, 4 to 5 mm long, with a slender short pedicel, the male somewhat smaller; both sexes with a cup-shaped calyx. Receptacle within the cup with a flattened tuberculate disc, the tubercules represented by minute fleshy structures that resemble short stalked colleters. Petals absent. Male flowers more numerous, each with an androgynophore arising from the centre of the disc and consisting of stamens fused together around a central column representing the pistillore; stamens 4 or 5 (–6), anthers minute and forming a ring around the central column, dehiscent extrorsely by 2 longitudinal slits to release the small amount of pollen. Female flowers less numerous and usually terminating distal panicles of branches; disc scarcely developed, staminodes minute below the carpels. Carpels 4 to 5(–6) sessile, united loosely, each laterally compressed and with 1(–2) basal ovule, styles as a short extension of the carpel, stigmas minute and recurved. Fruits maturing in pendulous clusters, usually only 1 developing from a single carpel in each flower; always 1-seeded, 6 to 8 cm long, 5 to 6 cm wide, ellipsoidal, flattened on 1 side (ventrally) and with a raised (dorsal) keel 5 mm (but sometimes up to 2 cm) high on the opposite side, epicarp woody, mesocarp fibrous, endocarp hard. Embryo represented by fused cotyledons, radicle directed ventrally.

(57) *Heritiera fomes* Buch–Ham.

This species is characteristic of the forest of the Sunderbans of the Ganges Brahmaputra Delta (Khuma District) and in the Irrawaddy Delta of Burma. The Sundarbans is said to get its name from this tree (vernacular 'sundri'), which is characteristic of the region.

It is distinguished from *H. littoralis* in its fruit and is described by Parkinson (1934) as having the ability to produce "erect pointed pneumatophores like those of *Xylocarpus gangeticus* [= *X. mekongensis*], but more numerous and longer". Kostermans refers to these pneumatophores as "blind rootsuckers" and "perpendicular shoots from its roots", but is seems clear that they are morphologically root structures. Watson (1928) makes the same mistake in describing pneumatophores. Parkinson compares these 2 species of *Heritiera* with the 2 mangrove species of *Xylocarpus*; in both genera there are contrasted root structures in different species. From the available descriptions of the habitat of this species, it would probably be classified as a mangrove if it were more widely distributed. It is said to grow "down to the edge of the tidal creeks" in Burma (Kostermans, 1959).

Heritiera fomes is considered to be the best of the 2 commercial *Heritiera* species in India; it has properties superior to those of teak. The difficult access, which makes it uneconomical to exploit, may be an important factor in its conservation.

(58) *Heritiera globosa* Kostermans

This species is known from Sarawak to Sabah and adjacent Indonesian Borneo but may have a wider distribution. It is distinguished from *H. littoralis* by its globose fruit, which has a shallow ventral crest extended at the distal end into a beak or vestigial wing, with the wing always slightly recurved in a characteristic manner. The seed apparently retains undigested endosperm at maturity, and the embryo has 2 distinct cotyledons but an obscure plumule and radicle. Details of germination are not known. It is described as occurring behind the tidal zone of the mangrove belt, but has been collected 70 km from the sea; even here, although the water is fresh, there is a tidal fluctuation. The buttresses are said to be well developed, snakelike, and extending 2 to 4 from the base of the trunk.

Chapter – 5

PATTERN AND PROCESS IN MANGROVE ECOSYSTEM (A Case Study of Mangrove Forests of Andaman Islands)

Geographical Situation of Andaman Islands

Millions of years ago, there was a continuous hill range from the Arakan Yoma of Burma to the Achin head of the Indonesia. But gradually, due to geological activities, these hills were being submerged by the surrounding sea water keeping up only the peaks and ridges. These standing peaks and ridges are presently known as Andaman and Nicobar Islands consisting of 319 Islands.

This group of Islands is situated in the Bay of Bengal between 6° to 14° North latitude and 92° to 94° East longitude. It has a coastline of nearly 1,500 km which is 33% of the total coastal area of the Indian Union. This isolated archipelago forms a chain of Islands lying in northernly and southernly directions. The most northernly point is Landfall Island which is 193 km away from the South of Cape Negaris and 901 km from the mouth of the river Hoogley in West Bengal. The southern most point is Pygmalian situated in Great Nicobar at a distance of 146 km from Lape Pendra of Sumatra.

The group of Islands is divided into two districts, the Andaman and the Nicobar. The Andaman group, generally known as Great Andaman Group, consists of four group of Islands (Fig. 9).

(i) North Andaman
(ii) Middle Andaman
(iii) South Andaman and
(iv) Little Andaman

The land area of two districts is about 8,383 sq.km. where the Andaman contributes a land area of about 6,430 sq.km. in the form of 291 Islands. The Nicobar covers 28 Islands and spreads-up in a land area of 1,953 sq.km. These two are separated as 10° NL by a deep sea channel, about 150 km across, which is also called as "Ten Degree Channel".

Port Blair, the capital of these group of Islands, is situated at a distance of 1,255 km. from Calcutta, 1,190 km from Madras and 1,200 km from Vishakhapatnam. According to latest census (1981) the population of Andaman is 1,58,287 where as the total population of two districts is 1,88,741 having a population density of about 23 per sq.km. on land basis.

These islands present varied natural panorama and are endowed with virgin lush evergreen dense tropical forests, extending from sea coast to top of the hills, covering 2,912 sq.km. as the Reserved Forest and 4,232 sq.km. as the Protected Forest Area.

TOPOGRAPHY

The terrain is mostly hilly, though some flat lands along the sea coasts are seen. Saddle peak, the highest point in North Andaman is 732.12 m high above sea level. Some other important mountains are Mount Diavolo (514.19 m), Mount Angelica (465.42 m) of the Middle Andaman, and in some South Andaman Mount Koyoh (480.20 m) the highest point of the Harrite range, the general range of Mount Harrite is 365 m. Another few mountain ranges of South Andaman are Mount Eord (434.6 m) on Rutland Island, Mount Caddle (324 m) and Mount Cholunga (327.6 m). The mountain ranges are steeper in tal Eastern areas than the Western sides. In general the elevation of these hills is between 76 and 381 m and in most of the cases they descend steeply to the tide water.

The coastline is deeply indented, irregular and broken at several points and forms several small creeks, narrow channels and well sheltered natural harbours. The coastal area are generally rocky and muddy but in some cases sandy also. By precipitation of silt brought down to these coastal sides muddy flats are formed on which mangroves profusely grow. These Islands completely lack fresh water springs and river systems. The coral reefs at certain places extend far away from the shoreline an provide ideal situation for mangorve development.

GEOLOGY

The geologist consider the formation of Andaman Islands similar to that of several parts of the coasts of the Burma. These Islands form a part of lofty range of submarine mountain, 1,125.6 m long, running from Cape Negaris in the Arakan Yoma range of Burma, through Sumatra and Java to the Lesser Sunda Islands and the Moluccas. These groups of Islands project as long narrow submarine range with only the peaks visible above sea level. The rocks, in most of the cases consist of mainly non calcareous, grey, sandsone and imbeded shells with occasional nests of poor coal, conglomerates and pale grey limestone. The coastline rocks of Ritchies archipelago consist of soft lime stone formed of coal and shell while the sand is made up of calcareous sandsone and white clay with occasional conglomerates. Volcanic fragmentary rocks, apparently younger, occur in main Islands especially on east coasts of South Andaman. Narcondam and Barren Islands are of volcanic origin : the extinct volcano of Narcondam belonging to the Regu group and the Barren Island to the Sunda group. The older rocks belong to early tertiary or late cretaceous but there are no fossil records to give correct picture of the age of these Islands.

SOIL

The type of forest establishment and proliferation depend upon the nature of underlying soil. In Andaman soil is generally soft and deep sandy loam varying from fine texture of alluvial flats to the gravel strewn soil of the low hills. The hilly regions consists of hard clayey soil with micaceous sand stone formation and conglomerates below. The different five types of soils have been described by Parkinson (1923) as under –

(i) Saline low lying land, the sub–soil of which is usually of alluvial formation. This is inundated at regular intervals by the rise and fall of the tides and is usually occupied by mangroves.

(ii) Alluvial land out of the reach of the sea, lying along the sea coast and in the valleys and consisting generally of deep fertile clayey or sandy loam. This is usually occpuied by luxuriant tall evergreen forests.

(iii) Low undulating ground of a somewhat poor and hard pebble or sandstone formation, where the forests are usually of a leaf shedding character and where many of the most valuable Andaman timber trees grow.

(iv) Hills consisting of a yellow–brown and rather stiff clayey soil of a micaceous sandsone formation usually covered by tall evergreen forests.

(v) High and steep hills such as Saddle peak and Mt. Ford, with a hard red brown infertile soil and an underlying rock often of an intrusive serpentine usually found clothed with dense, scrub–growth, bamboos and small hard wooded trees.

VEGETATION

In general, the vegetation is directly related to the climatic conditions and soil factors. Andaman has short latitudinal and longitudinal geographical extent, hence climatically, the Islands may be considered more or less uniform throughout and hardly influence the distribution of vegetation in Islands. The varied habitat in these Islands, however, has resulted in widely different types of vegetation with great diversity of plant communities.

Based on habitat factor, Parkinson (1923) recognized five main types of vegetation in Andaman group :

1. Mangrove forest, generally found on creek and protected coastal areas.
2. Beach forest, confined on the open coasts extending a little interior some times.
3. Evergreen forests, at lower elevation in alluvial land and vallies.
4. Moist deciduous forests.
5. Moist evergreen forest.

CLIMATE

These islands have warm, humid climate with perceptible touch of cold during the December and January. On account of its short latitudinal and longitudinal extent, there is not much difference in rainfall and temperature from north to south and east to west. These islands are subjected to heavy rainfall and little temperature fluctuations throughout the year which is most favourable for mangrove forests. The year may be divided into two seasons the summer season and rainy season, though, a touch of cold is also found for 2 months.

RAINFALL

Rainfall is very significant in mangrove communities because it directly affects the salinity of surface soil by adding rain water in huge quantity and also by leaching soil salts. The climate of the islands is governed by the south-west monsoon, which is almost uniform and least diversified. April to December months are of rainfall but May, June, July and August are of heavy rainfall. From January to March or mid March show a dry period with scanty or no rains. During 6 years the highest rainfall has been recorded in the month of May, when the rainfall was 798.30 mm. But generally the highest rainfall is recorded in the month of July. The average rainy days for a year recorded in the month of July. The average rainy days for a year are 128 and average is of 10–11 days per month. Rainy season extends from April to December where it reached its maximum in July and afterwards decreased steadily from September to December. On an average, these islands receive rainfall (3,180 mm $year^{-1}$) which is much higher than the other areas of the world (Table 23).

Table 23. Rainfall Data for Last Six Years of Andaman

Months	1995	1996	1997	1998	1999	2000
January	124.9	008.0	000.0	019.7	00.64	008.4
February	032.3	000.0	000.0	007.4	00.10	000.0
March	000.0	010.8	000.0	000.0	00.00	000.0
April	041.2	074.8	008.2	041.2	061.4	000.0
May	440.5	153.6	798.3	302.9	269.4	217.7
June	335.5	366.0	586.8	571.4	439.1	369.1
July	647.1	404.0	529.0	252.9	439.2	610.6
August	297.3	106.5	531.6	611.4	336.0	662.1
September	333.4	336.2	343.4	507.2	490.4	334.0
October	305.4	149.5	150.5	314.5	103.2	423.8
November	099.6	030.3	099.1	308.7	277.7	243.9
December	006.1	001.5	709.7	266.7	022.9	084.0

TEMPERATURE

Temperature plays very important role in mangrove distribution and occurrence. The mangroves are restricted to tropical and warm temperate regions only. The restriction to these regions is conditioned by the sensitivity of mangroves to forest. Davis (1940) has produced some evidence of minimum temperature. He concluded that below 25°C temperature are fatal to mangrove communities. Temperature of the Andaman is very favourable for these swamps and no forest has been recorded since last 30 years. Islands have the average temperature of minimum 22.5°C and maximum 29.9°C. The months with higher temperature are February, March and April and coldest months are December and January (Table 24).

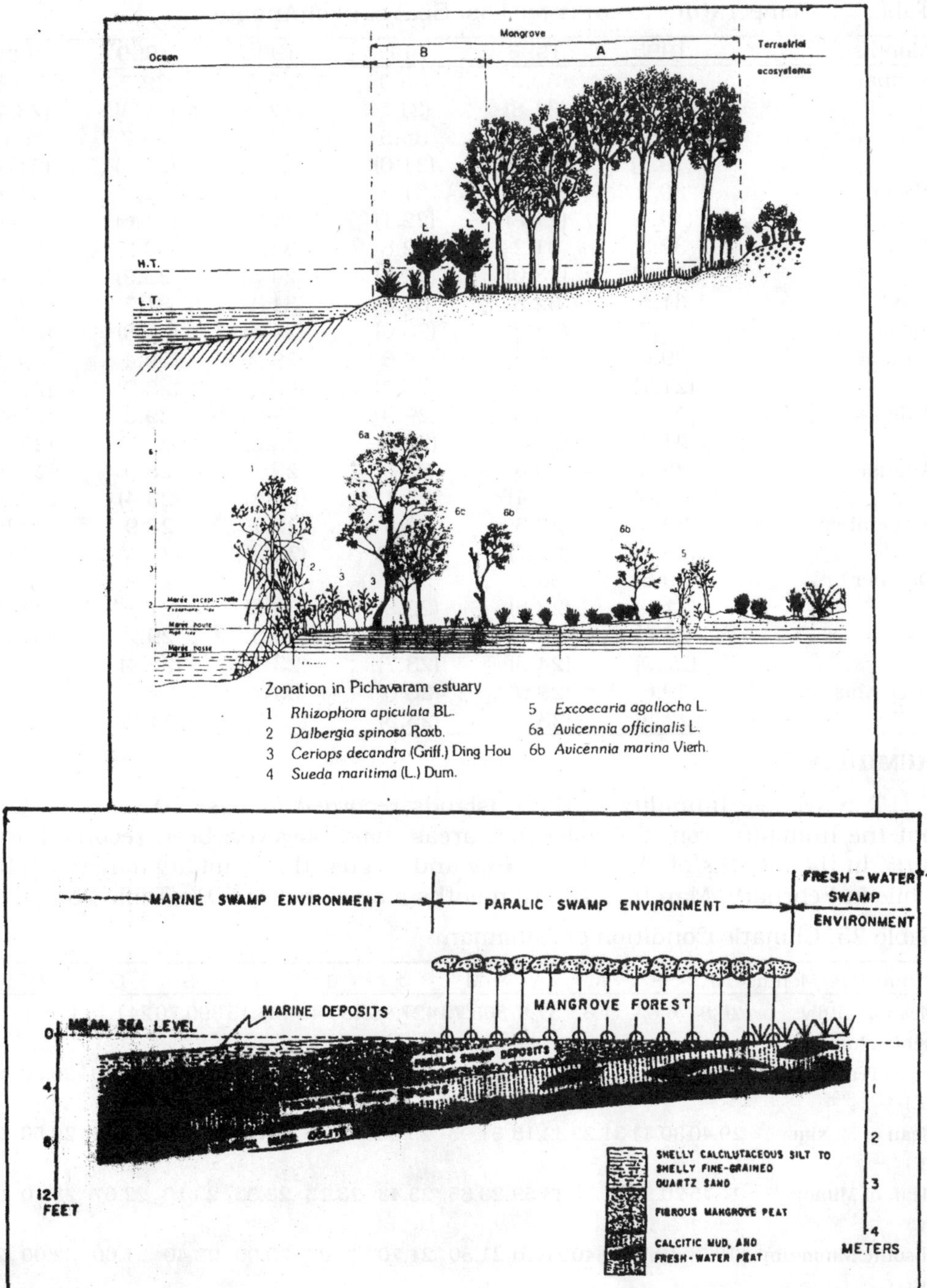

Fig. 23. Zonational patterns in mangroves

Table 24. Temperature Records for Last Six Years of Andaman

Months	1995	1996	1997	1998	1999	2000
January	29.0	30.0	39.0	29.2	29.0	29.3
	(18.7)	(21.8)	(21.5)	(22.0)	(21.3)	(23.4)
February	29.9	30.6	30.5	30.0	29.7	30.0
	(19.5)	(20.5)	(21.6)	(21.9)	(20.8)	(21.4)
March	30.9	31.5	31.3	31.4	31.2	31.1
	(19.4)	(21.8)	(22.1)	23.6)	(21.6)	(22.1)
April	32.3	31.2	32.8	32.1	32.0	32.7
	(21.1)	(23.0)	(24.8)	(24.5)	(23.9)	(24.5)
May	31.5	32.8	31.3	31.4	31.7	32.9
	(21.7)	(22.2)	(23.6)	(24.5)	(23.8)	(24.5)
June	29.4	29.5	29.6	29.2	29.4	30.1
	(21.7)	(24.9)	(23.7)	(23.1)	(23.3)	(23.9)
July	28.4	29.3	29.9	29.4	29.0	27.4
	(21.3)	(24.0)	(23.3)	(23.9)	(23.0)	(23.3)
August	28.6	29.6	29.8	29.1	28.9	28.9
	(23.0)	(24.4)	(23.1)	(23.2)	(23.4)	(26.5)
September	29.0	29.6	29.8	29.1	28.9	28.9
	(22.4)	(23.7)	(23.7)	(23.3)	(22.5)	(23.0)
October	29.5	30.1	30.0	29.7	29.9	29.1
	(21.6)	(23.6)	(22.8)	(23.5)	(21.8)	(22.7)
November	29.6	29.7	30.0	29.3	29.7	28.7
	(22.0)	(24.5)	(23.7)	(23.3)	(22.4)	(22.7)
December	29.6	29.9	28.8	28.7	29.0	28.7
	(22.8)	(23.8)	(23.5)	(22.7)	(21.6)	(22.6)

HUMIDITY

The average humidity in these islands recorded for last 30 years is 77.6% but the humidity from the mangrove areas, has, however, been recorded about 80%. In the months of May, June, July and August the humidity may reach 90% while in February, March and April months may go below 70% (Table 25).

Table 25. Climatic Condition of Andaman

Parameters/Months	J	F	M	A	M	J	J	A	S	O	N	D
Mean monthly rainfall (mm)	26.87	6.63	1.8	37.8	363.73	427.98	480.46	426.15	390.76	241.15	176.55	187.87
Mean Temperature (°C)	25.42	25.53	26.49	27.88	27.64	26.48	26.01	26.56	26.16	26.19	26.30	25.97
Mean of Maxima (°C)	29.40	30.11	31.23	32.18	31.93	29.53	28.90	29.20	29.22	29.72	29.50	29.12
Mean of Minima (°C)	21.45	20.95	21.76	23.59	23.35	23.43	23.13	23.93	23.10	22.67	23.10	22.83
Absoute Minimum (°C)	18.7	19.5	19.40	21.10	21.50	21.70	21.30	23.00	22.40	21.60	22.00	21.60
Absolute Maximum (°C)	30.00	30.60	31.50	32.70	32.90	30.10	29.90	29.80	29.80	30.10	30.00	29.90
Mean relative humidity (%)	71.50	71.00	67.00	67.50	69.00	82.50	84.50	86.50	85.00	86.00	78.00	77.00

WIND SPEED

During last six years in the Andaman Islands the average wind speed was 10.21 km hr^{-1}. The greater speed has been recorded during rainy season in the months of May to September. During this period wind speed may go up to 26.2 km hr^{-1} which has recorded in August.

5.1. STRUCTURAL PATTERN OF MANGROVE FORESTS

Mangrove, which is generally low forest or bush–wood, a dense type of vegetation is very characteristic due to its general physiogonomy, production of special types of aerial roots and viviparous seedlings. Plants give bright green colour and have bushy tree *dimension* as opposed to deep green lofty trees of hinterland forest. There are a few herbaceous forms associated with the shrub and tree species. Apart from these terrestrial forms, there are some epiphytes including *Hydnophytum formicarum*, *Dischidia major* and *Dischidia nummalaria*, several orchids e.g. *Eulaphia–andamanensis*, *Cymbidium–aloifolium* and semi parasites such as *Dendropthoe falcata* etc. Some lianas like *Derris scandens*, *Derris andamanica*, *Ventilago maderaspatana*.

Study Sites

On the basis of extensive survey of these islands following sites have been selected for detailed study due to differences in substratum, topography, habitat etc (Fig. 39).

Site – 1. Alexandra Island : Situated nearly 36 km from Port Plair. Here the substratum is sandy and mangal formation is dense and mostly species of *Rhizophora dominate*.

Site – 2. Wandoor : Situated at about 36 km. from Port Blair very near to site–1. The substratum is muddy and sandy and vegetation is almost similar to that of the 1st site with dense population of animals, e.g. mudskippers and snails.

Site – 3. Carbynscove: Located towards the east coast of South Andaman, 15 km. from Port Blair, the substratum is clayey and muddy, with lot of human disturbance.

Site – 4. Chiriya–Tapu : On the south coast of South Andaman, 42 km. from Port Blair. There is open formation of rocky and stoney shore. Here mangrove formations are found on rocky substratum.

Site – 5. Bambooflat : This site is along in inlet stream and is muddy with marked mangal formation. Some of the area is under acute human disturbance.

Site – 6. Wrightmyo : The site is along creeks and is marshy and muddy with dense mangal formation.

Site – 7. Oralkatcha : Situated a distance of 35 km. from Port Blair. This is an island having a sea creek and inlet streams. Here the best developed mangrove was noticed.

Site – 8. Nilamboor : An island having a sea creek and inlet streams. Here also very good growth of mangrove was noticed. The soil is muddy and sandy and vegetation is almost similar to previous one.

Site – 9. Gandhi Jetty : This site is situated along long creeks and mangrove formations are found on large flat areas.

Site – 10. Kandamtala : Situated in the Middle Andamans at distance of 60 km from Port Blair. Mangrove formations are found on large flat areas along the creek.

Floristic Composition

The tidal zones of tropical seas are frequently lined with great mangroves. The outstanding features of these plants are adaptation to grow in sea water and establishing themselves in estuaries, creeks and lagoon areas. The coast line of the islands is irregular and deeply indented thereby giving rise to a number of tidal creeks. These creeks are densely populated by this peculiar plant community. Here the mangroves are well protected against heavy tidal action and some what sheltered from high wind. Most of the islands have muddy, sandy and to some extent muddy sandy soil type. These factors provide ideal conditions for mangrove development and establishment. Extensive survey of various islands of south and middle Andaman was done and we came to the conclusion that the mangrove community of these islands is greatly dominated by the family Rhizophoraceae and the seaward line is always occupied by *Rhizophora* spp. From Table 26 it is seen that some species e.g. *Rhizophora apiculata, R. mucronata, Ceriops tagal, Bruguiera gymnorhiza* and *Avicennia officinalis* are common to all sites. Sites 3 and 7 are very rich in floristic composition contributing 68% and 64% of the total mangrove species respectively.

Similarly, Index similarity of among 10 sites were analysed (Table 27). From the table it is seen that maximum similarity is present between mangals of Site–3 and Site–7 and minimum between those of Site–6 and Site–7.

Table 27. Similarity Index of different Mangroves Sites

Sites	Site 1	Site 2	Site 3	Site 4	Site 5	Site 6	Site 7	Site 8	Site 9	Site 10
Site 1	1.0	58.82	48.79	57.14	43.75	78.28	50.00	48.22	43.75	50.00
Site 2		100	51.28	46.15	53.33	46.15	42.85	51.85	46.66	53.33
Site 3			100	54.54	69.45	47.05	87.14	52.94	64.84	59.45
Site 4				100	50.00	38.09	68.88	67.12	58.35	41.66
Site 5					100	50.00	45.00	64.00	64.28	42.85
Site 6						100	33.33	54.84	48.00	64.00
Site 7							100	59.45	60.00	50.00
Site 8								100	72.00	72.00
Site 9									100	64.28
Site 10										100

Complexity Index

This is a very simple mathematical expression for the quantitative characters of the community in term of complexity index. The coefficient was used originally only for comparing the floras of larger areas in relation to rainfall. Holdridge (1967) gave the idea of complexity index in case of mangrove forest. We have determined the complexity index of all the 10 sites. Maximum

Table 25. Floristic Composition of Mangrove Forest at Various Sites

S.N.	Name of the species	Occurrence of species at different sites											Percent
		1	2	3	4	5	6	7	8	9	10	Total %	Class
01.	*Acanthus ilicifolius*	—	X	X	—	X	X	X	—	—	—	50	3
02.	*A. ebracteatus*	—	X	X	—	—	—	—	X	—	—	30	2
03.	*Avicennia marina*	—	—	X	—	—	—	X	X	X	X	50	3
04.	*A. officinalis*	X	X	X	X	X	X	X	X	X	X	100	5
05.	*Aegiceras corniculatum*		—	—	—	—	—	—	X	—	X	'20	1
06.	*Bruguiera gymnorhiza*	X	X	X	X	X	X	X	X	X	X	100	5
07.	*B. parviflora*	—	—	X	—	X	X	—	—	—	—	30	2
08.	*B. cylindrica*	—	—	X	—	X	X	—	—	X	X	50	3
09.	*B. sexangula*	—	—	—	X	—	—	—	—	—	—	10	1
10.	*Ceriops tagal*	X	X	X	X	X	X	X	X	X	X	100	5
11.	*Rhizophora apiculata*	X	X	X	X	X	X	X	X	X	X	100	5
12.	*R. mucronata*	X	X	X	X	X	X	X	X	X	X	100	5
13.	*R. stylosa*	—	—	X	X	—	—	—	—	X	—	30	2
14.	*R. lamarckii*	—	—	—	—	—	—	—	—	—	X	10	1
15.	*Excoecaria agallocha*	—	—	X	—	—	—	X	—	—	—	20	1
16.	*Lumnitzera littorea*	—	—	X	—	X	—	X	X	X	—	50	3
17.	*L. racemosa*	—	X	X	—	—	—	—	—	—	—	20	1
18.	*Nypa fruticans*	—	—	X	—	X	—	X	X	—	X	50	3
19.	*Phoenix paludosa*	—	—	—	—	—	—	X	—	X	X	30	2
20.	*Scyphiphora hydrophyllacea*	—	X	—	—	—	X	—	X	X	X	50	3
21.	*Sonneratia alba*	—	—	X	X	—	—	—	—	—	—	20	1
22.	*S. apetala*	—	X	—	—	—	—	—	—	—	—	10	1
23.	*Xylocarpus granatum*		X	—	—	—	—	X	X	X	—	50	3
24.	*X. moluccensis*	X	—	—	—	—	—	—	—	—	X	20	1
25.	*Heritiera littoralis*	X	—	X	X	X	—	X	X	X	—	70	4
	Total number of species at various sites	8	10	17	9	12	9	16	11	13	12		
	% Contribution	32	40	68	36	48	36	64	44	52	48		

complexity index has been noted at Oralkatcha (Baratang Island), the value is 330.82 and minimum is at Carbynscove the value is 9.52. The high value of Complexity Index at Oralkatcha is due to high rainfall and the undisturbed condition of the mangrove forests (Table 28). There is no felling, while at other sites of South Andamans complexity–index is generally low, this may be due to the low rainfall and high human interference in these islands. At Sites–1 and 2, felling in many places is so common that whole area is devoid of vegetation and there is no natural regeneration, resulting in change of the basic structural pattern of mangrove forest. A comparative complexity index has been given in Table 29, from the table it is seen that mangroves of Andaman Islands are richer and healthier in comparison to mangroves of Costa–Rica and Peurto–Rico.

Table 28. Complexity Index of Various Sites of Andaman Islands

S.N.	Site	No. of species	No. of trees above 10 cm diamere	Height (m)	Basal Area (m^2)	Complexity Index	Rainfall (mm)
1.	Alexandra Island	8	90	8.7	1.51	9.45	2286
2.	Wandoor	10	102	10.0	1.40	14.28	2286
3.	Carbynscove	17	80	7.7	0.91	9.52	2286
4.	Chiriya Tapu	9	120	7.7	4.93	40.99	2212
5.	Wrightmyo	12	160	9.3	2.43	43.3	2417
6.	Bambooflat	9	157	9.0	1.75	22.25	2417
7.	Oralkatcha	16	175	17.0	6.95	330.82	3429
8.	Nilamboor	11	131	12.7	5.64	103.21	3429
9.	Gandhijetty	13	180	12.6	3.15	92.87	3429
10.	Kadamtala	11	163	12.0	4.68	100.69	2810

Table 29. Comparative Structural Pattern of Mangroves of World

Country	No. of species	No. of trees above 10 cm diameter	Height (m)	Basal Area (m^2)	Complexity Index	Rainfall (mm)
Costa Rica (Main river)	4	118	9.53	16.0	72.0	3300
Costa Rica (Santa Rosa)	2	80	2.22	10.0	3.5	1800
Puerto Rico (Mona Island)	2	109	2.97	15.0	15.9	1631
Puerto Rico	2	17	0.17	4.8	1	860
Andaman South	9	120	7.7	4.93	40.99	2286
Andaman Middle	16	175	17.0	6.95	330.82	3429

5.2. STRATIFICATION PATTERN IN MANGROVE FOREST

Stratification

The structure that results from the distribution of organism and their interaction with their environment can be called pattern. Many different kinds of arrangement in the standing crop of organisms contribute to pattern diversity in the community as for example : 1. Stratification pattern (vertical layering), 2. Zonational pattern, 3. Activity pattern, 4. Food-web pattern, 5. Reproductive pattern etc. In all, stratification pattern is most important pattern for the study of any forest. In a forest the two basic layers the autotrophic and heterotrophic strata-that are characteristic of all communities are frequently and distinctly stratified into additional layers.

Table 30. General Stratification Pattern of Species in the Mangrove Forest of Andaman Islands

	Stratum 1	Stratum 2	Stratum 3
	Trees or tall shrubs abundant and contributing to the physiognomy of the association	Shrubs or small trees which may be locally abundant	Low shrubs or lianas frequently present in edaphically restrictive situation
	Outer		
1	**Inner Zone**		
	Rhizophora mucronata	*Ceriops tagal*	
	R. apiculata	*Bruguiera parviflora*	
	R. stylosa	*Aegiceras corniculatum*	
	Ceriops tagal		
	Avicennia marina		
	A. officinalis		
2.	**Middle Zone**		
	Bruguiera gymnorhiza	*Excoecaria agallocha*	*Acanthus ebracteatus*
	B. parviflora		*A. ilicifolius*
	Lumnitzera littorea		
	Xylocarpus granatum		
	X. moluccensis		
3.	**Distal Zone**		
	Heritiera littoralis	*Phoenix paludosa*	*Nypa fruticans*

In the case of the mangrove forests of Andaman there is layering but the layering is not as clear as is found in the beach forests or evergreen forests of the Andamans. In some places where the mangrove forest is well developed and there is no interference of biotic factors, e.g. Kadamtala, Gandhi Jetty and Oralkatcha, the height of many trees, e.g. *Bruguiera gymnorhiza, Lumnitzera littorea, Avicennia marina, Rhizophora mucronata* and *R. apiculata* reaches upto 30 to 45 meters.

We have noted three strata in case of mangrove forests of Andamans:

Upper stratum

Occupied by tall mangrove species, e.g. Bruguira gymnorhiza, Lumnitzera littorea, Rhizophora mucronata, R. apiculata etc.

Middle stratum

Occupied by shrubs and small trees which are abundance in middle zone, e.g. *Ceriops tagal, Excoecaria agallocha, Aegiceras corniculatum, Bruguiera parviflora* etc.

Lower stratum

It is occupied by small shrubs and fern, e.g. *Acanthus ebracteatus, A. ilicifolius, Nypa fruticans* etc.

5.3. ZONATIONAL PATTERN

A significant amount of work has been done so far on floristics and zonational pattern of world mangroves. Some authors consider the zonation as the result of the successional sequence, while many others (Egler, 1952; Thom, 1967; Lugo and Snedaker, 1974) consider that zonation does not always represent a successional sequence. They stressed the role of soil type, frequency of tidal inundation, salinity changes and drainage in the existence of different communities in mangals. Thom (1982) also supported the same concept who stated that mangrove zones are responsive to geophysical, geomorphic and biological components. Inter–specific competition also plays an important role in determining the species diversity and distribution pattern in a given area.

In India, zonational studies have been done by various authors (Untawale, 1973; Rao, 1972, Blasco, 1977, Singh *et al.*, 1986) in different regions.

The zonational studies were carried out with the help of several belt transacts across the tidal flat from the seaward to landward mangal area to see the trend of species settings. The distribution of plant species was recorded along the continuous transacts. The soil salinity was measured following Jackson (1958) after 10 m, interval at each site to note the pattern of zonation in accordance to salinity gradient.

In mangal zonation, the role of tidal flooding, soil salinity, and different soil properties has been often emphasized. One major underlying factor controlling physical, chemical and to some extent biological features in an environment is geomorphology. Besides these common factors, the authors assume that anthropogenic factor also plays a vital role in mangal zonation.

Keeping in mind, the above causes of mangal zonation studies were carried out on the muddy, sandy and rocky tidal flats and also on some sites under acute human disturbance.

Based on the study of the structure and composition of mangrove forest at different sites, the following 7 major types of mangrove communities have been recognized –

1. *Rhizophora mucronata* – *Rhizophora apiculata* Community
2. *Bruguiera gymnorhiza* – Ceriops tagal Community
3. *Rhizophora mucronata* – *Bruguiera gymnorrhiza* Community
4. *Lumnitzera littorea* – *Avicennia officinalis* Community
5. *Bruguiera gymnorhiza* – *Avicennia officinalis* Community
6. *Heriteria littoralis* – *Pongamia pinnata* Community
7. *Acanthus ilicifolius* – *Acrostichum aureum* Community

Rhizophora species grow typically on the outer seaward fringe of the swamp, where the water is saline. The distribution of species in such habitat is largely controlled by salinity of substratum, frequency and duration of flooding by tidal water, and moisture content of substratum. Zonational patterns of mangrove of all sites have been studied and is given in Table 31. From the table, it is seen that mangroves of all sites have three zonation, namely proximal zone, middle zone and distal zone.

Table 31. Ecological Distribution patter of Mangroves at different sites in relation to Habitats

S.N.	Sites	Habitat	Proximal Zone – A zone of pure patch of mangrove	Middle zone- A zone of mixed mangrove	Distal zone – A zone of border line mangrove
A. South Andaman					
1	Alexandra Island	Sandy	Length : 10-20 m patch of mangrove *R. apiculata, Ceriops tagal, Avicennia officinalis*	Length : 40-50 cm zone not clear occupied by *Ceriops tagal, Bruguiera gymnorhiza, Xylocarpus granatum, X. moluccensis, Clerodendrum inerme*	Length: 10-50 m *Heritiera littoralis, Hibiscus tiliaceus, Pandanus tectorius, Acrostichum aureum*
2	Wandoor	Sandy muddy	Length : 10 m *Rhizophora mucronata, R. apiculata, Sonneratia apetala*	Length: 30-60 m *Ceriops tagal, Bruguiera gymnorhiza, Scyphiphora hydrophyllacea, Lumnitzera racemosa, Guettarda speciosa*	Length: 20-25 m *Acrostichum aureum, Acanthus ilicifolius, Pandanus tectorius, Caesalpinia nuga*
3	Carbynscove	Sandy clayey	Length : 5-15 m *Rhizophora mucronata R. apiculata, R. styl-osa, Sonneratia alba*	Length : 30-60 m *Avicennia marina, A. officinalis, Bruguiera gymnorhiza, B. parviflora, B. cylindrica, Aegiceras corniculatum, Lumnitzera littorea, L. racemosa*	Length: 20 m or greater *Hibiscus tiliaceus, Acrostichum aureum, Nypa fruticana, Acanthus ilicifolius, Heritiera littoralis, Excoecaria agallocha*
4.	Chiriya Tapu	Rocky Stony	Length: not clear *Rhizophora mucronata, R. apiculata, R. styloa, Ceriops tagal, Bruguiera gymnorhiza, Sonneratia apetala*	Length: 20-40 m *Ceriops tagal, Bruguiera gymnorhiza, Avicennia officinalis, R. stylosa*	Length: 5-10 cm *Hibiscus tiliaceus, Heritiera littoralis, Pandanus tectorius*
5.	Bamboo flat	Muddy	Length: 30-40 m *Rhizophora mucro-nata, R. apiculata*	Length: 60-90 m *Avicennia officinalis, Bruguiera gymnorhiza, B. cylindrica, B. parviflora, Ceriops tagal*	Length: 15-20 m *Acanthus ilicifolius, Nypa fruticans, Heritiera littoralis, Caesalpinia nuga, C. bonducella*

6.	Wrightmyo	Muddy sandy	Length: 25-40 m *Rhizophora mucronata, R. apiculata, Avicennia officinalis, Ceriops tagal*	Length: 50-70 m *Bruguiera parviflora, B. cylindrica, B. gymnorhiza, Xylocarpus granatum, X. moluccensis, Ceriops taga, Dolichandrone rheedii*	Length: 15-20 m *Acanthus illicifolius Acrostichum aureum, Heritiera littoralis*
B. Baratang Islands					
7	Orakatcha	Muddy	Length: 50-60 m *Rhizophora mucronata, R. apiculata*	Length: 100-150 m *Rhizophora mucronata, R. apiculata, Avicennia officinalis, A. mania, B. gymnorhiza, Ceriops tagal, Scyphiphora hydrophyllaceae, Aegiceras comiculatum, Afzelia bijuga, Lumnitzera littorea, Cerbera floribunda, Excoecaria agallocha, Xylocarpus granatum, X. moluccensis*	Length: 20-35 m *Liacula spinosa, Phoenix paludosa, Heritiera littoralis, Pongamia pinnata, Thespesia populnea, Pandanus tectorius, Acrostichum aureum, Dolichandrone rheedii, Clerodendrum inerme, Hibiscus tilliaceus*
8	Nilamboor	Muddy	Length: 40-60 m *Rhizophora mucronata, R. apiculata*	Length: 80-130 m *Bruguiera gymnorhiza, Ceriops tagal, R. apiculata, R. mucronata, Scyphopohra hydrophyllacea*	Length: 20-35 m *Nypa fruticans, Heritiera littorea, Acrostichum aureum*
9.	Gandhi Jetty	Muddy	Length: 30-35 m *Rhizophora mucronata, R. apiculata, Avicennia marina*	Length: 60-120 m *Bruguiera parviflora, B. gymnorhiza, B. cylindrica, Ceriops tagal, Avicennia marina, A. officinalis, cyphiphora hydrophyllacea, Aegiceras corniculatum, Dolichandrone rheedii*	Length: 20-25 m *Heritiera littoralis, Phoenix paludosa, Acrostichum aureum*
C. Middle Andaman					
10	Kadamtala	Muddy	Length: 15-25 m *Rhizophora apiculata, R. mucronata*	Length: 72-130 m *Avicennia officinalis, A. marina, R. mucronata, B. gymnorhiza, B. parviflora, B. cylindrica, Ceriops tagal, Xylocarpus grantum, X. moluccensis, Aigiceras corniculatum*	Length: 15-20 m *Phoenix paludosa, Acrostichum aureum, Nypa fruticans, Acanthus ilicifolius*

Proximal Zone

This is the seaward zone where frequency of inundation is maximum. The species in this zone are *Rhizophora apiculata, R. mucronata, Ceriops tagal, Sonneratia alba* etc. Salinity is in neighbourhood of that of sea water.

Middle Zone

This zone is towards the interior, soluble salts are more than in sea water. The common species are *Bruguiera* spp., *Lumnitzera littorea, Aegiceras corniculatum* etc. usually they attained great height and girth.

Distal Zone

This zone is towards landward fringe, where salinity may be high. The common species are *Heriteria littoralis. Acrostichum aureum, Acanthus* species, etc. This type of zonational pattern is called Metabolic zonation. From outer edge to inland, a pattern of change in soil salinity and corresponding zonation of the mangrove species is usually observed. Each species occupies a salinity zone to which it is best adapted; more of the energy goes to growth and less to maintenance (i.e. respiration) as compared to potential competitor species. This adaptation to salinity regime is known as the metabolic zonation. Mangroves such as *Avicennia, Aegiceras* and *Aegiallis* have salt secreting glands on their leaves. The sap which passes up their xylem contain 0.2–0.5% Sodium Chloride. Mangroves such as *Rhizophora, Bruguiera* and other lack the salt secreting glands and the concentration of salt in their sap is only about 1/10 of the above group. They have an ultra–filtration mechanism in their roots enabling selective absorption of ions.

In some species, e.g. *Sonneratia* and others have been found to have excessive amount of ions in their organs and thus with the absorbed and accumulated ions the leaves become quite fleshy. In all cases the osmotic pressure of cell sap is known to near that of sea water. The normal process of transpiration accounts for flow of water through the plant. The main difference between mangroves and other plants is that the mangroves have an usually high osmotic pressure in the leaf cell sap. It seems that physiologically salt excluding membrane system is more efficient than the system having salt secreting gland on their leaves. The occurrence of *Rhizophora* spp. in proximal zone or at seaward side at various study sites supports this idea. Other species like – *Avicennia, Aegiceras* and *Lumnitzera* spp. are always recorded in the middle zone of the mangrove forest, where soluble salts are more.

From the zonational study it seems that habitat it sandy and not flat has less species composition and less zonational pattern. Those areas which are flat and muddy have greater number of species composition and broad zonational pattern.

A well revealed impact of anthropogenic pressure on zonation is noticed on some Islands e.g. Carbynscove, Manjaree and Wandoor. The illicit cutting for the purpose of fuel, fodder and urbanisation changes the zonational pattern of these islands. Preferential cutting is common of several species e.g. *Avicennia marina, Bruguira* and *Xylocarpus* for electric poles, huts, etc. *Rhizophora* for fodder and fuel, while the *Nypa fruticans* and *Phoenix paludosa* for thatchery. This cutting alters the soil type, texture and salts of that area which gives the chance to develop new mangals and disturbs the original zonation.

In general, muddy habitats show everywhere a good picture of zonation, because it provides ideal environment for the growth and development of mangal. Sandy substratum represents only two zones the eroding proximal zone and distal zone while rocky habitat does not show the clear picture of mangal zonation, because, it pertains the large tidal flat which is subjected to high tidal amplitude, daily inundation and rocky nature substratum. Here only a few species can be established which can tolerate external factors.

The different mangrove zones can be viewed as steady state response to their respective tidal levels (Thom, 1967). In other words, each constituent species finds, within its tolerance range to salinity, its place in the environmental gradient resulting from the substrate level and the tidal range. This environmental sorting selects the species which are capable of growing in particular zones and intra and inter–specific competition determines the fine zonation.

5.4. ARCHITECTURAL PATTERN

The structure and especially the microscopic anatomy of mangroves have received considerable attention which is summarized by Chapman (1976). Information about form is necessary to an understanding of functional processes, particularly those that are specialized and adapt plants to mangal environments. The interpretation of observed structures in terms of function is, however, often based on assumption rather than experimental verification. Physiological investigations of mangroves are few, which is remarkable when one considers that it is the suite of functional characteristics that allows mangroves to survive in tidal environments.

Vegetative survival and competitiveness of trees are dependent on a number of factors, including an efficient display of foliage and the ability to response to environmental changes and stress (e.g. mechanical damage). The result of the response to these factors can be translated into crown shape. The principles of the responsive processes are well understood, although an appreciation that crown form is a dynamic expression of deterministic and opportunistic processes is slow in developing. A useful descriptive background for tropical trees, where most principles are enunciated is provided by Halle *et al.* (1978). They provide a convenient series of references points in the set of 23 "architectural models" they define. These models are abstract, but dynamic, they constitute for each species the deterministic component of form; the visible expression of the architectural model of a tree at any time is referred to as its "architecture" (Fig. 24). Trees may conform more or less precisely to their architectural model (i.e. they may or may not be 'model conforming'). They can deviate considerably from the idealized model because the architecture is modified by environmental influences (mechanical damage, predator attach, shading) that divert the normal course of development. Shaded mangrove seedlings remain suppressed, with little or no branching and short internodes. The most usual response in the form of the branched tree to other environmental.

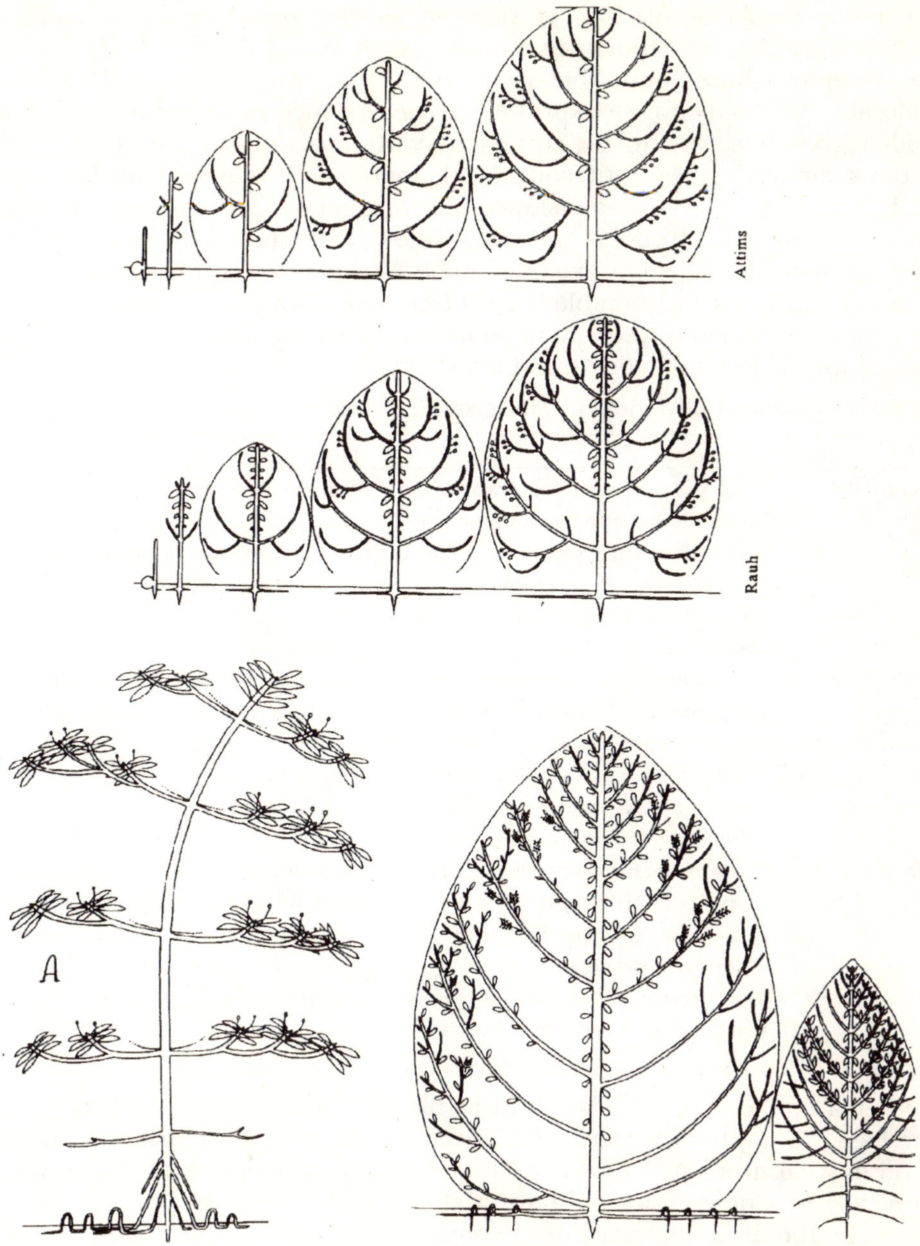

Fig. 24. Architectural patterns in mangroves

The range of architectural models exhibited by mangroves is small; this may simply reflect floristic poverty, but since the observed models are structurally similar, a more likely explanation is that architecture is constrained by the environment. Halle *et al.* (1978) comment on the high incidence of Attims's model in mangrove. It is found, for example, in all *Rhizophoraceae* (except some species of *Bruguiera*).

Attims's model (Table 32) is defined by the presence of a trunk with continuous growth, all orthotropic branches with lateral flowers, and continuous or discontinuous branching. This seems an all–purpose, generalized model with considerable plasticity in its expression. The absence of pronounced rhythmic (episodic) growth seems to be correlated with the relatively uniform growing conditions for mangroves (but capable of appreciable seasonal modification in strongly seasonal climates, as shown for *Rhizophora* by Gill and Tomilison, 1971). Inherent orthotropy of all axes allows for the rapid substitution of damaged leaders by adjacent existing shoots. Branch axes are readily converted into trunk axes as is notable in *Rhizophora.* Branch expression is very opportunistic, vigorous trees may show continuous branching but usually branches appear in loosely organized tiers.

Table 32. Architectural Models in Mangrove

Architectural models	Characters	Mangrove species
Attims's model	Polyaxial tree vegetative axes differentiated into trunk and branch. The architecture is determined by axes with continuous growth differentiated into monopodial trunk and equivalent branches.	*Rhizophora mucronata, R. apiculata, R. stylosa. R. lamarckii Lumnitzera racemosa, Avicennia officinalis. Ceriops tagal, Sonneratia alba, S. apetala.*
Scarron's model	The architecture is determined by an orthotrophic rhythmically active terminal moristem which produces an intermediate trunk bearing branches	*Lumnitzera littorea, Xylocarpus granatum, X. moluccensis, Hibiscus tiliaceus*
Prevosts model	Trunk and branches essentially orthotrophic with spirally arranged leaves	*Excoecaria agallocha*
Aubreville's model	The architecture is determined by monopodial trunk with rhythmic growth and spiral or deccussate *Phyllotaxy.* Branches plegiotropic.	*Bruguiera parviflora, B. gymnorhiza, B. cylindrica, B. sexangula*
Schoute's model	The massive creeping axis burried in mangrove mud with erect leaves and axillary inflorescence	*Nypa fruticans*

Other architectural models found in mangal may be described as variants on the basic *Rhizophora* pattern. Rauh's model, which has rhythmic branching, is found in *Xylocarpus* and possibly *Heritiera.* Massart's model differs in that the rhythmically produced branches have a strong endogenous plagiotropy; *Myristic* is probably the best example. It is notable that these three genera are all elements of the back mangal.

Pettit's model is common in species of the front mangal; it differs from Attims's model only in the presence of terminal inflorescence so that the development of branches is sympodial by flowering. *Avicennia* and *Sonneratia* approach this, but much of their branching is independent of flowering and corresponds Attims's model. *Lumnitzera* shows contrasted architecture in its two species : *L. racemosa* (Attims's model with lateral inflorescence, and *L. littorea* (Pettit's model with terminal inflorescences. Here the difference in flower position may relate to contrasted pollinators, small insects that easily penetrate the crown in the former species and birds that do not easily penetrate the crown in the second species.

Aubreville's model (Table 32) is well expressed in the large flowered species of *Bruguiera* (e.g. *B. gymnorhiza*). It differs from Attims's model (e.g. *Rhizophora*) in the pronounced plagiotropy by the apposition of its branches, developed abruptly, resulting in branching of the familiar *Terminalia* type, best represented by *Terminalia catappa* (Fisher, 1978). The developmental difference between Aubreville's and Attims's model is somewhat subtle; apposition growth is achieved gradually by branch complexes in *Rhizophora*, which are initially more or less orthotrophic. *Terminalia* branching, whether by correlative control of the *Rhizophora* type of by some endogenous mechanism of the *Bruguiera gymnorhiza* type, has been interpreted as functionally beneficial in minimizing self–shading within a tier (e.g. Fisher and Honda, 1979 a,b). The hexagonal pattern of structural units that supports the leafy rosettes itself represents a minimum path length arrangement.

Interpreting crown structure in terms of self–repeating units that maximize photosynthetic efficiency as far as light interception is concerned but a preliminary step in understanding competitiveness in mangroves. Nevertheless, the characteristic dense canopy of tall mangrove seems understandable in terms of leaf arrangement. In undisturbed mangroves little or no understorey develops except for suppressed, unbranched seedlings.

5.5. SUCCESSIONAL PATTERN

Succession

In general, in areas of large mangals, five geographic belts can be distinguished (West, 1956) : (i) a belt of shore water and midflats along the coast, (ii) a series of discontinuous sand beaches, variable in size, which are interrupted by tidal inlets and mudflats, (iii) a zone of mangrove forest, usually one–half to three miles wide, (iv) a fresh–water swamp, and (v) equatorial rain forest. Although the beach zone is frequently absent, this zonation was described in the Malaya Peninsula (Watson, 1928), western Africa (Grew, 1941), the Congo (Pyneart, 1933) and in Gujana (Martyn, 1934). Within the mangrove belt, there is usually a seral succession of vegetation in relation to hydrological and climatic conditions. Day *et al.* (1953) held the salinity gradient to be of great importance to distribution in South Africa, and described correlation between rainfall, evaporation, upflow of salt water from sea, and change.

There have been several attempts to classify mangrove vegetation according to physical characteristics of the environment. Watson (1928) described two general classifications : (i) mangroves that grow on accretive shores and (ii) those that grow on sand. Watson also related species to the tidal cycle and described five classes : (i) inundated by all high tides, (ii) inundated by medium high tides, (iii) inundated by normal high tides, (iv) inundated by sparing tides, and (v) occasionally inundated by exceptional or equinoctial tides.

Stevensons and Tandy (1931), working at Low Isles in Australia, described the mangrove habitat as (i) dense woodland, (ii) muddy glades, and (iii) shingle tongues. At present, mangrove types are some times considered to be related to the type of soil present. Troll and Drangedorff (1931) and Walter and Steiner (1937) described mud and reef mangroves; Chapman (1944a) added the categories of sand and peat mangroves. These four types are generally recognized today.

Tansley, Watt and Richards (1939) suggested that mangrove vegetation can be considered as formation type on a world-wide basis. They recognized two subformations : (i) the New World subformation, including Western Australia, and (ii) the Old World subformation. Chapman (1944a) recommended that a third subformation, the *Australisian*, be recognized because the species of *Avicennia* are very distinct in their distribution and segregate into these three geographical regions.

Davis (1940) said that the mangrove formation is composed of seral communities. Although reef and sand communities appear to be climax stages, the statement of Davis is generally true. The order of zonation varies considerably even in geographically related areas. For example, in the Old World subformation, *Rhizophora* is the pioneer species along river banks and in the more protected regions along oceanic shores, whereas *Avicennia* or *Sonneratia* pioneer on shores of greater wave and tidal action. In Jamaica, *R. mangle* pioneers along protected shores, while *Laguncularia racemosa* pioneers on sand spits where wave action is greater (Chapman, 1944a).

Chapman (1940) pointed out that the presence of sea grass in submerged areas accelerates the seaward extension of mangrove because it raises the height of the sea bed, allowing *R. mangle* seedlings to grow.

The seral nature of mangrove vegetation in the Indo–West–Pacific region was described in details by Macnae (1968), who recognized succession in every mangal he visited. Macnae stated that variation in development was often found, succession being complete only where the amount of available fresh water exceeded that lost through evaporation and transpiration. When losses through evaporation and transpiration exceed the income from rain and rivers, the soil becomes hypersaline and zonation is interrupted.

Macnae described the effects of freshwater imbalance in a zone of *Ceriops tagal*. This zone was located between a seaward fringe of *Bruguiera gymnorhiza* and a landward fringe of *A. marina*. Where freshwater loss exceeded gain, *C. tagal* became stunted. With increasing excess of evaporation, the *Ceriops* bushes died, forming a bare area which expanded both landward and seaward until only a few bushes grew near the *Avicennia* and *Bruguiera* fringes.

In Florida, Davis (1940b) was the first investigator to give a detailed account of succession in a mangal. He recognized seven principle communities : (i) The pioneer *Rhizophora mangle* zone. This seaward stage was composed of mangrove seedlings of various age growing in marsh soil below the level of low tide in shallow undisturbed water. *Thalassia testudinum* Koenig and Sims and *Cymadocea manatorum* Aschers grew in shoal areas near this zone, and *Spartina alterniflora* Loisel was present in some parts of Florida. (ii) Mature *Rhizophora* Consocies. This stage was composed of mature *R. mangle* with well–developed prop roots growing in mangrove peat. (iii) *Avicennia* salt marsh plants *Batis maritima* L., *Salicornia perennis* Mill, *Spartina alterniflora, S. spartinae* (Trin.) Merv, *Monanthochloe littoralis*, and *Sporobolus virginicus* (L.) Kunth. This mangrove salt marsh consocies grew on peaty soil and accumulated large amounts of organic and inorganic detritus. During dry periods, soil salinity was very high, whereas salinity was very low during rainy periods. In some places, *Avicennia* was more than 30 cm in diameter, but in other places was a small gnarled bush. (iv) Mature mangrove associes. This stage consisted of large tree of *R. mangle* and *R. nitida* growing together on peat soils in water of low salinity at approximately the mean high tide mark. (v) *Laguncularia racemosa* Consocies. This stage did not occupy a specific habitat, but was found with both the mature mangrove associates or between the *Avicennia* salt marsh associes and a *Conocarpus* associes when the natural mangrove associes was not present. (vi) *Conocarpus erectus* transition associes. This was the final stage of mangrove succession in Florida, as it was bordered on its landward age by sand dunes, upland tropical forest or freshwater marsh. Davis considered the *Conocarpus* associes to be an ecotone, not a definite seral stage. Chapman later (1940a) who reviewed the work of Borgeson (1909) and concluded that *Conocarpus* formed a true seral stage. Chapman later (1970) considered the *Conocarpus* community to be an ecotone between saline and freshwater communities. West (1956) described the final stage in seral succession in Columbia as dominated by *C. erectus* in the drier and less saline areas. (vii) Dwarf–form mangroves. Davis recognized a scrub–mangrove facies of dwarfed *Rhizophora, Avicennia* and *Laguncularia*, which grew above the high tide mark in freshwater. This dwarfed form was common in the everglades region.

Thorne (1954) listed many other plants in the mangals of Florida. Holdridge (1940) have an extensive description of the vegetative characters and general characteristics of *R. mangle, Laguncularia racemosa, C. erectus* and *A. nitida* and reported *Pterocarpus officinale* Jacq., *Anona glabra* L., *Bucia buceras* L. and *Drepanocarpus lunatus* G.F.W. Meyer living in the mangals of Peurto Rico.

Chapman (1944a) compared succession in the swamps of Jamaica with that in Florida (Fig. 25, 26, 27). In both cases *Rhizophora* was the pioneer form, with *Avicennia, Laguncularia* and *Conocarpus* inland. Asprey and Robbins (1953) stated that there were few associates with mangrove in Jamaica, a pattern similar to mangals in other parts of the world. *Batis maritima, Salicornia ambigua* Michx., *Acrostichium aureum* L., *Altermanthera ficocides* and *S. virginicus* occurred in the swamps of Jamaica. In other parts of the world, other genera and species occupy similar positions. For example, Taylor (1959) described the mangals of Paupa New Guinea, as follows–

1. Salt water swamps
    a. Tidal mangrove sequence
2. Brackish water swamps
    a. Brackish swamp sequence
    b. Estuarine sequence

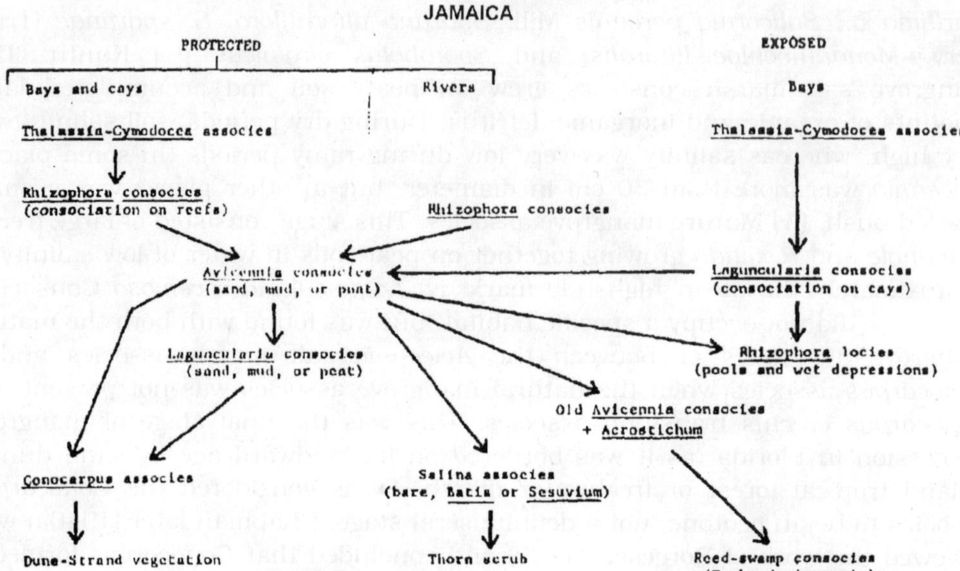

Succession in the mangrove swamps of Jamaica (Chapman 1944a).

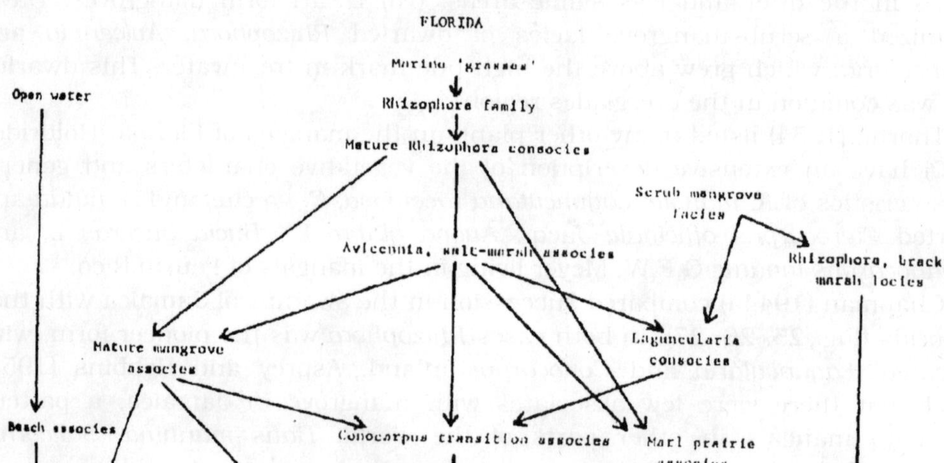

Succession in the mangrove swamps of Florida (Chapman 1944a).

Fig. 25. Successional Pattern in mangroves

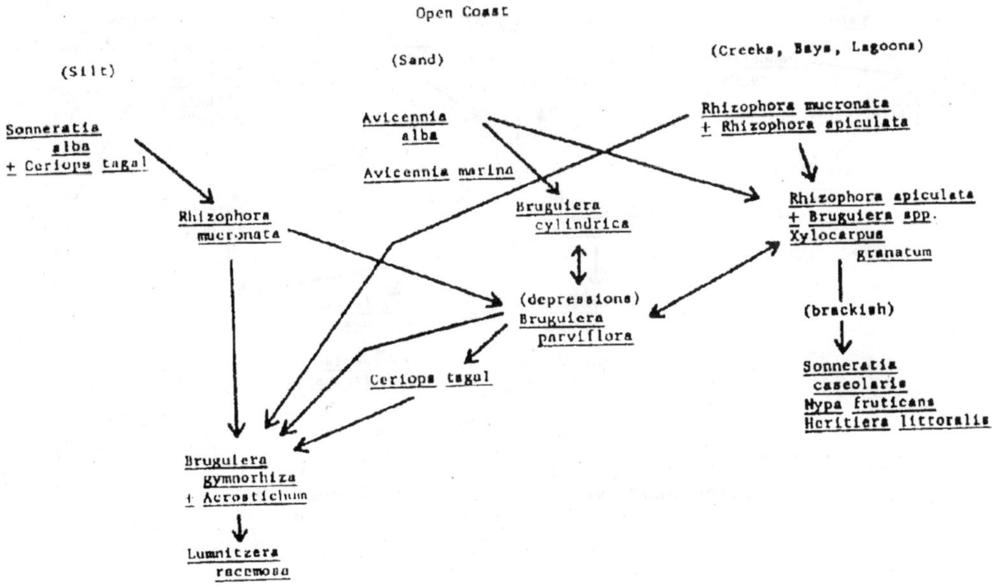

Succession in the mangrove swamps of Malaysia, Indonesia, and Borneo (Chapman 1970, Schema 7)

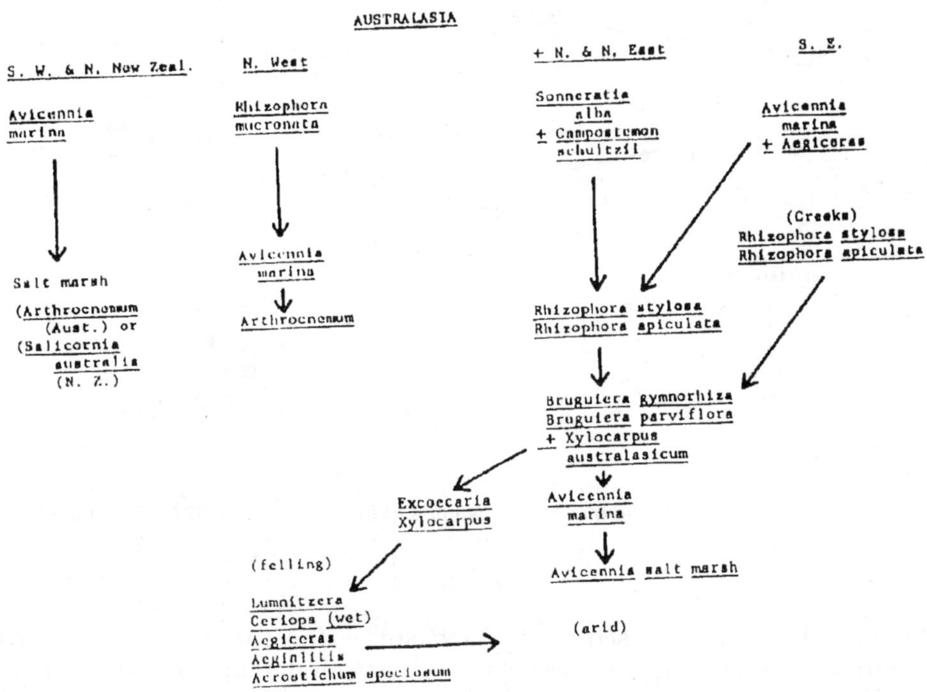

Succession in the mangrove swamps of Australia (Chapman, 1970; Schema 8)

Fig. 26. Succession in the mangrove swamps of Australia (Chapman, 1970; Schema 8)

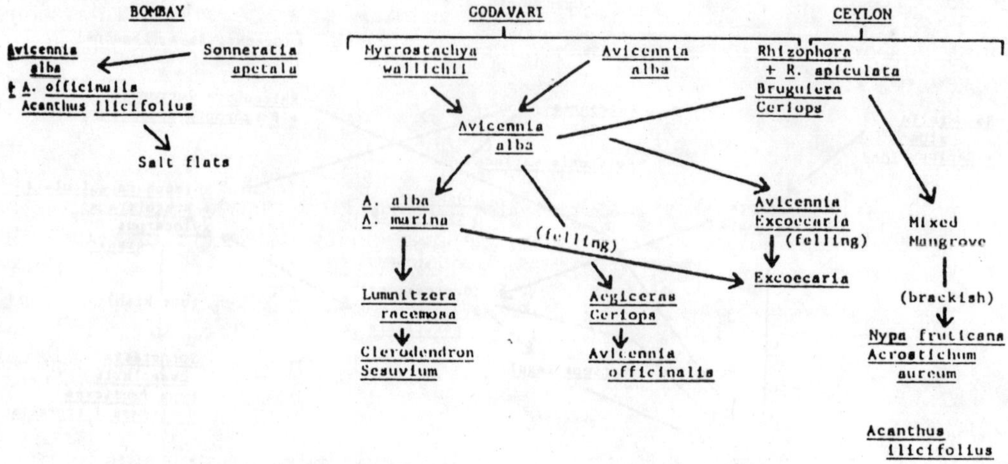

Succession in the mangrove swamps of India (Chapman, 1970; Schema 5)

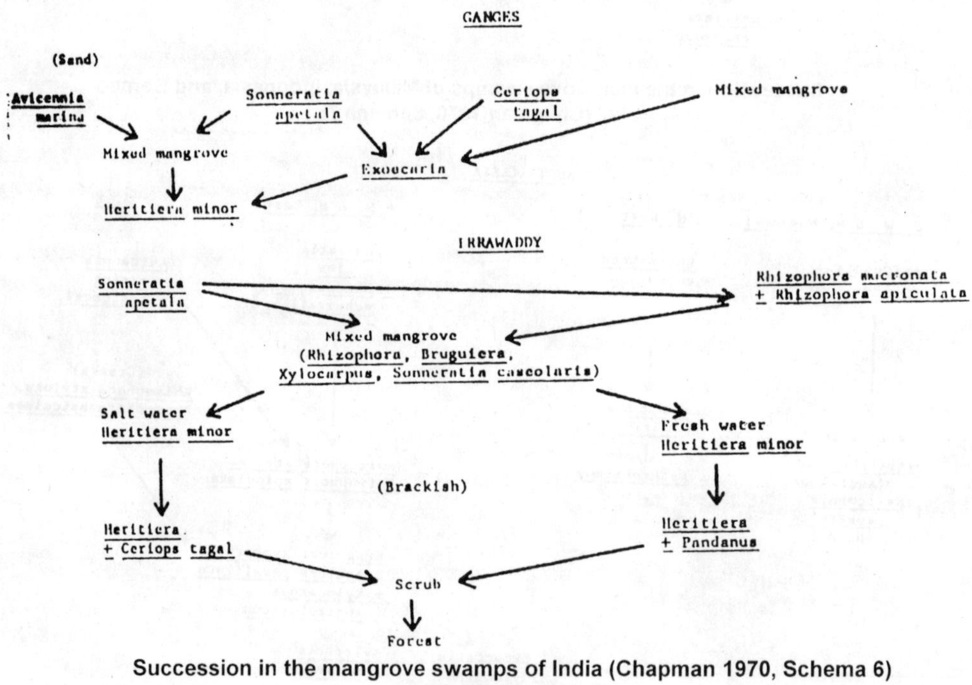

Succession in the mangrove swamps of India (Chapman 1970, Schema 6)

Fig. 27.

In the tidal mangrove sequence, each succeeding community occurred at sites with successively longer periods of tidal inundation. The shore pioneer species was. *C. tagal*, followed by a broad zone of *R. mucronata* and *B. gymnorhiza*. The final stage was dominated by *Heritiera littoralis*. The boundary between the *H. littoralis* zone and the rain forest was dominated by *Intsia bijuga* (Colebr.) O. Kunz. and was very sharp. All of the mangrove species occurred as scattered individuals in all the communities, although the dominant species comprised over 50% of the number of species, and there were sharp transitions between the zones.

In West Africa (Nigeria) Jackson (1964) recognized six groups of mangrove on the dual basis of range of habitat and dispersal of seeds.

Group I. Species restricted to the tidal areas and with specialized seed habits. This group included *R. racemosa*, the pioneer species at the water's edge and on the storm beach, and *A. nitida* dominant along the inner edges of closed lagoons. Boughey (1957) found *Rhizophora* only in open lagoons and *Avicennia* from lagoons in Natal. In the case of *Rhizophora* species, Keay (1953) considered *R. racemosa* to be the pioneer species, with *R. mangle* following on drier ground and *R. harrisonii* on wet ground. These were followed inland by *A. nitida*, *Laguncularia racemosa* and *C. erectus*. Gledhill (1963) pointed out that the propagules of *R. mangle* are approximately 20cm long carpus, *Phyllanthus* and those of *R. harrisonii* 30 cm long. Gledhill felt these were adapted for establishment on less heavily silted under more vigorous water current conditions.

Group II. Species found normally in tidal areas and with specialized fruits but normal seeds, or with buoyant seeds. Genera in this group were *Drepanocarpus*, *Delbergia*, *Ormocarpum* and *Hibiscus*. The seeds of these groups germinate in the water.

Group III. Species widely distributed along water courses and in swampy areas, usually with unspecialized fruits. The genera here were *Pterocarpus*, *Cynometra*, *Loinchocarpus*, *Phyllanthus* and *Phoenix*. The seeds of the first four are buoyant and are found floating in water. Seeds of *Phoenix* have never been found in the drift.

Group IV. Species not restricted to water courses in the forest areas and with no marked specializations. These are associated with high or well–distributed rainfall. The fruits and seeds are not associated with dispersal by water. This group included *Anthocleista*, *Elaies*, *Combretum*, *Alchornea* and *Paullinia*.

Group V. Cultivated plants. These are plants which occur in strand vegetation and have floating fruits which are dispersed by water, Genera include *Cocos*, *Terminalia* and *Anacardium*.

Group VI. Species whose seeds and seedlings are found in the swamp but with few or no mature individuals present. These include *Lochocarpus*, *Halomosia*, *Spondias*, *Cleistopholis*, *Dioclea* and *Entada*.

Blasco (1977) classified 6 types of Indian Mangals on the basis of phytogeographical position :

1. Bengal mangals,
2. Mangals of the Krishna and Godavari deltas,
3. Mangals of Cauvery delta,
4. Mangals of the basaltic coasts of Bombay,
5. Mangals of the Arid Zone Kathiawar,
6. Mangals of the Andaman–Nicobar archipelago,

but he has not given successional pattern of these mangals. First time, Naskar and Guha Bakshi (1987) have given phyto-succession in the Sunderbans. They have mentioned 5 phase in ecological succession of the mangrove swamps of the Sunderbans delta :

Phase 1 Swampy mangroves in intertidal mangrove zones
Phase 2 Tidal mangroves
Phase 3 True mangrove decline
Phase 4 Colonization of non–tidal species
Phase 5 Xerophytic non–mangroves and dry evergreen.

Authors have worked on the successional pattern of mangrove forests of Andaman from 1980 to 1985.

In Andaman, perennial rivers are absent and consequently no delta like substratum is formed. The case of marked accretion could not be noted and mangrove are along narrow straits (creeks), protected seashore and sea inlets and do not have considerable depth. The types of substrata noted are as follows:

1. **Rocky type** – On some sites of Chiria tapu and seashore side of Burma Nala mangrove species are present on rocky substratum, and roots of mangrove species enter in through elongated cracks. The density of plants is not high.
2. **Sandy type** – Along narrow strait (Creek of Parkinson) on Alexandra Island, some areas of Wandoor and Peel Island the substratum is sandy towards inner side, i.e. Landward side.
3. **Muddy type** – Along sea inlets, where washing of hinterland forest areas bring fine silt and clay, the substratum becomes muddy. Even in regions of sandy soil in the outer zone, where good terrestrial forests are present, the soil becomes muddy.

The substrata supporting mangrove forests of Andaman of Islands are somewhat different from those of the main land of India.

Geologically the Andaman Islands form a part of lofty range of submarine mountains, 1525–b km long running fro Cape *Nigarais*; in the Arakan Yoma range of Burma, through Sumatra and Java to the lesser Sunda Islands and the Moluccas and contain much that is geologically characteristics of the Arakan Yoma. In general, rock consists of principally non–calcareous, grey sand stone and embedded shales with occasional nests of poor coal, conglomerates, and pale grey lime stone. Volcanic fragmentary rock, apparently occur in the main islands, especially on the east coast of South Andaman. The older rocks are supposed to belong to early Tertiary or late Cretaceous.

Mangrove develop only a long protected shorelines as is evident from COE (1983). Protected shorelines are : (i) Protected shallow bays, (ii) estuaries protected or unprotected, the former have a much heavier vegetations, (iii) lagoons, (iv) leeward sides of peninsula of islands, (v) protected sea way (straits, usually narrow ones), (vi) behind spits (small narrow point of land running into a body of water) and (vii) along back waters affected by tidal waves.

Substratum most suitable for development of mangroves have been regarded to be muddy, but they also develop on gravelly, sandy, rocky and sometimes peaty substrata. Also there are some physiographic elements responsible for their development such as : (i) shores shelving gradually, and (ii) shores where rivers bring down eroded sediments.

In Andaman, we have recognized three basic categories of mangal and these are :

(1) **Pure stand** : Mangals consisting of pure stand have one mangrove species in the community. In Andaman in many places there is colonization of only single species of mangrove.

Rhizophora stands are mostly on seaward fringe, *Avicennia* stands are mostly on flate littoral zone. *Bruguiera* stands on flate littoral zone have been noticed. Frequently these pure stands from broad zones in their own right or else may occur as discrete patches within other associations.

(2) **Paired associations** – The more common paired associations are –

(i) *Rhizophora mucronata* – *R. apiculata* association
(ii) *Bruguiera gymnorhiza* – *Ceriops tagal* association
(iii) *Rhizophora mucronata* – *Bruiguiera gymnorhiza* association
(iv) *Lumnitzera littorea* – *Avicennia officinalis* association
(v) *Heritiera littoralis* – *Pongomia pinnata* association
(vi) *Acanthus ilicifolius* – *Acrostichum aereum* association

(3) **Mixed associations** – are the most complex mangrove associations. Mixed stands may contain upto 6 to 10 species. The species not commonly taking part in these associations are *Avicennia marina, R. stylosa, Bruguiera gymnorhiza, B. parviflora, Aegiceras cornculatum, Excoecaria agallocha, Ceriops tagal* and *Xylocarpus* species. Within a mixed stand one species may be dominant, and these are – *Rhizophora stylosa, B. gymnorhiza, Avicennia marina, Ceriops tagal* and *Lumnitzera littorea* species.

A detailed scheme of successional pattern of mangrove forest of Andaman is given in Fig. 28.

From Fig. 28 it is seen that *Rhizophora mucronata* and *R. apiculata* formed the proximal zone along creeks. These two species grew in all types of habitat. The substratum in the well developed Rhizophora community is always very soft and muddy due to entrapment of sediments between the prop roots. The first plant species to appear in the newly formed areas are *Acrostichum aureum* and *Acanthus ilicifolius* species. After stabilization of these species *Avicennia officinalis, A. marina, Ceriops tagal* and *Excoecaria agallocha* alongwith *Bruguiera gymnorhiza* and *X. moluccensis* appears one after another. The proximal zone finally is replaced by *Rhizophora mucronata* and *R. apiculata* because these species flourishes well in high tides. Nypa fruticans normally grows in middle zone where soil is muddy.

It is clear that mangrove swamps are not the simple communities some writers thought them to be. Macnae (1966) described in details a complicated succession in the swamps of Queensland, Australia. There, the pioneer tree was *A. marina* where the influence of saline water was strong. The *Avicennia* zone was composed of a row of mature trees. two or three trees deep, with thicklets of seedlings and samplings extending out onto a beach. Often, where the influence of freshwater was strong. *Aegiceras corniculatum* Blanco occurred in large numbers in the seedling and sampling thickets.

Ding Hou (1958) and Van Steenis (1962) held that *R. stylosa* was found only on sandy shores and coral terraces. Macnae stated that both *R. stylosa* and *R. mucronata* grew in mud, sand and on coral debris, and speculated that the two forms are actually variants of a single species. Whatever the taxonomic position may be, it is clear that in contrast to the New World genera the South–Eastern –

Asiatic and Eastern Africa forms of *Avicennia* are better adapted for pioneering than Rhizophorous forms.

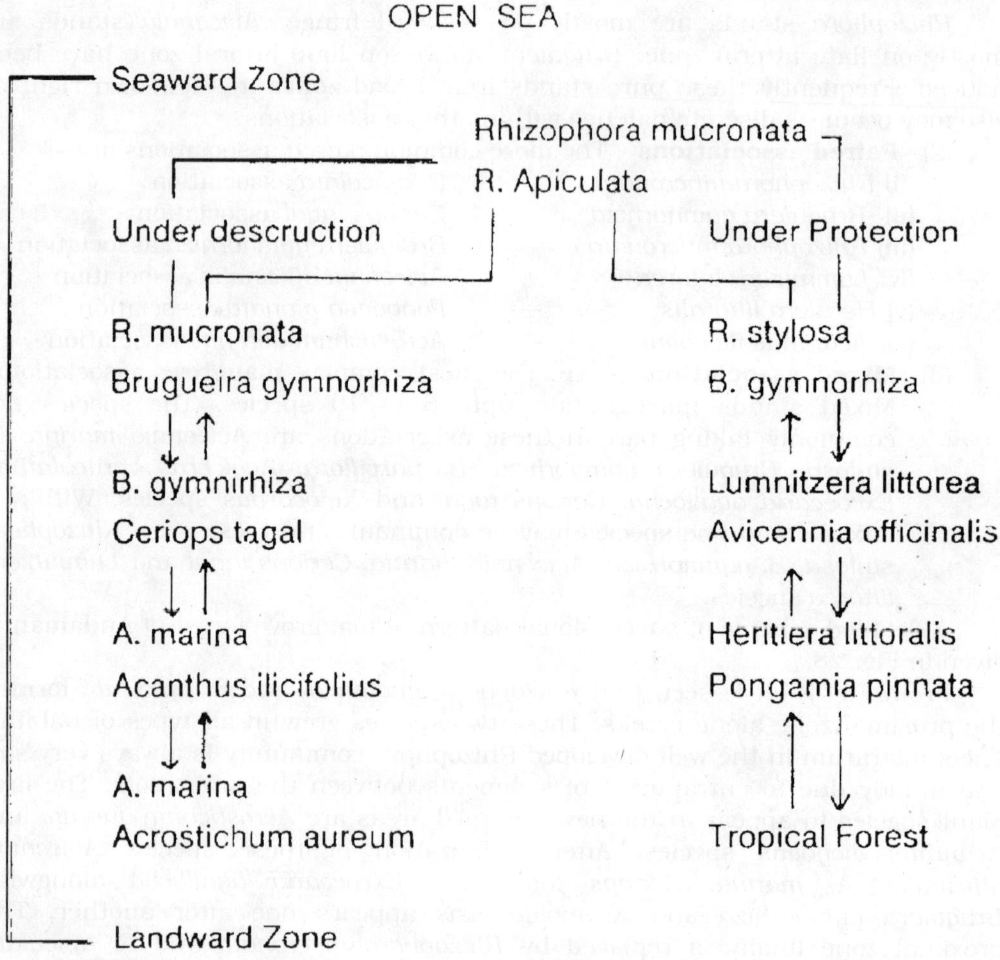

Fig. 28. Successional Pattern of mangrove forest of Andaman Islands.

Landward of the *Rhizophora* forest lay broad areas of either (i) thickets dominated by *C. tagal*, where the amount of rainfall was intermediate, or (ii) forests dominated by *Bruguiera exaristata* Ding Hou was sometimes subdominant. Occasionally, *A. marina, B. gymnorhiza, R. apiculata, R. stylosa, Xylocarpus granatum* Konnig, and *X. australasicum* Ridley were present. In areas of much rainfall, the *Ceriops* thickets were narrow and bordered by very dense forests dominated by either *B. parviflora* or *B. gymnorhiza.* In *B. gymnorhiza* forests, scattered specimens of *X. australasicum* occurred and the fern *A. speciosum* grew between the trees. The *Bruguiera* forests were the tallest of the Australian mangrove. Height of the trees appeared to depend upon the amount of freshwater available, with tallest trees in areas of highest rainfall.

The landward fringe of the Queensland mangals was the most diverse of all seral stages. *Avicennia marina* was the most abundant tree, but *B. exaristata* was common, and *C. tagal, C. decandra, L. littorea, R. apiculata* and *Excoecaria agallocha* were present. *Xylocarpus granatum* and *X. australasicum* were present occasionally. Where the landward fringe bordered a rain forest, many of the forest epiphytes grew on the mangrove trees. These included the orchids *Dischidia numularia* R.Br. and *Dendrobium* sp. the ant plant *Myrmecodia* (beccari Hook?) and the ferns *Drynaria rigidula* Bedd., *Platycerium* sp., *Polypodium acrostichoides* Forst. and *P.quercifolium.*

The above description of Macnae of seral succession in a mangal is, in general, typical of the large swamps of the Indo–Pacific region. This author (1966, 1968) described characteristics of large mangals in detail, and concluded that zonation of mangrove trees was due to the interaction of (i) frequency of tidal flooding, (ii) salinity of the soil water, and (iii) water logging of the soil. All three are modified by the presence of creeks, gullies, channels and rivers. The second and third depend upon rainfall and/or the supply of freshwater, evaporation, transpiration, and the nature and quality of the soil. Chapman and Trevathen (1953) stated that on muddy or sandy shores, distribution of organisms is related to the nature of the substratum which controls drainage, aeration, and penetrability.

The possible role of tidal flooding in relation to succession of Jamaican mangroves was shown by Chapman (1944b), who related vegetation types to the number of tidal submergence per year. His data showed decreasing tidal influence between the seaward *R. mangle* stand and the landward *C. erectus* stand.

Tidal flooding alone does not determine species compositions, zonation or succession in mangals. Clarke and Hannon (1967, 1969, 1970, 1971) studied the physical habitat of mangroves in Australia in great detail. They concluded that soils did not play a major role in control of plant distributional patterns and that plant reaction on the soil did not regulate seral change. Microclimate was important in providing conditions necessary for seedling development, determining soil characteristics and influencing competition between species. Clarke and Hannon showed that the holocoenotic complex was intricate and that variation in any of the components affected species distribution. The main factors were degree of tidal flooding, elevation of the land, and salinity of the soil water. Plant zonation was associated closely with elevation above mean sea level, seasonal patterns of soil salinity, and small differences in microtopography.

The authors stated that the sharpness of zonation depended upon the intensity of species interaction at ecotones. Slight environmental changes related to topography produced intense competition which made significant factors that were normally of secondary importance. Generally, the severity of the environment, including covering by mud and tidal scouring, determined the success of a species in advancing seaward, whereas landward extension was governed by ability to compete with other species in relation to salinity, availability of freshwater, temperature, light and humidity.

Chapman (1970), in a very important paper on mangrove phytosociology, compared succession in mangals throughout the tropics and gave eight schemata that depicted zonation. Chapman concluded that there is great similarity in the vegetational communities and suggested an extensive classification of natural associations according to the Braun–Blanquet system. The classification consisted of 8 alliances, 15 orders and 40 associations, but must be considered tenuous at this time because of lack of taxonomic and systematic data from many localitics.

Chapter - 6

FUNCTIONAL ASPECTS OF MANGROVE FORESTS - BIOMASS, LITTER FALL, LITTER DECOMPOSITION AND SOIL RESPIRATION

Mangrove swamps are considered one of the productive ecosystems of world, where the rate of primary productivity is quite high in stressed environment and contributes energy to detritus marine food webs (Heald, 1969; Odum and Heald, 1972; Boto and Bunt, 1981). Snedaker (1978) reviewed the importance of the mangrove swamps and suggested that "the value of mangroves is accepted and based largely on the production of organic matter as leaf litter fall that enters the estuarine system where it forms the basis for a complex food web". Mangrove forests reach their greatest structural and floristic diversity in the tropical and sub-tropical region and generally show higher rate of primary production and litter decomposition than the temperate forest ecosystems (Bray and Graham, 1964).

Andaman mangroves which floristically and structurally seems to be somewhat different from those of mainland. They occur along protected sea coasts, narrow sea channels, creeks and sea inlets; which do not penetrate deep due to higher slope gradient of hinterland. These islands lack the perennial river system. The functional aspects of following two types of mangrove forests are described :

(i) Monogenerie mangrove forest (MNMF) which is dominated by only *Rhizophora* species on larger areas of coastal fringe.

(ii) Mixed mangrove forest (MXMF) inwards dominated by *Bruiguiera gymnorhiza*, *Ceriops tagal* and other species.

Observations were made along a series of transects across the tidal flats of two types of forests. Transects were selected to cover a wide range of area and the composition of the forests were determined by following Mueller-Dombois and Ellenburg (1974).

Biomass was calculated by harvest technique as given by Snedaker and Snedaker (1984). Above ground mangrove biomass was calculated by felling of 10 trees from similar habitats of each forest and biomass was determined. These trees were sorted into trunk, branch, leaf and inflorescence components and fresh weight of each component was measured. Sub-samples were then dried at 80°C till constant weight and dry weight was recorded.

Litter production was usually monitored by collection of litter in litter fall traps made from Serlon cloth sewn onto steel wire frames 50cm x 50cm and depth 0.1 m and were suspended from convenient branches with PVC support and were randomly elevated, so that they did not get immersed during high tides. Litter was collected from litter fall traps at 15 days interval. The material was oven dried for 48 hours at 80°C.

The decomposition of mangrove leaf litter was studied using the litter bag techniques reported by Goulter and Allaway (1979). The techniques differs from that previously used by Albright (1976) in New Zealand, but can be more easily compared with results from terrestrial ecosystems. Decomposition studies were undertaken at sites on the creeks of the two types of forests. Samples of mangrove leaves weighing 50 gm were placed in bags, approximately 16cm x 16cm made from Serlon cloth (2 mm mesh size). At each site 10 litter decomposition bags were threaded on to Nylon line and tide between trees. The bags were rested on the substrate which were able to float during high tides. One bag was collected from each site after every 20 days. Its contents were oven dried at 80°C till constant weight.

Soil respiration was measured by the inverted box alkali absorption method (Walter & Haber, 1957) at 20 days intervals in both types of forests. In the field bottomless glass boxes having area of 50 sq km were kept sunk into the soil upto 10 cm depth. Boxes were completely covered with black cloth and CO_2 evolution from mangrove soils were absorbed by solution of 0.25N KOH for one hour in 6 cm petridish The amount of CO_2 fixed by the KOH solution was measured by titration (after addition of an excess of $BaCl_2$) with 0.25N HCl solution using methyl orange as an indicator and the rate of respiration was calculated.

Structural Composition of Mangrove Forests

During extensive survey of the various Islands of Andaman, it was found that there are two types of mangrove colonization –

(i) Colonization of *Rhizophora* spp. along the bank of creeks which attained the height of 5 to 10m and 20 to 40 m wide on the both sides of creek.

(ii) Colonization of mixed mangrove species on large tidal flats in the inner region. They are 20 to 35 m tall and form a band of 200 to 350 m wide near sea inlets and lagoons.

For qualitative and quantitative data, phytosociological studies were taken on larger areas of the two types of mangrove forests and their composition is mentioned in the Table 33.

Table 33 shows that mixed mangrove forests are richer in species composition than that of monogeneric forests. A detailed structural patterns of these two types forests are compared and same is given in Table 34.

From the Table 34 it is seen that MXMF are very rich and divers type of mangrove forests. This is edaphically climax vegetation under prevailing condition. In monogeneric mangrove forests diversity is very low and there is no zonational pattern.

Table 33. Differences in structural pattern of mangrove forests

Parameters	MNMF	MXMF
Species richness	Poor	Very h igh
Diversity of species	Very low	Very high
Stratification	Absent	2–3 strata
Zonation	Nil	Having three zones
Large stem diameter sp.	Nil	Many
Complexity Index	65.5	271.8

Table 34. Species composition and importance value index of each species of two types of mangrove forests

MNMF		MXMF	
Monogenric mangrove forests		Mixed mangrove forests	
Species	IVI	Species	IVI
Rhizophora mucronata	93.8	*Bruguiera gymnorhiza*	32.8
R. apiculata	84.3	*B. parviflora*	5.0
R. stylosa	40.4	*B. cylindrica*	3.4
		Ceriops tagal	30.8
		Lumnitzera littorea	13.8
		Avicennia officinalis	11.9
		R. mucronata	10.5
		R. apiculata	6.7
		Sonneratia alba	5.4
		Xylocarpus granatum	3.8

Mangrove grow in areas of high solar radiations and have the ability to use saline water so they are in good position to achieve high biomass. Biomass of two types of forests was determined at various places of Andaman and the same is given in Table 35. The Table 35 shows that the MXMF is more productive in comparison to MNMF. The value of the biomass was 213.8 t ha^{-1} in MXMF and 124.1 t ha^{-1} was in MNMF.

Table 35. Biomass production of two types of forests (t ha^{-1})

Plant Parts	Biomass (t ha^{-1})	
	MNMF	MXMF
Leaves	6.6	10.3
Fruits & Flowers	0.3	0.7
Stem & Branches	11.3	15.7
Wood	75.4	140.3
Prop roots	30.5	7.3
Pneumatophores		39.5
Total	124.1	213.8
SD ±	8.6	12.9

Litter Production

Litter production was observed throughout the year beneath two types of forests. There was a distinct seasonal pattern in leaf litter fall. The maximum litter fall was observed in between October to April (unpublished data). Leaves accounted the largest fraction of the litter fall throughout the year. In MNMF fruits were recorded during summer only while MXMF, fruits were recorded for longer period of the year. The data of litter fall of two types of forests are given in Table 36.

Table 36 : Estimate of annual total litter fall from litter fall traps (t ha^{-1} yr^{-1})

Litter fall trap No.	Monogeneric mangrove forests				
	Leaves	Twigs	Fruits	Inflorescence	Total
1	3.62	0.78	1.26	0.17	5.83
2	5.83	0.92	0.11	0.87	7.33
3	4.63	1.21	0.81	0.45	7.10
4	6.18	0.79	0.76	0.26	7.90
Mean	5.06	0.92	0.76	0.47	7.16
SD ±	1.23	0.23	0.13	0.11	1.26
	Mixed mangrove forests				
1	5.82	0.93	0.43	0.13	7.31
2	7.63	1.16	0.83	0.96	10.58
3	4.78	1.29	1.21	0.41	7.69
4.	5.91	1.43	0.77	0.52	8.63
Mean	6.03	1.20	0.81	0.50	8.55
SD ±	1.93	0.30	0.10	0.12	1.36

From the Table 36 it is clear that MNMF litter fall was estimated 7.1 t ha^{-1} yr^{-1}. This value was based on the average of 10 litter fall in mixed mangrove forests was estimated 8.5 t ha^{-1} yr^{-1}. This value is significantly higher than that of MNMF. The major contribution in litter fall was of leaves in both types of forests.

Litter decomposition rate

The decomposition of mangrove litter is an important aspect in nutrient cycling and the supply of organic matter to estuarine food web. It is a function of the availability of oxygen, substrate characteristics, animal and micro–organism activity (Lugo and Snedaker, 1974). Table 37 shows the ash free weight of leaf litter recovered from decomposition bags at two types of forests. Rapid decomposition of litter occurred during the first 10 weeks and decomposition was gradually slower thereafter. The response of litter decomposition of MNMF and MXMF was different. Decomposition rate was slow in MXMF after 60 days while in MNMF it was so after 80 days. The litter decomposition was assessed in different ecological condition e.g. high salinity, low salinity, aerobic and anaerobic mud. Different habitats were selected on the basis of concentration of soil salinity and dissolved oxygen, the rate of decomposition is given in Table 37.

Table 37. Litter decomposition rate in mg g^{-1} d^{-1} in two types of forest in relation to salinity and dissolved oxygen

Day	Low salinity 30–50‰ with aerobic mud D.O. 1.3–8.0 ppm		High salinity 60–80‰ with anerobic mud D.O. nil	
	MNMF	MXMF	MNMF	MXMF
20	1.7	2.3	0.8	1.1
40	1.9	2.9	0.9	1.9
60	2.6	3.5	1.1	2.3
80	2.9	2.7	1.9	1.4
100	1.8	2.4	1.7	1.1
120	1.1	1.6	1.5	0.8
140	0.7	1.4	1.1	1.2
160	0.2	0.9	1.1	1.0

It was found that fast decomposition rate was in aerobic muds where dissolved oxygen was 3.0 ppm to 8.0 ppm and salinity was 30.0/oo to 50.0/oo. The rate of decomposition was greater in MXMF in comparison to MNMF. Initially the decomposition rate was faster in MXMF afterwards a decline was noted. While in MNMF rate was comparatively lower.

Soil Respiration

Soil respiration of the two types of mangrove forests was measured in different islands where bags were put-up for decomposition. The average values are given in Table 38.

Table 38. The amount of CO_2 evolution (mg CO_2 m^{-2} hr^{-1}) in two types of forests

	MMNF	MXMF
CO_2 evolution rate	486.2	853.6
SD ±	16.3	29.4

Table 38 shows that the maximum value of CO_2 evolution was in MXMF. The value was 853 mg CO_2 m^{-2} hr^{-1}.

Andaman group of islands shelter one of the best and most important mangal area of the world. Mangrove forests throughout the world are fundamentally similar in their pattern, in space and time because similar process of succession and maintenance operate. But, they are dissimilar in biomass production, cycling of matter, productivity, litter fall, litter decomposition and soil respiration rate. Environmental factors such as tidal factor, water chemistry, topography, etc. do play a vital role in diversity of functional aspects of mangrove forest ecosystems.

Many authors, Golley *et al.* (1975), Golley and Golley (1972) and Rodin and Bazilevish (1967) have demonstrated the utility of working with absolute measure of organic production which is defined as standing–stock biomass. Not only do such data provide increasingly valuable means for making comparison (Rodin & Bazilevish, 1967) among ecosystems, but they also provide a basis for evaluating worldwide productivity pattern (Golley and Golley, 1972). The biomass production of mangrove forests of Andaman is significantly very high and is the second highest forest in biomass production (Productivity in mangrove ecosystem is given in Fig. 29).

Productivty in mangrove Ecosystem

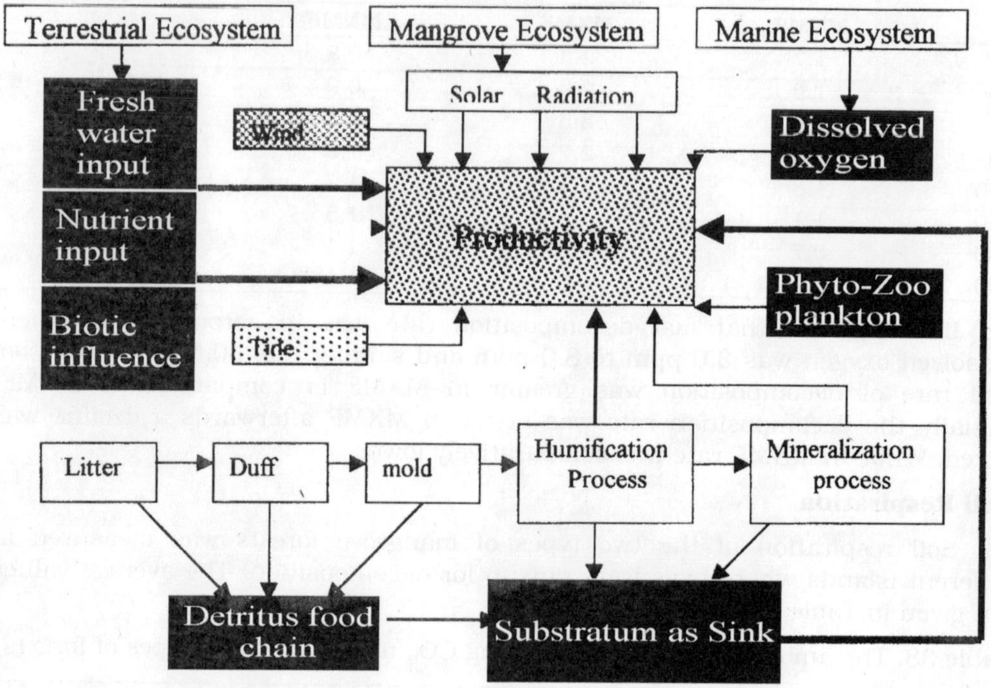

Fig. 29.

The variability in the biomass data may be attributed to age, stand history, structural differences and biotic interferences. The high biomass of Panamanian forests or Andaman forests is due to their age and maturity. Moreover, they are more protected from hurricane, human disturbance and other destructive factors. The forests in Florida had been influenced by hurricane few years back before the biomass sampling harvest. The difference among the Florida mangrove type estimates of biomass suggest that real variations exist among forest type.

Table 39. Estimates of mangrove biomass

Location	Mangrove species	Biomass production t ha^{-1}	References
Panama	—	279.2	Golley *et al.* (1975)
Puerto Rico	—	62.9	Golley *et al.* (1962)
Philippines	Mixed mangrove sp.	45.93	Dela Cruz & Bannag (1967)
Florida (Overwash)	*R. mangle*	129.64	Lugo & Snedaker (1974)
Florida (Riverine)	*R. mangle*	173.9	—
Thailand	*R. apiculata*	159.0	Christensen (1978)

Japan	*R. mucronata*	108.1	Suzuki & Tagawa (1983)
Australia Sydney	*Avicennia marina*	129.4	Briggs (1977)
W. Port Bay	*A. marina*	8.6	Clough & Attiwill (1975)
New Zealand	*A. marina*	104.1	Woodroffe (1982)
India			
Andaman Islands	Monogeneric mangrove forest	129.1	Mall *et al.* (1991)
	Mixed mangrove forest	213.8	Mall *et al.* (1991)

In comparison with other tropical forests, the biomass of mangrove forests is relatively low (Rodin & Bazilvich, 1967). The low biomass production associated with MNMF is due to result of structural differences, lower species diversity, age of the forest and complexity index.

Table 40. Comparison of litter fall estimate from Andaman Islands and different mangrove swamps of world

Location	Species	Litter fall t ha^{-1} yr^{-1}	Source
Thailand	*Rhizophora apiculata*	6.70	Christensen (1978)
Puerto Rico	*R. mangle*	4.75	Golley et al. (1967)
Florida USA	*R. mangle*	8.50	Heald (1969)
Florida	*R. mangle* (Scrub)	0.60	Pool *et al.* (1975)
Florida	*R. mangle* (Fringe)	7.58	Pool *et al.* (1975)
Florida	*R. mangle* (Fringe)	1.28	Teas (1979)
Florida US	*Avicennia germinans*	6.50	Lugo & Snedaker (1974)
Florida USA	—do—	2.85	Teas (1979)
Queensland Australia	*Rhizophora apiculata*	10.91	Duke *et al.* (1981)
Hinchinbrook Island	*R. lamarckii*	8.69	Duke *et al.* (1981)
Hinchinbrook Island	*Ceriops tagal*	7.18	Duke *et al.* (1981)
Hinchinbrook	*Bruguiera gymnorhiza*	7.99	Duke *et al.* (1981)
S. Africa	*Avicennia marina*	9.67	Steinke (unpublished)
N.S.W. Australia	*Avicennia marina*	5.80	Goulter & Allaway (1979)
New Zealand	*A. marina* Tall	8.10	Woodroffe (1982)
New Zealand	*A. marina* Low	3.65	Woodroffe (1982)
Andaman	Monogeneric mangrove forest	7.16	Mall et al. (1991)
Mixed mangrove	Forest	8.15	Mall et al. (1991)

Mangrove with greater tidal activity and water turnover generally have higher litter fall than mangrove in areas with stagnant water (Pool *et al.*, 1975), as has been observed for other forested wetlands (Conner and Day, 1976; Brown *et al.*, 1979 and Brinston *et al.*, 1980). Also, as tidal activity decreases, surface litter is less susceptible to export (Twilley, 1985), suggesting that *in–situ* litter decomposition should have more influence on litter dynamics and nutrient cycling. Storms are known to be an important factor in promoting litter fall both in mangroves and elsewhere (Pool *et al.*, 1975 and Goulter and Allaway, 1979). In Andaman storms are occasional but wind velocity is very high hence which promote litter fall in these islands. High amount of litter fall in mixed mangrove forests is result of high species diversity. A high amount of litter production in

MXMF results from differences in mangrove biomass, canopy height, density, nutrient availability and soil salinity. Andaman forests are very rich and relatively undisturbed and have good height, girth and stand maturity, than any mangrove forest of world, accounting more biomass and more litter decomposition. The values of litter fall of Andaman compared well with values of the other countries. Our data are very similar to data accounted by Duke *et al.* (1981) in case of Hinchinbrook Island (Table 40).

Organic matter dynamics is an important aspects of the decomposer sub–system in mangrove ecosystems. A major part of the carbon fixed by the plants is subsequently returned to the soil in the form of dead organic debris, which plays an important role in the evolution of the structure and function of mangrove ecosystems. The gradual decomposition of litter and its incorporation into the soil is responsible for the process of mineralization. The process of litter decomposition and the factors affecting them in case of terrestrial ecosystems have been critically reviewed by Dickinson (1974), but regarding mangrove ecosystems, the role of litter decomposition is largely governed by chemical composition of the litter, environmental condition, soil animals and micro–organisms (Singh & Gupta, 1977). But in mangrove ecosystems the rat of litter decomposition is mainly governed by the nature of substrate, soil oxygen availability, soil salinity, micro-organisms and scavenger activity. The rate of litter decomposition to aerobic and low saline substrates. Environmental conditions, mainly temperature and moisture play a role in litter decomposition in terrestrial environment. But in mangrove ecosystems these two factors are always constant for all seasons, thus variability occurs only in soil salinity and soil oxygen. These two factors are usually responsible for governing the rate of litter decomposition. In areas with a large tidal range most of the litter may be exported from the system (Boto & Bunt, 1981) and decomposition in sea water noted very slow but where the tidal impact is weaker a proportion of the litter accumulates on the substrate and degrades in situ. Present study includes that there is a little impact of the seasons on litter decomposition because of almost uniform climatic condition of Andaman. The low rate of decomposition in monogeneric mangrove forests is result of substrate conditions, chemical composition, decomposing materials micro-organisms activity and low soil salinity. In Andaman most of monogeneric mangrove forests are associated with sandy soil which are poor in micro-organism population and having low soil salinity. Hesse (1961) also mentioned that the rates of decomposition vary according to oxygen availability, substrate characteristics, micro–organisms and scavenger activity which are generally high in mangrove swamps. Our observations is that apart from aerobic bacteria, fungi and actinomyces in mangrove swamps, the nature of decomposing material is important factor in deciding the rate of decomposition as it was revealed from Table 38. Soil respiration is one of the earliest and still the most frequently used index of microbial activity in the soil (Lundegardh, 1927 and Haber, 1958). It is manifested by the magnitude of the life processes which take place in soil. Soil microorganisms mainly bacteria and fungi play an important role in releasing CO_2 and by metabolizing dead organic debris. Reichie *et al.* (19750 while studying an mesic deciduous forest have concluded that soil microflora contributed over 90% of the total decomposer respiration data is primarily due to microbial population and secondary due to habit composition.

Managed and Unmanaged Mangrove Forests

Mangrove grow in areas of high solar radiation and have the ability to extract fresh water from the sea water, so they are in an excellent position to achieve high primary productivity. Mangroves with greater tidal activity and water turnovers generally have higher productivity and higher biomass. Several other factors are identified as important regulators of mangrove productivity: (i) amount of dissolved oxygen, (ii) total salt content which governs the osmotic pressure gradient, (iii) macronutrient content of soil etc. These factors are responsible for physiognomic differences and functional attributes of mangroves (Lugo and Snedaker, 1974).

The functional attributes of mangrove forests in relation to environmental gradients have been studied by many workers e.g. Golley *et al.*, 1975; de la Cruz & Banaag, 1967; Snadekar & Lugo, 1973; Christensen, 1978; Suzuki & Tagawa, 1983 and Woodroffe, 1982 & 1985.

No data are available on functional aspects of mangrove forests of India, so in this chapter functional studies of mangrove forests of Andaman have been taken into consideration in relation to two types of mangrove forests.

(i) Forest naturally managed type, situated on fringe on flat littoral zone.

(ii) Forests are unmanaged type, situated along creeks with narrow belt in between tropical rain forest and sea.

Species richness and dominance

Mangrove forests of Andaman is the richest in species and most luxuriant and diverse vegetation in relation to physical environment. Species distribution and physiognomy of mangrove seems to be directly related to landform and substrate condition (Singh *et al.*, 1987). The most important substrate condition is substrate salinity, redox-potential and chemical composition. Davis (1940), Chapman (1970 and 1976), Macnae (1968), and other concluded that successional tendencies are limited to particular localities by the dynamic nature of external forces which are constantly including changes to environmental gradient and plant habitats. Thom (1967 and 1974) studied the ecology of mangroves and examined in terms ot the plant habitat change induced primarily by geomorphic process. A very diverse type of mangrove forests of Andaman is result of diverse type of landform.

A quantitative evaluation of two types of mangrove forests in terms of height, girth and IVI has been given in Table 41. Trees in the managaed forest were taller and attained greater girth than those of the unmanaged forest, though, the species diversity in unmanaged forest was more. The average community height (15.1 m) and tree girth (113 cm) in managed forest were more in comparison to average community height (6.7 m) and girth (65 cm) in unmanaged mangrove forest. Further, the size of contributing species were significantly very high in managed forest because of age and balance from all stresses, trees with large green leaves provide the protection and utilized maximum solar radiation for maximizes production and root in preserving soil erosion and organic matters are well known. The totality of all forests functions are as the biomass system that keeps the ecological balance needed for various physiological processes. For the better the clear quantification of mangrove forests, a complexity – index was accounted and same is given in Table 42. It is seen that the value of complexity index of managed forest is significantly very high in comparison to unmanaged mangrove forests of Andaman and also to mangrove forests of world.

Table 41. Composition, Height and Importance-Value-Index of Different Tree Species in the Two Forests

Name of the species	Managed Forest			Unmanaged Forest		
	Av. Height (m)	Av. Girth (cm)	IVI	Av. Height (m)	Av. Girth (cm)	IVI
Rhizophora apiculata	20.7	101	29.2	9.4	81	8.3
R. mucronata	11.2	109	15.3	7.6	69	6.4
R. stylosa	15.2	126	5.4	5.2	76	1.3
Ceriops tagal	10.0	132	12.6	6.2	60	8.3
Bruguiera gymnorhiza	27.8	189	17.3	6.4	75	16.3
B. cylindrica	5.3	59	1.0	—	—	—
Avicennia officinalis	25.6	118	2.3	10.5	121	32.4
Lumnitzera littorea	15.8	127	3.6	4.4	49	2.3
Sonneratia alba	12.6	83	1.3	—	—	—
Scyphiphora hydrophyllacea	—	—	—	3.8	51	1.0
Excoecaria agallocha	—	—	—	4.3	52	2.8
Aegiceras corniculatum	—	—	—	5.3	59	1.2
Xylocarpus granatum	10.3	86	2.1	6.5	42	1.3
X. moluccensis	8.4	76	1.2	6.3	49	1.1
Heritiera littoralis	—	—	—	6.4	69	1.8
Average height of community in m	14.8			6.4		
Average girth of community in cm		109		65.6		

Table 42. Complexity Index of Managed and Unmanaged Mangrove Forests and Comparison with other Mangrove Forest of World

Location	No. of species	No. of trees above 10cm diameter	Height (cm)	Basal area m^2	Complexity Index
Andaman Island Managed forest	11	185	15.3	7.4	226.0
Unmanaged forest	13	121	6.7	4.1	43.2
Costa Rica (Main rivers)	4	118	16.0	9.53	71.0
Costa Rica (Santa Rosa)	1	80	10.0	2.23	3.5
Puerto Rico (Mona Island)	2	179	15.0	2.97	15.9
Puerto Rico (Jabus Bay)			4.8	0.17	1.9

Many authors, Golley *et al.* (1975), Golley and Golley (1972) and Robin and Bazilevich (1967) have demonstrated the utility of working with absolute measures of organic production as defined by standing stock biomass. Not only such data do provide increasingly valuable means for making comparison (Robin and Bazilevich, 1967) among ecosystems, but they also provide a basis for evaluating worldwide productivity patterns (Golley and Golley, 1972). The biomass of managed forest and unmanaged forest was accounted and same is represented in Table 43. In managed forest the biomass was 239.4 t ha^{-1} in comparison to 69.4 t ha^{-1} in unmanaged forest. Mangrove of Andaman are highly productive as compared to another mangrove forests of world (Table 44). Maximum biomass 279.212 t ha^{-1} was accounted by Golley *et al.* (1974) in Panamian mangrove forest. In our study also we have noted very high biomass in managed mangrove forest, it is next to Panamian mangrove forest. From Table 44 it is seen that there is lot of variations in mangrove biomass from country to country. The variability in the biomass data may be attributable to age, stand history, structural differences and biotic influence (Woodroffe, 1982). The high amount of biomass production of managed forests of Andaman is representative of an older stand protected from human destruction. The managed mangrove forest of Malasia, the biomass was found to be 300 t ha^{-1} in case of 25 years old forest (Eong *et al.*, 1979).

Table 43. Estimate of Mangrove Biomass of Managed and Unmanaged forests in (t-ha^{-1})

Type of forest	Dry weight (t ha^{-1})			
	Leaves	Woods	Roots	Total Biomass
Managed Forest	22.4	182.7	34.3	239.4
Unmanaged Forest	7.4	49.6	12.4	69.4

Table 44. Total biomass of various mangrove forest of world and comparison with Andaman

Panama	*Rhizophora brevistylis*	30–40	279.2	Golley *et al.* (1975)
Puerto Rico	*R. apiculata*	08	62.9	Golley *et al.* (1962)
Phillippines	Mixed mangrove species	—	45.9	Cruz & Benaag (1967)
Florida	*Rhizophora mangle*	Overwash	124.6	Lugo & Snedaker (1974)
Florida	—do—	Riverine	136.0	—do—
Florida	—do—	Fringe	118.9	—do—
Florida	—do—	Scrub	7.9	—do—
Florida	Mixed Mangrove species	Island	49.0	—do—
Florida	*Laguncularia racemosa*	Succession	8.1	—do—
Thailand	*Rhizophora apiculata*	121	159.2	Christensen (1978)
Japan	*R. mucronata*	5–6	108.1	Suzuki & Tagawa (1983)
	Bruguiera gymnorhiza	5-6	97.6	—do—
	Rhizophora + Bruguiera	5-6	78.6	—do—
Australia Sydney	*Avicennia marina*	6–8	128.4	Briggs (1977)
W. Port Bay	—do—	2–4	86.0	Clough & Attiwill (1985)
New Zealand	*Avicennia marina* var.	2-5.4	104.1	Woodroffe (1985)
Andaman	Managed forest–mixed sp.	14.8	239.4	This Study
	Unmanaged forest mixed sp.	6.4	69.4	—do—

Litter Fall

Mangrove swamps reach their greate structural and floristic diversity in the tropics, and tropical forests generally have higher rates of litter production than temperate forests (Bray and Gorham, 1964). The annual average amount of litter fall of managed and unmanaged forests of Andaman is given in Table 45. From the table it is seen that the value of litter fall 15.22 t ha^{-1} a^{-1} in managed forest is significantly very high than the 8.12 t ha^{-1} a^{-1} of unmanaged forest. High amounts of litter fall in mangrove forests is depended upon the physical environment of mangroves and varies from place to place according to community physiognomy. Mangroves with greater tidal activity and water turnover have greater litter fall than mangrove in areas with stagnant water (Pool *et al.*, 1975) as has been observed for other forested wetlands (Conner & Day, 1976; Brown *et al.*, 1979 and Brinston *et al.*, 1980). Also as tidal activity decreases suface litter is less susceptible to export (Twilley, 1985), suggesting that in situ litter decomposition should have more influence in litter dynamics and nutrient cycling. Storms are known to be an important factor in promoting litter fall both in mangroves and elsewhere (Pool *et al.*, 1975 and Goulter and Allaway, 1979) and in other forest types in New Zealand (Danial, 1975). In Andaman storms are occasional but high wind velocity is very common, hence promoting litter fall in these islands. The high amount of litter fall in managed forest is result of greater value of complexity index, floristic composition, canopy height, density and altitudinal variability apart from other environmental factors. The litter fall data of Andaman is compared with other forests of world in Table 46. It reveals that litter fall data is significantly very high in comparison to litter fall of other mangrove forests of world. Maximum 10.91 t ha^{-1} a^{-1} is recorded by Duke *et al.* (1981) in Queensland, Australia of *Rhizophora apiculata* stand.

Table 45. Estimate of annual total litter fall from litter fall traps beneath managed and unmanaged mangrove forests

Type of forest	Dry weight of litter fall				
	Leaves	Twigs	Fruits	Inflorescence	Total
Managed mangrove forest	12.5	1.03	1.05	0.54	15.22
Unmanaged mangrove forest	5.8	0.93	1.07	0.29	8.12

Table 46. Comparison of litter fall estimate of various mangrove forests of world

Location	Mangrove Species	Litter Fall t ha^{-1}	Source
Puerto Rico	*Rhizophora mangle*	4.75	Golley *et al.* (1962)
Florida (USA)	*R. mangle* (Scrub)	0.60	Pool *et al.* (1975)
----do----	*R. mangle*	8.80	Heald (1969)
----do----	*R. mangle* (Fringe)	7.58	Pool *et al.* (1975)
----do----	*R. mangle*	1.28	Tease (1979)
----do----	*Avicennia germinans*	6.50	Lugo & Snedaker (1975)
Hinchinbrook Is Queensland	*Rhizophora apiculata*	10.91	Duke *et al.* (1981)
Hinchinbook Is. Queensland	*R. lamarckii*	8.69	----do----

----do----	*R. stylosa*	9.34	----do----
----do----	*Ceriops tagal*	7.18	----do----
----do----	*Bruguiera gymnorhiza*	7.99	----do----
----do----	*Sonneratia alba*	7.92	----do----
----do----	*Avicennia* species	8.01	----do----
South Africa	*Avicennia marina*	9.57	Stainke (unpublished)
NSW Australia	----do----	5.80	Goulter & Allaway (1979)
Thailand	*Rhizophora apiculata*	6.70	Christensen (1978)
New Zealand	*Avicennia marina* Tall	8.10	Woodroffe (1982)
New Zealand	*A. marina* low	3.65	----do----
Andaman Managed	Mixed mangrove species	15.22	This study
Andaman unmanaged	----do----	8.12	----do----

Litter Decomposition

Organic matter output in the mangrove ecosystem refers to the fascinating process of litter decomposition, a major portion of the energy fixed by the photosynthetic canopy of green plants ultimately finds its way into the detritus component. Decomposition of organic matter is one of the dynamic process in the marine ecosystems, which has a direct bearing on the energy flow of these ecosystems. In terrestrial ecosystems, the rate of litter decomposition is largely governed by chemical composition of the litter, environmental conditions, soil animals and microorganisms. But in mangrove ecosystem the rate of litter decomposition, apart from soils, animals and microorganisms, is mainly governed by the nature of substrate, amount of oxygen and soil salinity. The data of % litter decomposition on weight loss basis is given in Table 47. In the evaluation of litter decomposition kinetics, time series, weight loss data of the litter of managed and unmanaged mangrove forest has been employed. An exponential decay is the approximation of percentage weight remain of litter as a function of time, showing the initial great loss 10.34% to 22.13% and 12.0 to 25.23% in managed and 7.6 to 16.3% and 8.3 to 19.3% in unmanaged as compared to the final loss.

Table 47. Mean quantities of percentage litter weight loss from the surface litter bags and bags burried at 10 cm in soil depth alongwith the quantities of non-decomposable litter material

Sampling periods	% litter weight loss			
	Surface litter bags		Litter bags burried at 10 cm in soil	
	Managed Forest	Unmanaged Forest	Managed Forest	Unmanaged Forests
2nd September	----	----	----	----
3rd October	10.3	7.6	12.3	8.3
2nd November	12.8	8.9	13.2	8.6
3rd December	15.2	12.4	15.9	14.3
2nd January	22.1	16.3	25.4	19.5
1st February	16.1	15.7	10.1	13.4
2nd March	10.3	14.3	6.2	11.7
1st April	3.8	9.32	4.1	7.4

1st May	2.6	5.9	3.8	5.3
Total % Weight loss	93.2	90.3	91.0	88.5
and of non-decomposable litter	6.8	9.7	9.0	11.5

The results indicate that the major portion of the litter in the system generally remained in the soil. It is evident from Table 47 that the percentage weight loss of litter in the bags burried at 10cm soil depth was higher, as compared to litter bags kept at the surface. It is clear from data in the Table 47 that the non–decomposable litter matter was more in the surface litter bags in comparison to the litter bags burried at 10 cm soil depth. The decomposition of mangrove litter is an important stage in nutrient recycling and the supply of organic matter to the estuarine food web, and is function of the availability of oxygen, substrate characteristics, and animal and microorganism activity (Lugo and Snedaker, 1974). In areas with a large tidal range most of the litter produced may be exported from the system (Boto and Bunt, 1981), but where the total range is smaller, a proportion of the litter can accumulate on the substrate and degrade in situ. Rapid decomposition was slower thereafter. Similar studies from other parts of the world have looked at the initial period of rapid weight loss. In Thailand, leaves of *Avicennia* sp. decomposes much more rapidly than those observed in this study.

Chapter - 7

NUTRIENT STATUS IN MANGROVE ECOSYSTEMS

7.1. NUTRIENT AND NITROGEN CYCLING

In tropical and sub-tropical countries mangrove ecosystems are often extensive and occur on the margins of oceans and estuaries. Here they receive enrichment from land and sea and in turn enrich the coastal water. The major process by which coastal water are enriched is the form of litter into the adjacent aquatic system.

There are many other factors which may control or influence the nutrient status, productivity and diversity in mangroves. These include :

(1) Climate
(2) Geomorphology
(3) Tidal range
(4) Fresh water input and its physico–chemical properties
(5) Biological factors

However, the substrate characteristics must be considered to exert one of the most direct controls on these ecosystems. The best development of mangrove ecosystems occur where upper tidal areas are exposed to a continuous supply of freshwater. This occurs :

(1) Where rainfall exceeds evaporation throughout the year.
(2) Where hinterland swamps provide a continuous seepage of freshwater.
(3) Where large freshwater catchment allow strong dilution of estuaries waters, reducing salt addition to the soil.
(4) Where large freshwater catchment and heavy seasonal rains provide regular and prolonged freshwater flooding of the tidal zone.

In each of these cases, the effects of the freshwater is to remove salt from the soil by leaching and the maintain soil water content above the permanent wilting point. Leaching of salt is more rapid in case of sandy soil and become more difficult in clay soil. The nutrient pool of the ecosystem is regulated by 5 interactive processes :

(1) Inorganic mineral ions are transported in solution to the root zone by freshwater or tidal flooding thus adding to the nutrient pool.

(2) Inorganic mineral ions are transported on the clay fraction of sediments deposited in the intertidal zone after freshwater flooding, thus adding to the nutrient pool.

(3) Inorganic mineral ions are blown in across the coast by wind thus adding to the nutrient pool.

(4) Flooding by freshwater and especially by the regular action of the tides, depletes the nutrient pool by exporting organic matter to the adjacent water bodies.

(5) Microbial decay of organic material, assisted by the activities of larger benthic fauna, results in the release of nutrients in soluble inorganic form, providing for internal cycling of mineral ions within the nutrient pool.

Of these 5 processes only the last concerned with internal cycling of nutrients back to the living organisms contributes to the conservation of the nutrient pool.

Detritus Ecosystem

Mangrove leaf litter in estuaries provides an important nutrient base for foodwebs leading to commercially important food fishes and invertebrates. Although, poor in nutrients when they fall from trees, mangroves leaves become nutritious due to the microbial enrichment process (Odum, 1971).

The productivity of the mangrove trees is not among the highest, if compared with other arborescent angiosperms. However, the rate of leaf fall is probably among the highest (Pool *et al.*, 1975). Heald (1971) considered that only 5% of the mangrove leaves are consumed on the trees and that rest is recycled by detritus food chain. Only another 1–2% are lost to the sediment mostly in the form of mangrove peat. About 60% of the litter may be recycled during the one year.

Detritus food chains are present in every ecosystems. Their importance has been neglected or even overlooked during the last decades, because of the over emphasis on the grazing food chain. This being a relatively simple and linear food chain. However, it seems that the much more complicated and much less quickly effective (Fig. 30).

Fell *et al.* (1975) has described the sequence of breakdown of mangrove leaves, goes through a complicated and often repetitive sequence of bacterial and fungal decomposers —> meiobenthic bacteria and fungus grazers —> macro-benthic detrivores and predators.

The detritus particles covered by a film of decomposers and of meiobenthic animals, undergo several sequences of ingestion by detrivores —> faecal ejaction of the non-digested and further minced particles —> recolonization of the particles by detrivores. When the detritic particles have been broken down to sizes of about 0.85 mm they are suspended in the mangal waters and consumed by the filtratory sessile fauna and by the zooplankton.

The essential difference of the mangal as compared to other aquatic detritic ecosystem is its high efficiency of 'within ecosystem' recycling. Although sizeable amount of the detritic food are lost to the open sea–about 25% in the Florida mangal according to Snedaker and Lugo (1973). The mangal waters are perhaps the most important feeding grounds for the shrimp and fish fauna of the tropical sea.

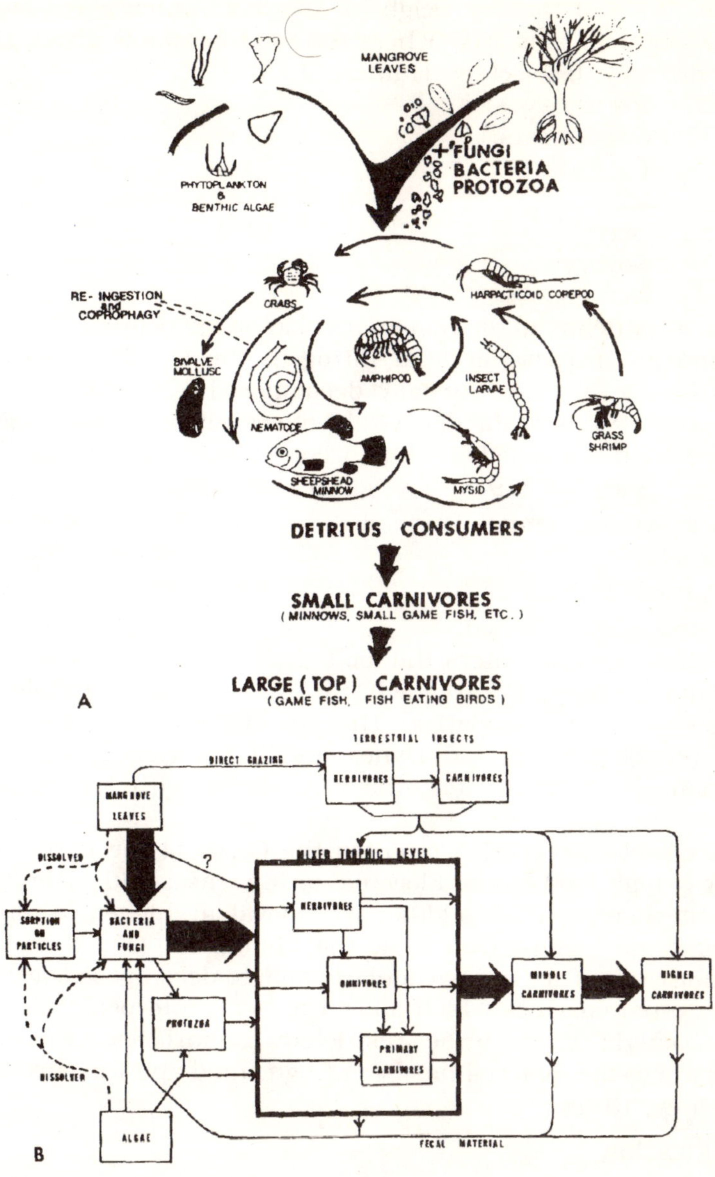

A detritus food chain based on mangrove leaves which fall into shallow estuarine waters of south Florida. Leaf Fragments acted on by saprotrophs and colonized by algae are eaten and re-eaten (copropgraphy) by a key group of small detritus consumers which, in turn, provide the main food for game fish, herons, storks and ibis. A picture model of the food chain is shown in A (upper diagram) and a "compartment model" in B (lower diagram) (Redrawn from W.E. Odum, 1970).

Fig. 30.

Carbon and Nitrogen Ratio

A convenient indicator of the nutritional value of a food is the ratio of carbon to nitrogen (C : N) content. The dietary requirement of protein for most animals is 16.5% of the dry weight of the diet, which corresponds to a C : N intake of about 17 : 1. In case where the C : N intake is above 17 : 1 there is a protein deficiency (Russel & Hunter, 1970). Examples of C : N ratios of agricultural crops are –

Sugar beats	48 :1
Potatoes	30 :1
Rice	31 :1
Beef	43 :1
Soybean	6 : 1
Diatoms	6 : 1

Nutritive enrichment of *R. mangle* leaf litter results from loss in carbon content and an increase in final nitrogen; the C : N changes from 120 in senescent leaves to 43 in partially decomposed leaves (Fell & Master, 1980), carbon represent 45% of the dry weight of senescent in *R. mangle* leaves; about half of the carbon is leached and half becomes particulate detritus.

Lugo and Snedaker (1974) have given some data –

1.	The average annual litter fall	896 g dry wt. m^{-2} y^{-1}
2.	Leached	224 g cm^{-2} y^{-1}
3.	Particulate detritus	224 g cm^{-2} y^{-1}
4.	Nitrogen contributed	4.5 g m^{-2} y^{-1}

The soluble carbon enters the food chain either as uptake by heterotrophs (Cooksey and Cooksey, 1974) which are than consumed by filler feeders or by direct ingestion after flocculation. The proportion of flocculates : solubles has not been reported. Lush and Hynes (1973) reported abiotic precipitation of organic material in freshwater; the rate varied with leaf species and water chemistry.

Wallace and Duce (1978) reported the C : N of the particles is 4.2 to 8.5 suggesting a high food value. Flocculation increases with salinity from 0 to 20 per ml, but there is little additional removal at higher salinities (Sholkoviz, 1976). Fungi may cause much of the change in C : N in some kinds of decomposing litter. They are particularly able to decompose leaves because, they produce external enzymes which allow the hyphae to penetrate in the litter. In contrast, bacteria tend to be restricted to surfaces and are particularly important where the material has been fragmented and the surface/volume ratio is high (Horley, 1971).

Nitrogen Fixation

Nitrogen is considered to be the major nutrient limiting production (Brown and Johnson, 1977; Mague *et al.*, 1974 and Ryther and Dumston, 1971), where concentration of nitrogen compounds are normally low, dissolved nitrogen gas becomes a potential source of inorganic nitrogen through biological nitrogen fixation. In recent years, considerable attention has been focused on nitrogen fixation in mangrove environment although, it is difficult to draw any broad conclusion regarding its overall significance. There is a consensus that nitrogen fixation may be of some importance in certain high productivity coastal environments such as coral reefs, salt marshes and sea-grass beds (Copne *et al.*, 1979; Jones, 1974; Webb *et al.*, 1975).

The sediments may mud, sands, peat or coral soils, often they are anaerobic and sometimes they have a pronounced smell of hydrogen sulphide. Details about mangrove soil are already mentioned in Chapter–3. Various reports indicate that mangrove sediments may have a high affinity for phosphorus (Clough and Attiwill, 1975; Hesse, 1962) as well as a considerable potential for denitrification (Nedwell, 1975; Pohm and Aspires, 1979).

Mangrove soils studied in the Tampa Bay region of Florida, were found to contain only traces of nitrate and nitrite, while ammonia was present at a concentration of between 0.1 to 0.3 ppm (Zuberer, 1976). Tidal water entering a marsh near a mangrove forest in Hawaii was largely depleted of nitrogen, phosphorus and oxygen on arrival in the forests yet within the mangroves, sediments removed between 10 and 100% of the remaining nitrogen and phosphorus (Walsh, 1967). Gotto and Taylor (1976) observed similar activity in the decaying leaves of *Rhizophora mangle*. Another supply of combined nitrogen can be found with nitrogen fixation by some mangrove leguminous plants, including the genera *Derris* and *Cynometra*.

Potential nitrogen–fixing non–phototrophic bacteria have been isolated from a variety of mangrove soils. A survey of microbial populations in marine sedimentary environments of parts of the Bahamas indicated that numerous bacteria, up to 17 x 10^6 sediment^{-1}, where associated with the calcarous surfaces silts and muds of mangrove swamps (Bavendamm, 1932).

The common potential nitrogen fixing–chemotrophic bacteria in the marine environment is given below :

Table 48.

	Genera reported	Environment	Locality	Author
1	Azobacter, Clostridium	Sediment	Baltic sea	Benecke & Kentner (1903)
2	Clostridium	Sediment	Estuary of Florida & USA	Brooks et al. (1971)
3	Azobactor, Clostridium Desulphovibrio	Sediment	Salt, Marsh, USA	Disker & Smith (1980)
4	Desulphovibrio, Enterobactor, Klebsiella	Sediment	Scotland Estuary	Herbert (1975)

Nitrogen is an essential macronutrient for all kinds of plants. On total dry weight basis it constitutes about 1.5 to 5%. Although about 4/5 of atmosphere is nitrogen, most higher plants are unable to utilize this nitrogen as they do not have adequate physiology and biochemistry for its uptake and assimilation. Plants growing in soil obtain nitrogen mostly in the form of nitrate and ammonium. Nitrate after entering in the plant tissues is reduced to nitrite and then to ammonia. This reduction may take place either in the roots or in the shoots (Srivastava, 1980).

The eventual products of N_2 assimilation are aminoacids, proteins, nucleotides, nucleic acids, chlorophyll, alkaloids, tannins and other secondary metabolites. Since development of the Acetylene reduction method, extensive assessment have been made of nitrogen fixation in natural materials rather than with cell masses of isolated nitrogen fixing organisms. Such studies have revealed that ecologically significant accretions of nitrogen may be found in diverse habitats including the rhizosphere of terrestrial plants, and rhizosphere of marine and freshwater angiosperms. The latter was of special interest because in marine system specially in mangrove ecosystems nitrogen fixations are practically unknown and less understood.

In mangroves, nitrogen is considered to be the major nutrient limiting production (Boto and Wellington, 1984; Stanley *et al.*, 1987; Singh and Garge, 1993), where concentrations of nitrogen compounds are normally low. Within recent years, considerable attention has been focused on nitrogen fixation in mangroves environment and although it is difficult to draw any broad conclusions, regarding its overall significance, there is a consensus that nitrogen fixation may be of some importance in certain high productivity coastal environments.

This work deals with experimental evidences of nitrogen assimilation in some mangroves and their significance. This work was conducted on mangrove forests of Andaman Islands, and South Japan.

For the determination of total nitrogen in different parts of plant species, samples were collected in different parts of mangrove forest, where mangroves are undisturbed and soils are very rich in humus contents.

Total nitrogen was measured by micro–kjeldahl method by Lang (1958). About 100mg of dry plant tissues were digested with 4.0 ml of concentrated H_2SO_4 by constant heating, when the digested solution became light yellow (after about 8 to 10 hr.), few drops of 20% H_2O_2 were added to get clear solution. The colourless solution was cooled and made up to 100 ml by distilled water. To 0.5 ml of this solution, 2–5 ml of distilled water was added. This was followed by the addition of 2.0 ml of Nesseler's reagent, dark yellow colour was developed. The OD was measured on a UV Spectrophotometer at 400 nm and total nitrogen content was calculated using standard curve prepared from acidic ammonium sulphate solution.

A mangrove forest can be viewed to consist of three quite distinct components –

(i) the mangrove sediment,
(ii) above ground vegetation and
(iii) mangrove trees.

Hence the fixation of nitrogen uptake and assimilation can be dealt in three parts (Fig. 31).

(1) Atmospheric nitrogen uptake by stomata

Numerous studies have indicated that atmospheric ammonia might be taken in either through the leaves, or directly through the soil. The foliar absorption of the gas is primarily stomatal, which accounts for up to 97% of the total uptake (Von Home *et al.*, 1989). The atmospheric level of NH_3 is the primary factor determining its uptake rate by the plants. Generally the uptake rate increases with increase in ammonium concentration (Porter *et al.*, 1972; Farquhar *et al.*, 1979). Atmospheric ammonia may combine with SO_2 and presence of humidity in mangrove environment to form $(NH_4)SO_4$ aerosol, which can then be absorbed by the plants. As in the case with gaseous ammonia, the absorption of $[NH_4]\ SO_4$ aerosol is also stomatal.

The source of SO_2 in mangrove environment is the sedimental sulphate present in sea water, which is reduced to sulphide in highly anaerobic conditions. Sulphide ion is highly toxic to plant and many intertidal mangrove and marsh plants have developed strategies to overcome sulphide toxicity, hence the mobility of metals in highly anaerobic soil is limited if sulphide is in excess. Over long–time periods and given the appropriate conditions in terms of iron concentrations and redox potentials, etc. iron pyrites is formed by anaerobic bacteria. Macnae and Kolk (1962) reported a mean increase from 34 to 54 mg H_2S per 100 gm dry sand. Now this H_2S reacts with atmospheric ammonia combine with SO_2, form ammonium sulphate aerosol (Fig.31).

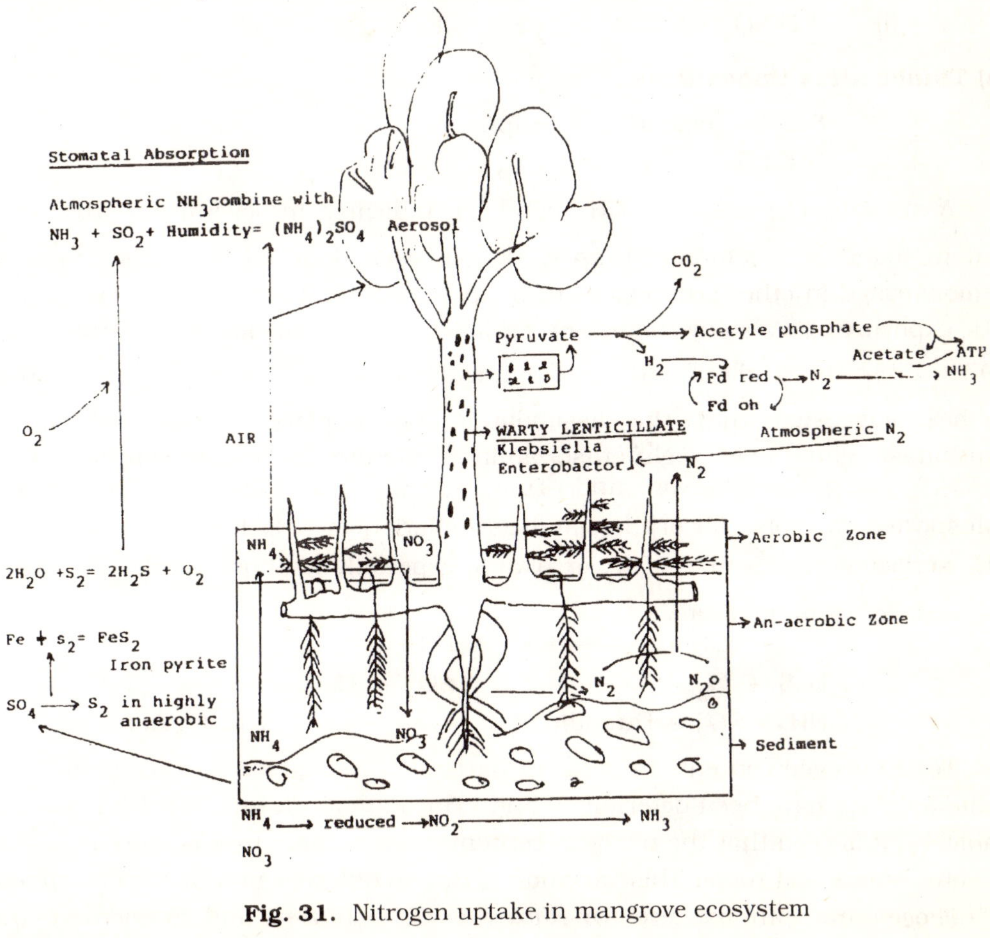

Fig. 31. Nitrogen uptake in mangrove ecosystem

There is only one report of sulphur oxidizer (*Thiobacillus thiooxidens*) found in the tidal mangrove soils of Sierra Leone. There is a general report on the ecology of sulphur cycle and a single study on the biological SO_4 reduction in Senegal.

In fact, in the primary swamps of mangroves one of the important processes take place is SO_4 reduction, which is evident by the smell of H_2S and the formation of Black sulphide in the region. The sulphur occur in various states of oxidation from –2 of sulphides to +6 of sulphates. The organisms capable of oxidizing inorganic sulphur compounds are both autotrophic and heterotrophic in nature. The most imortant genus being *Thiobacillus*. This genus contains strict chemoautotrophic organism capable of oxidizing inorganic sulphur (*T. thiooxidens*) and ferrous ions and sulphur (*T. ferrooxidens*) at acidic pH (pH 3.0 or below). They are obligate aerobes except (*T. denitrificans*). The reactions carried out by some important *Thiobacillus* are as follows –

(a) *Thiobacillus ferrooxidens* :

Fe^{+2} oxidation to Fe^{+3}

(i) $2\ FeSO_4 + 1/2\ O_2 + H_2\ SO_4 = Fe_2(SO_4)_3 + H_2O$

(b) *Thiobacillus thiooxidens* :

Sulphur oxidation to Sulphates

$S + 1.5\ O_2 + H_2O = H_2SO_4$

Assimilation of absorbed NH_3 has been indicated in several investigations. But in mangroves community no work has been done so far. This has been demonstrated in other group of plant by increased total nitrogen content during NH_3 exposure and also by the use of ^{13}N labeled or ^{15}N labeled NH_3 (Porter *et al.*, 1973). From the study of pattern of aminoacid labeling during $^{13}NH_3$ exposure, it has suggested that the assimilation takes place through glutamine synthatase, glutamine exogluterase, amino renferase (glutamate synthase) GS–GOGAT pathways (Hanson and Tully, 1979). The foliar NH_3 is readily transported to roots. Absorption and assimilation of NH_3 has the potential of NH_3 servicing as a source of N_2 or plants, especially in those conditions where soil is deficient in nitrogens.

$Fe + H_2S \longrightarrow FeS_2$

$H_2S + O_2 \longrightarrow SO_2 + H_2$

$NH_3 + SO_2 + \text{Humidity} \longrightarrow [NH_4]_2SO_4$ Aerosol

Total nitrogen contents in different parts of 13 mangrove species, growing in same habitats have been determined and values are given in Table 48. From the Table 47, it is seen that the nitrogen content in some mangrove is more in leaves in some stems and roots. This variation is due to different mechanism of uptake of nitrogen from atmosphere. In *Rhizophora, Lumnitzera* and *Sonneratia*, the nitrogen amount is more in leaves than roots and stems. In *Bruguiera* sp. and *Ceriops* sp. the nitrogen content is more in stems than roots and leaves. From the table, it is seen that the amount of nitrogen in sediment (substratum) is much lower than in plant body. The amount of nitrogen available to sediment is due to denitrification. In aerobic zone, nitrogen is available in form of ammonia, and this ammonia is easily absorbed by various types of nitrogen in root is due to absorption of sedimental nitrogen by various types of root systems.

(2) Atmospheric nitrogen uptake by pneumatophores and warty tree bark

Lenticellate warts on tree barks are common occurrence in broad leaved trees, especially those growing under wet temperature and tropical environments. Mangroves are not exception, many species have warty lenticellate cell, which take part in the biological nitrogen fixing activity. Yatazawa *et al.* (1983) revealed that warty lenticellate barks of many kinds of leguminous and non–leguminous trees reduced acetylene significantly and the amount of nitrogen fixation of barks could contribute considerably to the nitrogen economy of mangrove ecosystems. Mangrove species *Bruguiera gymnorhiza, B. parviflora, B. sexangula, B. cylindrica, Ceriops tagal* etc. are not exceptions. To confirm this, it seemed essential to obtain definite information on the microorganisms responsible for nitrogen fixation in the aerial part of the mangrove tree. Warty barks and non–warty of all the 13 species, growing in same habitats were collected from mangrove forest and materials were analysed for nitrogen and results were given in Table 50. It is seen that in some mangroves, the amount of nitrogen in warty–bark is significantly higher than non–warty bark. *Bruguiera* and *Ceriops* are examples.

The bacteria responsible for nitrogen fixation were isolated from the bark, and their capacity for acetylene - reduction and the incorporation of $^{15}N_2$ into bacterial cells was confirmed (Uchino *et al.*, 1984).

(3) Sedimental nitrogen uptake by various types of root systems

Nitrogen fixation in aerobic sediments usually ascribed to bacteria, is generally slow in comparison to fixation by blue–green algae (Brezonik and Harper, 1969; Howard *et al.*, 1970 and Kevin and Brezonik, 1971). Nitrogen can only be utilized in organic form NO_3 and NH_4. Severe losses of nitrogen can occur in flooded anaerobic soils via denitrification. Many mangrove flooded soils have aerobic layer overlying the more highly reduced soil. Ammonium in the aerobic zone can be reduced to gaseous nitrogen and nitrous oxide and lost from the soil.

The sedimental NH_4^+ also combine with SO_2 and humidity and formed ammonium sulphate aerosol, which is easily absorbed by the stomata. In majority of mangroves, the major input of nitrogen in the system is through the stomatal absorption (Singh, 2000). In a very few mangrove species (*Bruguiera* sp. and *Ceriops* sp.), input of nitrogen is through warty lenticellate cell, with the help of nitrogen fixing bacteria. Sedimental nitrogen and the nitrogen generated by the process of denitrification in anaerobic muds, the major parts of these nitrogen is not utilized by the mangrove plants. These nitrogen goes to atmosphere, where is absorb by mangroves. Sedimental nitrogen is assimilated by all the mangroves species but this nitrogen is not sufficient for all metabolic processes hence additional amount of nitrogen is essential to maintain the systems. In *Avicennia* and other mangroves species, which are having various types of numerous root systems in anaerobic and aerobic zones, they are more efficient and capable to absorb maximum nitrogen from habitats. This study needs further more experimental evidences and support which are in progress.

Table 49. Nitrogen contents in different parts of mangrove species in mg g^{-1} dw

Name of species	Total Nitrogen			
	Sediments	Roots	Stems	Leaves
Bruguiera gymnorhiza	4.9 ± 0.2	21.4 ± 2.1	49.3 ± 3.8	30.2 ± 2.4
B. parviflora	7.6 ± 0.4	19.1 ± 1.6	41.2 ± 2.2	39.4 ± 2.8
B. cylindrica	4.9 ± 0.1	24.3 ± 2.7	45.8 ± 2.9	32.6 ± 2.2
Rhizophora mucronata	4.2 ± 0.6	26.3 ± 2.8	20.6 ± 3.6	39.8 ± 3.8
R. apiculata	8.4 ± 1.1	22.4 ± 3.1	17.5 ± 1.9	37.6 ± 2.9
R. stylosa	6.2 ± 0.2	29.3 ± 1.9	19.3 ± 2.6	43.0 ± 3.6
Ceriops tagal	10.2 ± 1.3	18.8 ± 1.2	40.3 ± 2.9	34.4 ± 2.6
Avicennia marina	12.3 ± 1.6	39.4 ± 3.8	22.4 ± 1.9	32.6 ± 3.1
A. officinalis	6.8 ± 0.4	36.9 ± 2.6	17.5 ± 2.6	29.4 ± 2.9
Sonneratia alba	7.2 ± 0.8	18.4 ± 1.6	32.4 ± 2.8	34.6 ± 3.6
S. apetala	4.2 ± 0.1	16.3 ± 1.2	33.3 ± 3.1	38.3 ± 3.3
Lumnitzera racemosa	5.8 ± 0.2	15.6 ± 1.9	24.6 ± 2.2	30.3 ± 2.1
L. littorea	6.2 ± 0.1	18.3 ± 2.6	22.6 ± 2.6	33.4 ± 2.9

Table 50. Amount of nitrogen contents in mg g^{-1} dw present in barks of mangroves

Name of species	Bark with Lenticel	Bark without Lenticel
Bruguiera gymnorhiza	50.3 ± 2.3	14.3 ± 2.1
B. parviflora	63.3 ± 6.5	12.6 ± 1.6
B. cylindrica	58.8 ± 3.2	8.2 ± 1.9
Rhizophora mucronata	12.6 ± 1.7	10.3 ± 1.3
R. apiculata	13.4 ± 1.2	11.7 ± 1.4
R. stylosa	8.3 ± 0.6	10.5 ± 1.0
Ceriops tagal	49.4 ± 4.7	13.9 ± 2.3
Avicennia marina	15.3 ± 1.6	12.3 ± 2.1
A. officinalis	14.3 ± 1.3	15.8 ± 2.4
Sonneratia alba	34.5 ± 3.1	12.5 ± 1.8
S. apetala	27.2 ± 1.8	11.2 ± 1.4
Lumnitzera racemosa	18.7 ± 2.5	16.3 ± 1.5
L. littorea	19.3 ± 1.6	17.8 ± 2.3

7.2. NUTRIENT DYNAMICS IN MANGROVE

Most of the tropical coastline between 25°N and 25°S latitude is vegetated by forested wetlands called mangrove (McGill, 1958). These plant communities have received considerable botanical investigation because of their unique taxonomy and ovivipary (Tomlinson, 1986), and the diverse fauna that inhabit these coastal areas (Macnae, 1968; Chapman, 1976). However, the ecology of mangrove is poorly understood, particularly the significance of these ecosystems to the productivity and nutrient cycling of estuarine and adjacent coastal waters. It has been suggested that the high fishery yields of coastal tropical waters are due to the presence of these communities (Macnae, 1974; Turner, 1977; Jothy, 1984), yet there is no evidence of a cause and effect relationship for mangroves and fisheries (Macnae, 1974). Thus the function of these wetlands in supporting secondary productivity continues to be a complex issue. Mangrove may also influence the primary productivity of coastal waters by controlling the fate of dissolved nutrients and suspended sediments. Mangroves are considered a source of organic detritus yet a nutrient sink, contributing to the confusion of their role in coastal processes.

There is a critical need to understand the function of mangroves in tropical ecosystems because of the rate these intertidal areas are being converted to alternate land uses. Mangroves are exploited for forestry products including fuelwood for cooking, fenceposts, charcoal, tannins, pulpwood, chipwood and timber (Polumin, 1983). Areas inhabited by mangroves are also reclaimed for agriculture, aquaculture and residential development. For instance in the Philippines, the construction of ponds for the production of milkfish and penaied prawns has resulted in the loss of 189,000 ha of forest since 1920, which represents nearly 45% of the original total mangrove resource (Jara, 1984). Other major shrimp pond operations have developed in the intertidal zone of Ecuador, Thailand and Panama. In south–west Florida (USA) the reclamation of mangroves for urban development resulted in the loss of 24% of mangrove area in Macro Island from 1952 to 1984 (Patterson, 1986). A projected 200% increase in population in this coastal area by the year 2000 demonstrates the pressure to develop these natural resources. In contrast to forestry, which attempts to maintain some sustainable yield in mangrove ecosystems, reclamation activities such as urban development, agriculture and pond mariculture result in the loss of this resources from the coastal zone.

Conflicting management plans for the intertidal zone in many tropical countries result from desire to preserve mangrove as a natural resource for humans. Mangroves may indirectly support economically important fisheries by providing free services such as habitat, food and good water quality. Thus the loss of these plant communities would negatively impact industries that rely on productive coastal fisheries. However, to settle these conflicts, information is needed that more clearly describes the function of mangroves to better understand the importance of these systems to the fisheries of tropical estuaries. The purpose of this paper is to review our current understanding of the coupling of mangrove to the productivity and nutrient cycling of coastal waters. This analysis should help present a framework for the development of future mangrove research that will address the issue of outwelling from these intertidal ecosystems.

Mangrove forest structure and phytosociology have long been associated with tidal inundation patterns (Watson, 1928; Chapman, 1944; Chapman, 1976). Lugo and Snadaker (1974) classified the structure of mangroves into six community types that related forest physiognomy with their hydrology and geomorphology. These six groups can be combined into three types (riverine, fringe and basin) using the grouping by Brown *et al.* (1979) for forested wetlands based on frequency and amplitude of water inundation. The hydrologic energy of riverine mangroves is high since it is dominated by river flow and tidal inundation, while fringe mangroves are influenced mainly by frequent tidal inundations. Basin mangroves have less hydrologic energy since they are located inland of fringe or riverine communities and as a result are less frequently inundated by either tides or river floods. This continuum of hydrologic energy has been associated with distinct types of mangrove forest structure (Lugo and

Snedaker, 1974; Pool *et al.*, 1977). It is less clear if the function of these systems is also specific among these types of mangroves along this tidal continuum. The objective of this chapter is to review information on the ecology of mangrove dominated estuaries for evidence that hydrologic energy controls the exchange of materials from mangroves to estuarine and coastal waters.

Organic Matter Dynamics

Leaf litter on the forest floor represents a major source of organic matter and nutrients for outwelling from mangroves to adjacent estuarine waters. Thus the balance of litter productivity, decomposition and export influence the exchange of these materials at boundary of the mangrove subsystem. Litter productivity values for mangrove forest worldwide range from 1.20 t ha^{-1} yr^{-1} for scrub mangrove in south Florida to 23.4 t ha^{-1} yr^{-1} for a 20 yr old managed forest in Malaysia (Twilley *et al.*, 1986). It has been suggested by Pool *et al.* (1975) that litter production rates in mangroves are a function of water turnover within the forest and the rank of the means of litter production in Fig. (riverine > fringe > basin > scrub) supports this hypothesis. In southwest Florida the apparent inverse relation of litter production among five basin mangrove sites and average soil salinity further supports the argument that environments with higher hydrologic energy support higher litter productivity (Twilley *et al.*, 1986). Fewer tides result in higher soil salinity (Cintron *et al.*, 1978) and/or the accumulation of toxic substances (eg. hydrogen sulfide; Carlson *et al.*, 1983; Nickerson and Thibodeau, 1985), which can result in increased stress on these inland mangrove forests (Hicks and Burns, 1975; Lugo *et al.*, 1981). Wharton and Brinson (1979) suggested that the production of forested wetlands was dependent on water movement, not only as a source of silts and clays, but also a supply of nutrient and aeration for optimal growth. Increased hydrologic energy in intertidal areas seems to increase the potential for litter production in the estuary.

Export from mangroves may also be associated with the amount of hydrologic energy within the intertidal zone. Rates of organic carbon export from basin mangroves was dependent on the volume of tidal water inundating the forest each month, and accordingly export rates were seasonal in response to the seasonal rise in mean sea level (Twilley, 1985). Rainfall also increased organic carbon export from mangroves (Twilley, 1985), especially dissolved organic carbon (DOC), similar to results for salt marshes (Harriss *et al.*, 1980). Total organic carbon (TOC) export from infrequently flooded basin mangroves in southwest Florida was 64 gC m^{-2} yr^{-1}, and nearly 75% of this material was DOC (Twilley, 1985). Particulate detritus export from fringe mangroves in south Florida was estimated at 186 gC m^{-2} yr^{-1} (Heald, 1969), compared to 420 gC m^{-2} yr^{-1} for a mangrove forest in Australia (Boto and Bunt, 1981). Estimates of tidal amplitude per tide in these three mangrove forests were 0.08 m, 0.05 m and 3 m, respectively. At low tidal amplitude in the basin forests, only 20% of litterfall was exported (Twilley, 1985), compared to 45% in the fringe forest, and at the upper tidal range leaf litter on the forest floor was negligible (Boto and Bunt, 1981). These results indicate that as hydrologic energy increases, both the magnitude of litter produced within mangroves and the proportion of this litter that is exported also increase. Both of these processes enhance the transport of organic matter across the mangrove boundary to estuarine waters.

Organic carbon exchange in mangroves can also be evaluated by looking at the net ecosystem productivity of these forested wetlands. Net ecosystem production (NEP) of mangroves is the balance of inputs from in situ production and allochthonous sources, minus losses from community respiration and export. Mangrove research in southwest Florida and in Peurto Rico have made significant contributions to the development of organic carbon budgets for mangroves in microtidal environments (Table 51).

Table 51. Mass balance of carbon flow (gC m^{-2} yr^{-1}) in mangrove forests in Florida and Puerto Rico

	Rookery Bay [1,2,3,4]		Puerto Rico [5]	Fahkahatchee Bay[6]		
	Fringe	Fringe	Fringe	Basin	Fringe	Fringe
GPP						
Canopy	2055	3292	3004	3760	4307	5074
Algae	402	26	276			
Total	2457	3318	3280			
Respiration (plants)						
Leaves, Stems	671	2022	1967	1172	1416	3084
Root-AG	22	197	741	146	182	215
Roots-BG	?	?	?	?	?	?
Total	693	2219	2708	1318	1598	3299
NPP	1764	1099	572	2442	2709	1775
Growth		186	153			
Litterfall		318	237			
Respiration (heterotrophs)		197	135			
Respiration (Total)		2416	2843			
Export		64	500			
NEP		838	-63			
Burial		?	?			
Growth		186	153			

1. Lugo *et al.*, 1975; 2. Twilley, 1985; 3. Twilley *et al.* 1986; 4. Twiley, 1982; 5. Golley *et al.* 1962; 6. Carter *et al.* 1973.

Nutrient Exchange

The idea that intertidal wetlands may be a nutrient sink is particularly confusing since it contradicts the "outwelling" concept of detritus exchange established above for mangroves. Indications that mangroves may be a nutrient sink come from a study by Walsh (1967) where nutrient concentrations in waters moving through a mangrove in Hawaii decreased in concentration. However, this study did not directly measure nutrient flux across the mangrove boundary, and very few studies have attempted to budget flows of nutrients in mangrove ecosystems. It has been argued for other plant communities that nutrient recycling may be of greater significance than inputs to maintaining productivity. Such processes include resorption or retranslocation of nutrients prior to leaf fall (nutrient use efficiency; Ryan and Bormann, 1982; Vitousek, 1982), the immobilization of nutrients in leaf litter during decomposition (Brinson, 1977; Twilley *et al.*, 1986), and nitrogen fixation. In mangroves these patterns in nutrient recycling processes may influence the exchange of nutrients at the boundary of mangroves, which would determine whether mangroves serve as either a nutrient source or sink to estuarine and coastal waters.

There is some indication that nutrient resorption can vary within a mangrove species in different environments. The mass/nutrient ratio of nitrogen in leaf litter relative to nitrogen return to the forest floor via litterfall is an index of nutrient–use efficiency. High efficiencies of *Rhizophora* occur in trees located in either riverine or fringe forests. In riverine and fringe mangrove forests were leaf export is high, higher nutrient recycling efficiency in leaves prior to leaf fall is an advantage to increasing nutrient conservation. Trees with lower efficiencies in nutrient utilization, which are represented by nitrogen loss less than 20 kg ha^{-1} yr^{-1} and mass/nutrient ratios less than 200, are located in more inland sites. Also, nutrient use ratios of *Avicennia*, which are also located in more inland sites, are lower than for the other mangrove species. Further testing will have to be done to see if in fact nutrient use efficiencies in mangrove species are an adaptive strategy to sites with varying hydrology or fertility.

Increases in nitrogen concentration of leaf litter during decomposition may immobilize nitrogen in basin mangrove forests, effecting within system nutrient conservation (Brinson, 1977; Vitousek, 1984). Nitrogen enrichment of leaf litter includes absorption and adsorption processes by bacterial and fungal populations (Fell and Master, 1973; Rice and Tenore, 1981; Rice, 1982), as well as nitrogen fixation (Zuberer and Silver, 1978; Gotto *et al.*, 1981). These inputs of nitrogen result in an absolute increase in nitrogen in decomposing litter, and particularly for *Rhizophora* leaf litter, recycle nitrogen on the forest floor (Twilley *et al.*, 1986). Inland mangrove forests have higher decomposition and mineralisation rates of leaf litter and along with infrequent tides, have lower rates of detritus export. Thus high litter turnover may increase the availability of nutrients for reabsorption by roots and prevent the loss of nitrogen from the system.

Organic Matter

A mass balance of organic matter in Rookery Bay, a mangrove–estuarine ecosystem in southwest Florida, indicates that mangrove detritus may be a significant source of energy for secondary productivity. Mangroves contribute nearly 345 gC m^{-2} yr^{-1} to the estuary (per m^2 of estuary) which is 83% of the allochthonus inputs. Since litter export from fringe mangroves is nearly double the per area rates for basin mangroves, they are considered important sources of detritus in south Florida. However, the area of basin mangroves is nearly double the area of fringe forest, thus the contribution of organic carbon to Rookery Bay is nearly the same for both types of communities. Thus the much larger land mass occupied by the inland mangroves establishes these systems as very important sources of detritus to estuarine ecosystems. Allochthonous inputs of organic carbon together with NPP of the water column represent the pool of organic matter available for secondary productivity. Allochthonous inputs accounted for 47% of this organic pool in Rookery Bay and mangroves alone accounted for 39% of the total organic matter supply (fringe = 21% and basin = 19%). During the summer nearly 52% of the organic carbon available for secondary production was allochthonous, indicating the significance of mangrove to this estuary.

A mass balance of organic carbon has also been estimated for a mangrove–estuarine ecosystem 25 km southeast of Rookery Bay (Carter *et al.*, 1973). The primary productivity of Fahkahatches bay was estimated at about 251 gC m^{-2} yr^{-1}, and phytoplankton and benthic plants contributed about 73 and 27% of the total, respectively. Input of organic detritus from freshwater discharge to this estuary was not intensively studied, so it was assumed that 40% of the litter produced in wetlands was exported, for an estimate of 338 gC m^{-2} yr^{-1} (per m^2 of estuary). Based on this estimate of export, mangroves contribute about 52% of the organic matter supply that is available for secondary productivity in this system. Thus the importance of mangrove to the organic matter budget of Fahkahatchee Bay is similar to results for Rookery Bay.

Table 52. Primary productivity and chlorophyll concentrations of Waters in Mangrove Dominated Estuaries and Coastal Waters

Location	Distance Offshore (km)	Chlorophyll (mg/m^3)	Primary mgC m^{-3} h^{-1}	Productivity gC m^{-2} yr^{-1}	Reference
			Backwaters		
Cochin, India	0	7.3-20.0		124 (NPP) 280 (GPP)	Quasim *et al.* 1969
Pichavaram, India	0	3.2-30.2	300-600		Krishnamurthy *et al.* 1975; Sundaraj and Krishnamurthy, 1973
Kollur, India	0	0.1-1.8	1.9-7.5		Untawale *et al.* 1977
Goa, India	0		11.4-91.1		Verlencar and Quasim, 1985
			Estuarine Waters		
Vellar Estuary, India	0	0.4-10.5			Santhakumari, 1971
Biscayne Bay Florida	0	0.2-4.2		13-46	Roman *et al.* 1983
Rookery Bay Florida	0			251	Twilley, 1982
Fahkaatchee Bay, Florida	0			183	Carter *et al.* 1973
			Coastal Waters		
Lacadive Sea	7-15	0.8			Shah, 1973
Goa, India	10	1.9		342	Verlancar and Quasim 1985
Phangha Bay Thailand	1			468	Wium-Anderson, 1979

Net productivity of phytoplankton in Rookery Bay was 251 gC m^{-2} yr^{-1} which is similar to productivity values in Fahkahachee Bay and the backwaters of Goa, India; yet much higher than the productivity of Biscayne Bay in Southeast Florida (Table 52). In Rookery Bay, phytoplankton production was highest during the summer when allochthonous inputs were also greater. This is in contrast to the mangrove backwaters of India (Qasim *et al.*, 1969; Krishnamurthy *et al.*, 1975; Verlencar and Qasim, 1985) and Brazil (Teixeria *et al.*, 1969). In those backwaters phytoplankton production was minimum during freshets when increases in turbidity apparently limited photosynthesis. Several investigators have observed a stimulation of primary productivity in the water column by the addition of DOC from mangroves (Prakash, 1971; Prakash *et al.*, 1973; Cooksey and Cooksey, 1978; J.W.Day Pers. Comm.). Prakash *et al.* (1973) demonstrated that there was an upper limit to this stimulation, above which the negative effects of light absorption limited primary productivity. Information in Table 52 suggests that primary productivity in coastal waters offshore from mangrove dominated estuaries is high for continental shelves, due to enhancement from both mangrove detritus and presence of upwelling along the coast. More studies are needed to determine if mangrove may not only contribute to the organic pool of coastal waters via export, but also by enhancing primary productivity.

Table 53. Community metabolism and production : respiration ratios for mangrove dominated estuaries

Station	Date	Production	Respiration	P/R	Reference
		($gO_2.m^{-2}$ d^{-1})			
Rookery Bay	Annual	6.25	5.76	1.08	Twilley, 1982
Florida	April	5.12	6.40	0.80	
	May	6.40	7.20	0.89	
	June	7.52	4.40	1.71	
	June	6.68	5.76	1.16	
	July	7.04	4.00	1.76	
	December	5.88	6.79	0.70	
Coots bay	July	7.50	14.10	0.53	Tabb *et al.*
Florida	October	1.10	4.97	0.22	1962
Station 3e	January	1.50	1.30	1.15	
	April	2.44	4.68	0.52	
Station 5e	January	0.81	1.66	0.49	
Whitewater Bay Florida	April	8.22	7.03	1.11	Tabb *et al.* 1962
Laguna Madre Texax	Annual	4.30	5.60	0.76	Odum and Hoskin 195
Kollur, India		2.24	2.90	0.77	Untawale *et al.* 1977

The ratio of gross productivity (Pg) to 24-h community respiration (R24) indicates the ability of autochthonous production to meet the heterotrophic demands of the system. The Pg/R24 ratio of Rookery Bay was 1.08 based on six diurnal surveys of changes in dissolved oxygen concentrations (Table 53).

During the summer, when net production was higher, Pg/R24 ratios were >1.75 on two occasions; afterwards ratios were <1.0 with the lowest value of 0.70 in December. Seasonal differences in Pg/R24 ratios have also been observed in other mangrove dominated estuaries, reaching as low as 0.22 in October at Coots Bay in south Florida (Table 53). Untawale *et al.* (1977) measured a metabolic P/R ratio of 0.77 for a backwater area of Kollur estuary in India. These mangrove dominated estuaries are basically heterotrophic systems except for periods of peak primary production and rely on imported organic matter to support what organic matter is available for net ecosystem production (NEP). In Rookery Bay the sum of gross production and allochthonous inputs minus losses due to respiration was equal to 482 gC m^{-2} yr^{-1}. It is still unknown how much of this excess material is utilized directly by higher trophic levels, and how much is exported as detritus to adjacent coastal waters.

Nutrients

Preliminary estimates of nutrient dynamics in mangrove–estuaries can be made using mixing diagrams of nutrient concentration along a salinity gradient. Nixon *et al.* (1984) compared mixing diagrams in a mangrove dominated estuary to an estuary where mangroves had been removed. This comparison was to determine the linkage of mangroves to estuarine waters. Curves exhibiting non-conservative behaviour of nutrients suggest that mangroves are a source if the curves are convex, and a sink if the curves are concave. A conservative nature of nutrients indicates little net exchange (Liss, 1976). This analysis has several shortcomings since changes in concentrations of inorganic species other than by dilution may only represent transformation within the water column and not a loss or input to the system. Surveys of total nitrogen and phosphoric, and suspended matter by Nixon *et al.* (1984) in the Sangga river estuary, Malaysia, during the dry season generally exhibited conservative behaviours in the estuary. Concentrations of these materials were much lower in the natural estuary than in the estuary where mangrove had been reclaimed. Suspended particulate materials were especially higher in the estuary that has been diked for agriculture (Nixon *et al.*, 1984). A nutrient survey of the Gambia River estuary found that total nitrogen and suspended sediments also exhibited conservative concentrations with salinity during October and November. These surveys may approximate the mass balance of nutrient flux in mangrove-dominated estuaries. However, because of their limited application, much more work is needed to determine the function of mangroves in nutrient dynamics in tropical estuarine ecosystems, and the exchange of nutrients with coastal waters. The extension of estuaries unto continental shelves in areas of high freshwater discharge are more obvious examples where materials from mangrove–dominated estuaries may be transported to coastal waters. Such is the case of extended estuaries along the west coast of India that are strongly influenced by southwest monsoons that occur from June to September (Qasim and Gupta, 1981). Characteristics of the coastal and estuarine waters of Goa which are surrounded by 1800 ha of mangrove (Untawale *et al.*, 1982), can be identified into time zones of premonsoon, monsoon and postmonsoon. This temporal sequence has been described as periods of stability, change and recovery, respectively (Table 54). In response to monsoons, salinities in coastal waters 10 km offshore dropped from 36.27 ppt during premonsoon to 31.09 ppt, and the 1 ppt isohaline moved seaward from 65 km upriver to about 15 km upriver. Shah (1973) also found an influence of mangrove backwaters on coastal waters 7–15 km offshore in the seas off Cochin, west India. In these areas coastal water salinities dropped from 35 ppt to values <30 ppt during peak discharge. Associated with this increased discharge was increased stratification, higher nutrient concentrations and turbidity, and lower primary productivity rates (Table 54). Following the monsoon, primary productivity of estuarine and coastal waters were stimulated relative to premonsoon conditions. These sequence of events associated with high freshwater discharge are also characteristic of mangrove backwaters in Cochin (Qasim, 1973).

Table 54. Characteristics of Goa estuaries dominated by mangroves in relation to monsoon season

Characteristic	Station[3]	Pre-monsoon	Monsoon	Post-monsoon
Time[1]		Feb-May	Jul-Sept.	Oct.-Jan.
Description[1]		Stable	Change	Recovery
Physics/Salt[1]		Well-mixed	Stratified	Stratified/Mixed
1% Isohaline (Distance upriver)		65 km	11-20 km	
Currents[1]		Flood dominated	Ebb-dominated	Flood dominated
Salinity[1]	S	36.27	31.09	34.45
	M	30.49	5.73	23.64
	Z	30.43	7.97	25.70
"F" (Freshwater fraction)[1]	M	0.16	0.82	0.31
Dilution time [1]	M	6.20	1.20	3.2
	Z	6.20	1.30	4.0
Residence time[1] (Days)	M	50	5-6	
	Z	(longer)	(longer)	
Phosphates 1 (μmol/1)		Low	0.42-2.50	
Silicates [1]		Low	High	
Prim. Prod. [2]	S	938.1	----	938.1
($mgC\ m^{-2}\ d^{-1}$)	M	300-1200	0-600	400-1600
Chlorophyll[2]	S	1.94	---	1.94
(mg/m^3)	M	1.8-6.0	0.5-50	6-11
Detritus[2] (% POC)	M	76	75.9	46
POC : CHL[2]	M	59.40	107.9	74.2

1. Qasim and Gupta, 1981; 2. Verlencar and Qasim, 1985; 3. S=Offshore station; M = Mandovi River estuary; z = Zuari River estuary.

The significance of detritus from mangrove dominated estuaries on the composion of particulate organic materials (POC) in coastal waters is indicated by seasonal shifts in POC/Chl ratios. Chlorophyll was 77.3% of the POC pool in estuarine waters based on annual averages of values. Peak POC/Chl ratios were >4400 followed by minimum values in postmonsoon when phytoplankton production was maximum. In the coastal region, the percentage of detritus was greater than in the estuarine region suggesting that this coastal area may act as a sink for materials transported from these mangrove–dominated estuaries (Verlencar and Qasim, 1985). In postmonsoon, most of the POC is from phytoplankton; however, in the monsoon season, allochthonous materials originating from surface runoff are dominate (Verlencar and Qasim, 1985). In the Zaire river estuary in southwest Africa, 50% of the material transported from the river was deposited in the estuary and the other 50% transported to the shelf (Eisma and Kalf, 1984). Other investigators have acknowledged that the coastal waters of the Bering sea and Northern coast of Taiwan (Tanoue and Handa, 1979; Tsu Chang *et al.*, 1981) could be influenced by detritus in areas subjected to land drainage (Verlencar and Qasim, 1985). Rao and Rao (1975) also observed maximum POC/Chl ratios (Chlorophyll based of Harvey pigment units) during monsoon season at a station 5 km offshore from Waltair (east coast of India) and minimum values occurred during postmonsoon.

The exact source of this detrital material is unknown, and is not likely from mangroves in areas of extremely high currents such as the Goa estuaries (Untawale *et al.*, 1982). There are no mangroves at the mouth of this estuary apparently due to the rocky substrate that, along with strong currents, prevent the establishment of mangrove seedlings in this area (Untawale *et al.*, 1982). The mangrove; estuary area ratio of this system is only 0.3, and thus freshwater discharge or in situ production may dominate as sources of organic material to coastal seas. Yet in more deltaic estuarine areas where the geomorphology is more suitable for extensive mangrove habitats, such as in northeast India or Southeast Asia, exchange of materials from mangrove to coastal waters may be significant. Mangroves in areas of monsoons present a particular case where the offshore coupling may be substantial. Even in microtidal areas with about 1500 mm of rainfall such as in south Florida, detritus from mangroves is significant in coastal waters. More studies are needed to document the magnitude of organic matter exchange in coastal ecosystems under various hydrologic energies.

The objective of this chapter is to review information on the flux of organic matter and nutrients between mangroves, and estuarine and coastal waters to focus future research in mangrove ecology. It seems that processes involving the exchange of materials from mangroves varies among tropical estuaries depending on hydrologic energy (tides and/or runoff) and geomorphology. Hydrologic energy influences the per area productivity and flux of materials in mangroves, and together with geomorphology, these two factors also determine the areal extent of mangroves surrounding an estuary. The importance of this total flux of detritus from mangroves to the organic matter budget of estuarine ecosystems depends on the relative size of both systems. It has been suggested that the ratio of wetland area to the surface water area of estuaries may be indicative of the importance of wetlands to the productivity of aquatic ecosystems (Mann, 1975; Welsh *et al.*, 1982). Macnae (1974) and Turner (1977) have both related shrimp productivity to the areal coverage of mangroves, suggesting the importance of these wetlands to coastal fisheries. Mass balance of organic matter and nutrients are needed from a variety of geographical regions representing high river discharge, high tidal amplitudes, and combinations of these hydrologic energies. Studies on sites other than the low and medium tidal environments reviewed in this chapter would further test the hypothesis that hydrology and geomorphology control the function of mangroves in estuarine and coastal ecosystems.

The concept of outwelling from mangrove has been based on less data than for salt marshes, yet it seems more conclusive that there exists a net flux of detritus from these forested wetlands. One reason for this may be related to the greater tidal amplitude and runoff at the sites used to study mangrove export compared to temperate intertidal wetlands. For instance, tidal amplitude for mangroves in Australia is greater than 3m, and rainfall in south Florida is more than 1500 mm. Also litterfall in mangrove is continuous throughout the year and for *Rhizophora*, this leaf material is very buoyant. These factors lead to a greater potential for exchange in mangrove ecosystems. However, studies of detritus flux in mangroves are few and are based on relatively simple hypsographic analysis of export. More studies on mangroves are needed that represent sites with different hydrologic energies and geomorphology. Also, experimental designs using flume techniques designed for salt marshes should be investigated to further test the hypothesis that the contribution of detritus from mangroves to estuarine and coastal waters depends on hydrologic energy.

The function of mangroves may be strongly related to forest structure such that processes may be specific to riverine, fringe and basin mangrove forests according to their respective hydrologic characteristics. This conclusion is based mainly on organic matter exchange in mangroves, although there are indications that nutrient recycling may also very along a continuum in hydrology. There are many processes that need to be studied before generalization of mangrove functions can be made. Studies on the importance of nitrogen fixation, denitrification, fertilization, sedimentation and exchange on nutrient budgets would greatly expand our understanding of nutrient dynamics in mangrove ecosystems. The contradiction that tidal wetlands contribute detritus to coastal ecosystems yet also serve as nutrient sinks has been explained by transformations of nutrients that occur within the wetland. The question of whether mangroves are nutrient sinks has received little attention, and processes associated with these transformations may be important in understanding this function.

Evaluating the importance of mangroves to sustaining secondary productivity in estuarine and coastal ecosystems depends on both the transport and utilization of organic matter from mangroves. Based on mass balances, some mangrove–estuarine ecosystems transport substantial amounts of organic material to estuarine and coastal waters. However, the mass balance approach does not indicate directly the utilization of this material by higher trophic levels. Also, mangroves may benefit several fisheries by providing habitat during certain stages of their life cycle rather than serving only as a source of food. Several excellent analysis of trophic food webs in mangrove ecosystems, such as that by Odum and Heald (1972), have demonstrated that mangroves are utilized for food and habitat. Evidence from stable isotope studies suggest that the utilization of mangrove detritus may vary depending on its quantity relative to other sources of food (Macko and Zieman, 1983), similar to conclusions based on mass balance studies. There are several controversies concerning the interpretation of stable isotope information, yet more research on experimental designs using this technique to determine if mangrove detritus is utilized are warranted. We are still far from establishing a cause and effect relationship between mangroves and fisheries.

Chapter – 8

LOSS OF MANGROVE BIODIVERSITY : CAUSES AND CONSEQUENCES

8.1 INTRODUCTION

Global biodiversity loss is a crisis whose importance is now well understood and widely accepted. What is also accepted is that a problem with such a global scale requires a global response. The International community's response has been to rely on a structural approach common to the resolution of international problems. The Convention on Biological Diversity (CBD) has developed a legal framework while the Global Environment Facility (GEF) has provided a financial framework. The combined framework continues to provide the rationale and standards for how the international community should address the issue of biodiversity loss. As such, it has not only played a role in maintaining the visibility of the issue on the international agenda, it has also provided the impetus for activities and programmes that have directly contributed to the conservation of biological diversity around the world. As a global organization committed to the conservation of biological diversity. WWF has been directly involved in the push for, and the development of, this global response, which have supported and encouraged the work of the multiplicity of institutional, governmental, and non–governmental responses that have emerged to address the problem.

Inspite of many positive and encouraging developments that have arisen in this area over the last ten years, it is our belief that the approaches and responses have failed to stem the loss of biodiversity in any meaningful way. Indicators of loss in terrestrial and marine ecosystems continue to show that species and habitats are disappearing. A recent WWF report estimates that the world has lost 30 percent of its natural wealth in the span of one generation from 1970 to 1995.

In fact an increasing amount of research in the environmental community has focused on why biodiversity loss continues to exist as a global problem; this despite the many initiatives and innovations that have been devoted to the issue over the years, and which have pushed the geographical and conceptual boundaries of what needs to be addressed to conserve biodiversity ever outward. The Root Causes project is contribution to this emerging school of thought. One of the assumptions underlying research is that biodiversity loss will continue unabated until its indirect or root causes are understood and addressed.

The root causes project is driven by a need to focus attention on what we believe to be the new requirements necessary to stem and halt the loss of biodiversity. Its four basic premises are that:

1. Biodiversity loss, as a global phenomenon, is accelerating in scale and scope.
2. The solutions proposed and activities undertaken to date have not been sufficient to address the problem.
3. This basic failure stems from an inability to understand and articulate the basic conflict that exists between the promotion of growth and consumption on one side, and activities and incentives promoting sustainable development and conservation of biological diversity on the other. We believe that this conflict and the large bias that continues to exist in favour of growth and consumption are the root causes of biodiversity loss.
4. New conservation approaches must begin with an understanding that conflict and be based on the expectation that the roots of that conflict must be addressed if they are to succeed in the long run.

The Sustainable Programme Office (SPO) designed the Root Causes project to test the basic validity of those premises and has generated threė distinct and related outputs. First, to help us understand how to carry out a root causes analysis and what it might tell us about the nature of the conflict driving biodiversity loss, developed an *Analytical Approach* designed to facilitate the analysis of the root causes of biodiversity loss. Second, to look at the status of biodiversity, and what the effectiveness of current approaches has been carried to case studies around the world to look at the root causes of biodiversity loss in specific instances. And third, to help define the development of new conservation approaches, offer recommendations on how the international community can develop new ways of thinking about biodiversity loss and how it should be addressed.

In the lead–up to the 1992 United Nations Conference on Environment and Development (the Rio Summit), the traditional aid donors were under pressure to provide financial resources to address what were acknowledged to be the major 'global' environmental problems : loss of biological diversity, climate change, loss of ozone layer and degradation of the marine environment. Beginning in the 1980s, these issues were capturing increasing scientific and public attention and a fledging consensus on how to address them was emerging by the time of this summit.

These issues represented, in economic parlance, problems of the 'global commons', i.e. problems whose manifestation might be local or national but whose consequences would be global in scale. As Timothy Swanson has noted, describing the global nature of the biodiversity issue, 'the global stocks of biological diversity generate a flow of services to all societies on Earth' (Swanson, 1997). The particular logic that followed this definition of the problem reasoned that the costs of inaction would be global in scope so the responsibility for addressing the problems – even if its manifestation was highly localized – should be borne by the global community. Two leading environmental economists, David Pearce and Dominic Moran, used the term 'global appropriations failure' explaining that '....if the country in question received no financial or other resources to pay for these global external benefits, it will have no incentive to look after the biological resources' (Pearce and Moran, 1994). As to who would actually pay for solutions to the problems, a debate raged over claims made by developing countries and the environmental scope of activities needed to ensure effective conservation. The fact that the CBD recognized 'sustainable use' reflected the emerging importance of these new approaches.

8.2 AN EMERGING CONSENSUS

It is belief that on the basis of the evolution in thinking that has occurred in the various organizations that are concerned with biodiversity loss, elements of a new consensus on how to address the loss of biodiversity are beginning to emerge. This emerging consensus rests on two assumptions :

1. That conservation strategies must be multi–disciplinary in nature to encompass activities that range from the creation and management of protected areas to the integration of sustainable development principles into international economic policy–making. These strategies must be based on analyses of the complexity of factors that drive biodiversity loss and must seek to involve many different actors, with their differentiated comparative advantages, in the identification and implementation of solutions.
2. The paradox of biodiversity loss is that it is considered a global problem but that its actual occurrence is a highly localized phenomenon. The conservation of biodiversity therefore needs to be an undertaking that occurs simultaneously at a variety of scales.

The rapid loss of biodiversity and habitats around the world is occurring at a local level as a result, for example, of farmers clearing new fields, timber companies opening new forests for logging and hunters producing for city markets. The explanation for these activities, however, is often found in socio–economic forces that arise not at the local level but far from the sites of biodiversity loss. As discussed earlier, conservation work has rarely understood the importance of these socio–economic factors, with the result that biodiversity loss continues to accelerate despite years of efforts to protect habitats and species. Conceptual model of mangrove forests and biodiversity loss is given in Fig.32.

(i) Socio–Economic Root Causes

(a) Social change and development
(b) Public policies
(c) Markets
(d) Macro–Economic Policies
(e) Poverty inequality
(f) Demographic change

(ii) Proximate Causes

(a) Climate change
(b) Pollution
(c) Over harvesting
(d) Habitat loss and alternation

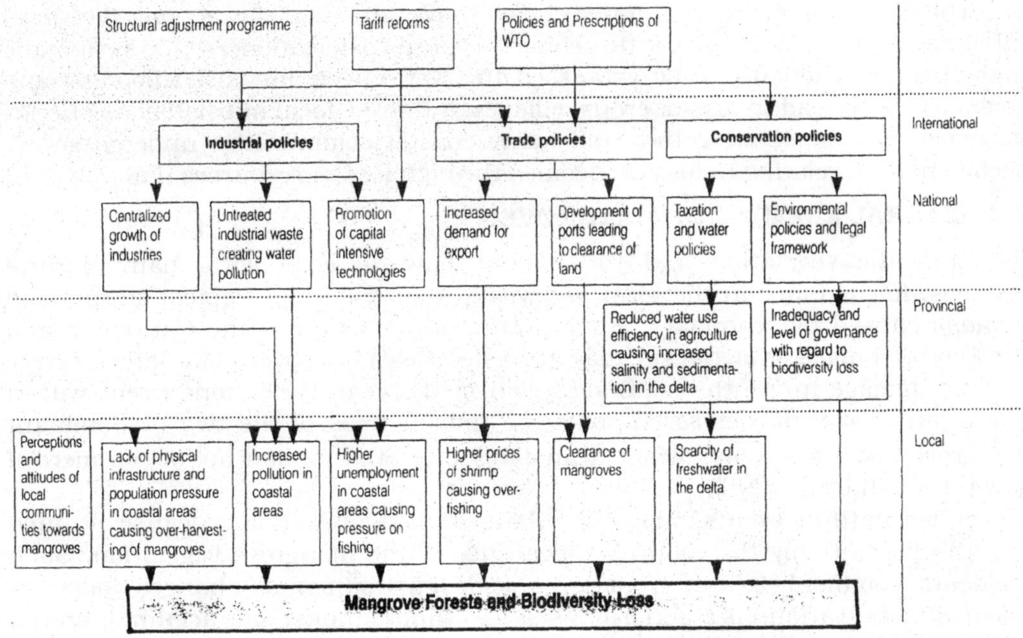

Fig. 32. Conceptual model for the loss of mangrove biodiversity

Mangroves form one of the most extraordinary ecological formations occurring in the coastal lowlands of the tropics. Neither land nor sea, they consist of trees with flying buttresses, and are adapted to brackish waters and unstable, often highly saline soils. This type of vegetation provides fish nurseries, rich fishing grounds and a vast number of both timber and non–timber products.

Mangroves are, one a global scale, exposed to many single destructive events. But the cumulative effects of natural disasters, clear–cutting, overlogging, fish– and shrimp farming, industrial and domestic pollution, dredging and industrial and agricultural land reclamation as well as fragmentation threaten their continued existence.

The world–wide damage to tropical tidal forests is a process which has received little public attention, despite the fact that the loss of mangrove wetlands has already reached alarming dimensions. On the whole, mangrove destruction is perceived with a similar in difference to the global concerns raised about tropical forest degradation a few decades ago. At that time, too, there was little enthusiasm for the concerns of ecologists and foresters.

Mangrove forests often die a slow death, a process that sometimes lasts for many years and goes largely unnoticed. Coastal communities often only realise that something is wrong when fish catches reach low levels or when dramatic events like mass fish death occurs. Such events would then catch the attention of politicians and administrators, but too late. It is rather the rule than the exception that these events pass and the voices of local people and concerned environmentalists continue to go unheeded.

The loss of mangrove forests can have drastic effects on the local inhabitants for many years to come. Typical consequences of systematic mangrove destruction can be witnessed the world over. Increased flooding and coastal erosion lead to loss of crops, villages and lives, local subsistence fisheries decrease or collapse altogether and many coastal communities undergo severe social changes relating to loss of incomes through loss of resources (Fig. 32).

8.3 NATURAL ENEMIES OF MANGROVES

Both the vegetation and the habitat have to cope with many natural enemies, for instance in the form of devastating events like tropical storms and *tsunami* caused by seaquakes. Such events are common in the Caribbean and the Bay of Bengal and forests may take decades to recover. The latest severe cyclone damage in Southwest Bangladesh in the year 1988, concurrent with a storm surge of 5 m, destroyed some 8.5 million trees of the two predominant mangrove species – a loss which translates into 66.3 million m^3 of commercial saw timber (Karim, 1995).

Other natural events, more of a long–termed nature, may cause stunted growth or eventually the total disappearance of the mangrove vegetation. Such incidents commonly result from large–scale environmental changes following macroclimatic variations : abnormal sea level fluctuations, salt accumulation in soil or massive sediment deposits after heavy rainfall and flooding. *Sundri (Heritiera fomes)*, the most important mangrove tree species of the Sundarbans, contributing some 60% of the area's marketable saw timber, has been severely affected during the past three decades by a disease described as 'top dying'. This disease kills the foliage and twigs of one or more branches, then moves down to the main stem and eventually causes the tree's death. The latest Bangladeshi forestry inventories documented some 45 million *sundri* trees – about 20% of the entire stand – affected by top dying (Hussein and Acharya, 1995). The possible causes for top dying discussed in a recent workshop include : increased soil salinity due to reduced water flow, reduction of the crucial level of periodic inundation, excessive sedimentation and waterlogging, nutrient imbalances and toxicity events in the soil, stress induced by cyclones and an outbreak of pathogenic gall cankers on the branches.

Biological pests can also have severe impacts on both individual plants as well as whole forest systems. Epidemics of bacteria, fungi, spiders, leaf–mining, herbivorous or word–destroying insects and invertebrates such as wood–boring mussels can all alter entire forest landscapes. For instance, throughout Australia caterpillars of the mangrove borer (*Poecilips rhizophorae*) are parasites to the fruits of *Rhizophora* species, inhibiting their ability to germinate (Hutchings and Saenger, 1987). Such epidemics can cause enormous losses in production forests. Fortunately, most insect attacks on mangroves are less severe. In fact, a whole host of insects naturally feed on mangrove foliage in different ways and in any healthy mangrove forest millions of insects will be nibbling and mining leaves without any visible overall effect on the forest.

Non–insect pests are also numerous. Young plants may be infested by growths of barnacles on aerial roots in such large numbers that respiration is reduced significantly. Leaf–eating crabs, mostly members of the *Sesarmid* family, chew the leaves and bark of young *Rhizophora* trees and seedlings and can devastate entire silviculture plots. In other situations over–grazing by domestic and wild mammals depletes vegetation cover severely and may even hinder mangrove development.

Sizeable forest damage is sometimes caused by grazing sheep, goat and camels in arid coastal zones of Asia and Middle East. In those habitats the leaf production can only take place when grazing is prevented until the mangrove vegetation recovers. Similar happenings are reported from Northern Australia where kangaroos and buffaloes graze intensively on mangrove foliage (Hutchings and Saenger, 1987).

8.4 ANTHROPOGENIC MANGROVE DESTRUCTION

Without doubt most of the recent mangrove destruction around the world has been due to man. Historically, the basis for mangrove utilization schemes has been economic rather than ecological (Paw and Chua, 1991). As a consequence, the decline of mangrove resources has reached alarmingly high rates all over the world, as illustrated by the figures in Table 55.

Table 55. Examples of the (estimated) world–wide decline of mangrove forests in recent times (Mastaller, 1996)

Country/Region	Period of record	Estimated original mangrove area (ha)	Estimated present mangrove area (ha)	Percentage
Cuba	1969–1989	476000	448000	94
Bangladesh	1963–1990	685000	587000	86
Guatemala	1965–1978	58000	50000	86
Peninsular Malaysia	1979–1986	113000.	89000	79
Ecuador	1966–1989	235000	177500	76
Thailand	1961–1993	300000	219200	73
Vietnam	1969–1990	425000	286400	67
U.S.A.	1958–1983	260000	175000	67
Colombia	1976–1989	480000	307000	64
Indonesia	1969–1986	4220000	2176000	52
Philippines	1968–1995	448000	140000	31
Singapore	1922–1989	700	180	26
Puerto Rico	1930–1985	26300	3000	11
South India (Kerala)	1911–1989	70000	250	04

Human interactions of various natures are responsible for the degradation of mangrove wetlands. In most regions the problems listed under specific headlines are similar, and cases quoted as typical examples could well be encountered in almost any country with mangrove resources in one way or another. A generalized view of loss of mangrove biodiversity is given in Fig. 33, 34.

This survey of the global status of mangrove forests has found that vast areas are being destroyed either intentionally or as a secondary result of other activities. Each example of destruction has, as a common basis, policy decisions relating to economic development by both government and private sectors. In each case, it appears that a decision was made which either ignored the value of the mangrove resource, or which placed a significantly higher value on the alternative land or resource use. In essence, short-term exploitation for immediate economic benefits has taken precedence over the long-term generation of benefits which have both economic and natural values. It is believed that, in the majority of the cases, had the full value of the mangrove resource been taken into account, the subsequent action would have been modified or altered so as to protect those values for sustainable use.

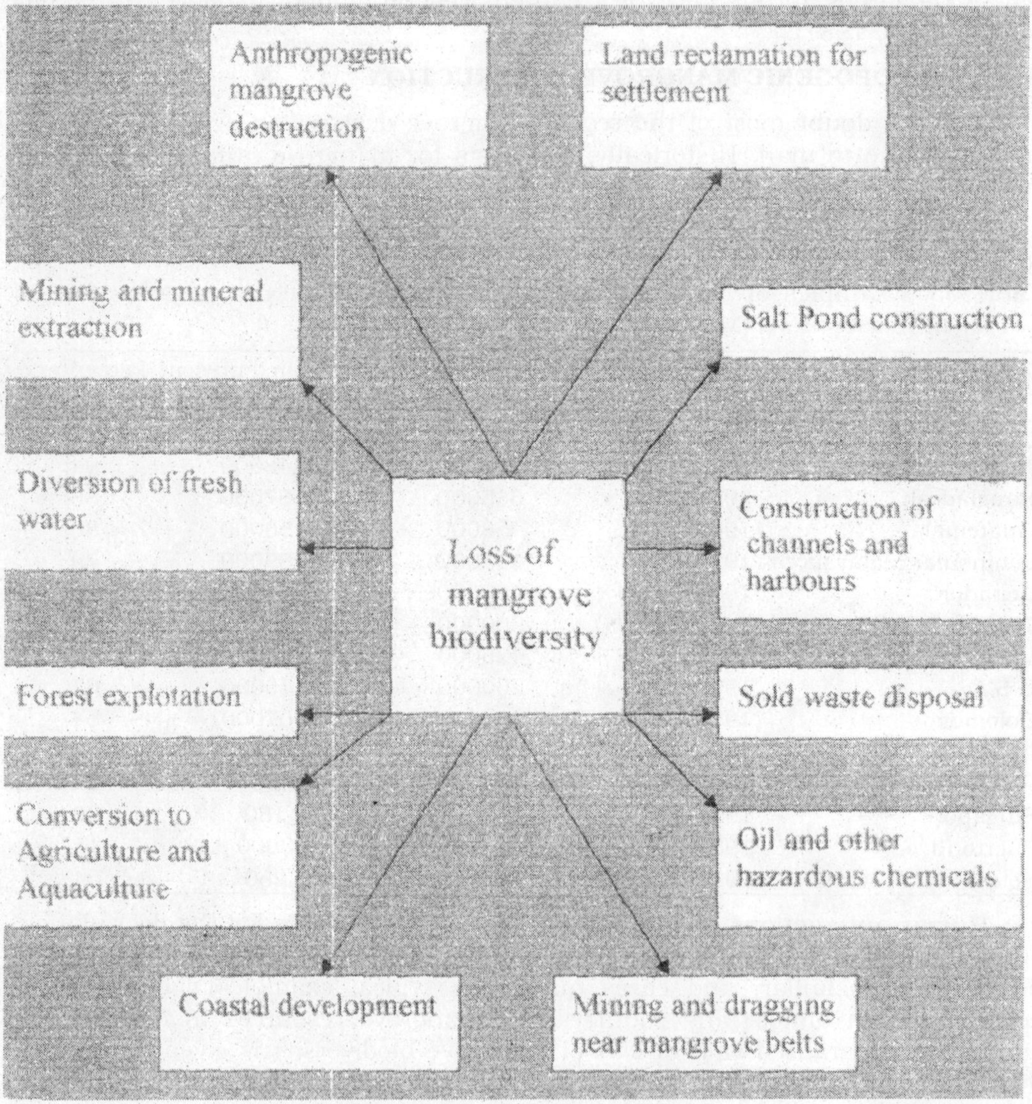

Fig. 33. Loss of mangrove biodiversity

Some of the dominant reasons for the destruction of the mangrove ecosystem are described in this section. The causes of mangrove destruction as currently underway in the World can broadly be distinguished as:

(i) over–exploitation by traditional users; and

(ii) destructive action resulting from activities generally unrelated to sustained uses of mangroves.

It is apparent in this classification that repeated or simultaneous actions in a region increase the total impact. For example, one traditional exploiter in insignificant, however 10,000 exploiters focusing on one area would have a significance impact. Also, a combination of actions imposed in a local area would have a cumulative impact on the total system.

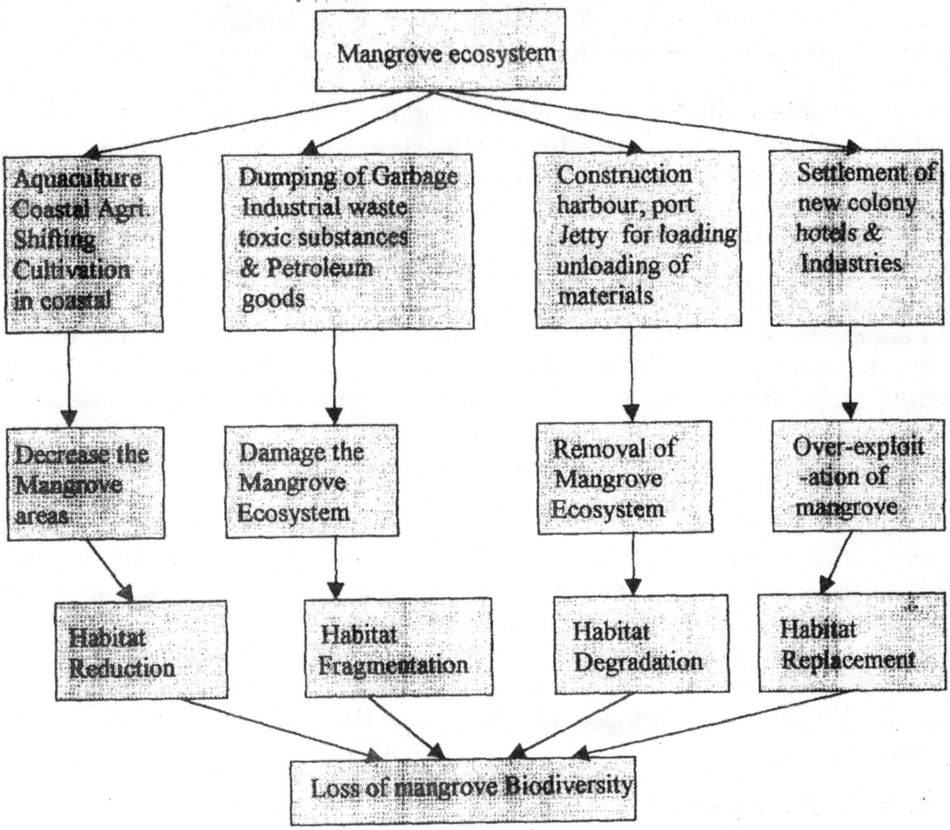

Fig. 34. Indicate the causes of loss of mangrove biodiversity

(a) Mining and Mineral Extraction

As the scarcity of industrial minerals increases, there is a corresponding increase in the mining of alluvial mineral deposits in the tropical coastal zone. Indeed, in many areas of the world there are rich alluvial deposits of tin and chromium, as well as other minerals associated with them such as titanium.

The exploitation of ore bodies in the coastal zone takes place upstream, downstream and within the mangrove ecosystem. Mining within the system results in its complete destruction, whereas mining in adjacent areas causes variable destructive effects. The dominant effect is the deposition of overburden materials which are transported to and within the mangroves by surface water. Excessive sedimentation is detrimental to mangroves through its blocking role in the exchanges of water, nutrients and gases within the substrate and between the substrate and overlying water. When this exchange is totally blocked, the death of mangroves occurs within a period measured in days. Partial cessation of exchange imposes a stress on mangroves which is manifested in reduced productivity and reduced survival as a result of being made significantly more susceptible to any further stress.

The turbidity and increased siltation caused by dredging and overburden disposal also results in destruction of local corals and sea-grass meadows, and their associated faunas. Further, mangrove detrital-based food webs are disrupted and overall there may be a reduction in fishery yield.

Mining activities also are frequently associated with various refining processes to further concentrate the ore for economical shipment. Such procedures usually involve such steps as crushing, washing, chemical flotation, screening and dewatering. As partial refining takes place close to the mine area, wastes are often disposed of in the coastal zone subject to local sedimentation and/or transport to adjacent areas.

In some countries drilling for and production of oil occurs in mangroves, as on the Mahakam River in the Handil Dua area of East Kalimantan. The individual wells are not so much a destructive impact as are the associated pipelines and roads which alter drainage, and the almost invariable oil leaks which accompany the activity. Storage areas and shipping facilities may add to the impact.

As would be expected, the potential short-term economic gains from the ore generally far exceed the short-term economic or natural value of the mangrove forest, and there are few examples of mining being deferred, or mining procedures altered, in deference to the mangroves. However, in contrast to several of the examples presented here, mining could be viewed as a temporary land use if efforts were made to stimulate recovery and regeneration. Such efforts, though are rarely being made in spite of the increased mining activities in the coastal zone.

(b) Diversion of Freshwater

It is a popular misconception that mangroves are salt-demanding plants when, in fact, mangrove development is best in areas which have significant input of freshwater runoff. Accordingly, mangroves in arid, semi-arid, and/or seasonally arid environments are highly dependent on periodic inputs of freshwater. However, it is also in these types of environments, many of which are heavily populated, that the dry-season demand for freshwater for human use is highest. To decision makers who view freshwater runoff into the ocean as a waste of a valuable resource, it is almost inevitable that decisions are made to capture and use this resource in the terrestrial environment rather than see it 'washed'. In considering the major rivers which have been dammed (e.g. Nile, Indus, Ganges etc.) the downstream effects on both coastal vegetation and nearshore fisheries, will become readily apparent.

Freshwater inflows into mangroves may be altered by various upstream activities in the catchment area. Changes in agricultural and forest land use (e.g. logging) may change the amount. Minor changes will cause adjustment which may not be severe. Roads, if located perpendicular to the surface flow patterns, and not adequately provided with culverts, can severely disrupt the mangrove system.

Of major destructive effect, however, are large scale reductions of freshwater inputs caused by human water uses such as irrigation or large water supply diversions out of the catchment. Moreover, large changes in volume and constancy of stream–flow through such activities as flood control or power generation can also have destructive impacts. These problems usually result from major engineering works such as dams and barrages.

The diversion of freshwater from the mangrove ecosystem affects mangroves and associated fauna in a variety of ways. The actual mechanisms for these effects are not fully understood.

(c) Forest Exploitation

Throughout the World, firewood and domestic fuel are decreasing in supply, while at the same time the demand is rapidly rising. To offset the deficit, mangrove forests are used directly by country people for firewood (e.g. India and much of Africa) or the wood is turned into charcoal for domestic or small industrial use. In those areas where the annual extraction is less than the annual regrowth of wood, the mangrove forests serve as a sustained–yield resource and could exist in perpetuity in that state. In many areas, however, the extraction far exceeds the annual regrowth and the forests are rapidly being degraded or are disappearing.

The other form of timber exploitation occurs on a very large scale and appears to be based wholly or in part on the cash returns from the sale of timber (minor objective) and reoundwood for chips and pulp (major objective). In several instances, it seems that the decision to exploit was based on a sudden recognition of economic gain for what had been considered a worthless swamp forest (e.g. Venezuela) or on the opening–up of new lands for colonization and development (e.g. East Kalimantan, Indonesia).

In some cases large–scale exploitation has resulted in a complete loss of certain areas, mainly due to unsuccessful natural regeneration and the great demand for conversion to other forms of land use. Although attempts have been made to regulate exploitation of these areas on an aerial basis (as in Sarawak) or though the retention of seed trees.

(d) Conversion to Agriculture and Aquaculture

Coastal swamp forests in saline, anaerobic environments have traditionally been considered to be marginal, or totally unsuitable for agriculture and / or aquacultural production. However, with improvements in cultigen hybridization and conversion of saline, acidic soils, and increasing demand for arable land, the mangrove environment is being viewed as a major alternative for the global increase in agricultural production including both small grains and aquatic animal protein. In fact, it is one of the objectives of the International Rice

Institute in Los Banos, Philippines, to find ways to utilize this 'marginal' resource which includes not only mangrove–dominated lands but also contiguous areas of tidal flats, beach ridges, intercreek saline open flats and reclaimed lands. It is not apparent that any of the natural values of mangroves are ever taken into account in decisions to convert this self–maintaining coastal ecosystem into an industrially subsidized agricultural or aquacultural factory.

(i) *Agriculture*

In many areas of the World, for example in Asia and Africa, the pressure on arable land has led to effort to convert mangrove lands into agricultural lands. This usually involves the digging of a narrow canal (parallel to the shoreline) and piling up the diggings to form bunds on one or both banks of the canal. A number of canals are also often dug perpendicular to the first canal and these lead to the sea to encourage draining (at low tides). The mangroves on the landward side become drained and eventually the salts leached by freshwater runoff and rains. The bunds thus prevent seawater intrusion, and gates in the bunds allow excess water to be drained at low tides.

Such a reclamation can lead to extensive loss of mangrove areas and their high productivity and to adverse effects on the fisheries of surrounding coastal areas. In addition, the canals cause a change in the freshwater regimes of the unreclaimed seaward mangrove and can have deleterious effects on the system.

One of the most unfortunate aspects of reclamation of mangrove areas for agriculture is the high failure rate of many schemes. The main problem is not due to saline conditions as many believe, but to the development of acid sulphate soil conditions. Many mangrove areas have soil containing large amounts of pyrite sulphur (FeS_2). When exposed to the air, oxidation releases sulphuric acid. The soils become extremely acid and very high in soluble salts. This condition often leads to nutrient and fertilizing problems, resulting in crops (e.g. rice) not growing or growing with very reduced productivity.

Large scale reclamation involves thousands of hectares and, should acid sulphate conditions develop, the loss (financial and environmental) can be extensive. It is thus important that the potential problem be made known so that potential users are made aware of the risks involved.

Since reclamation eliminates the mangroves, the question is not how it affects mangrove, but rather its adjacent ecosystem. Here, the question of its economic impact on the coastal fishery and social impact and benefits on the people the fishery supports needs consideration. Will the reclaimed land bring in more than the loss to the coastal fishery ?

(ii) *Aquaculture*

Impoundment of mangrove areas for aquaculture ponds is not a new activity. However, in recent years the number and scale of aquaculture projects has increased rapidly. In 1977 it was estimated that 1.2 million hectares of mangrove forests in the Indo–Pacific region had been converted to aquaculture ponds.

Aquaculture operators clearfall the mangrove forest and build small dykes with sluice gates to retain water at all stages of the tide. While individual ponds generally only range from 0.1 to 1.0 ha in size, from hundreds to thousands of hectares of continuous fish ponds may be built in an area.

Much of the mangrove flora and fauna in the areas surrounding the ponds may be destroyed as well, because of major changes in drainage conditions, in tidal inundation frequency, and in nutrient availability as well as because of the toxicity of runoff waters from the ponds and drainage channels in areas of acid sulphate soil.

Highly pyritic soils also cause serious difficulties for the operators of aquaculture ponds. The efficiency of phosphate fertilizers is severely inhibited, resulting in inadequate growth of algae on which fishponds depend for high yields. High acidity and high aluminium concentrations will kill the fish or, in less severe cases, weaken them so that they are vulnerable to diseases or parasites. During rain the sudden influx of these toxins from the sides of the dykes is commonly lethal to a large proportion of the remaining fish. Finely divided ferric hydroxide subsequently appears in the pond water and clogs the gills of the survivors, killing some and weakening the remainder. The prawns grown in aquaculture ponds appear to be even more vulnerable to the toxins from acid sulphate soils.

Aquaculture ponds sites on acid sulphate soils tend to be uneconomic to operate. To be economically viable ponds are best situated within areas of moderate tidal range with suitable land elevation and soils.

While small–scale pond construction would probably have little significant adverse effects on the mangrove ecosystem and its adjacent coastal ecosystem, this is not the case with large–scale impoundment. The conversion of large areas of mangrove to ponds may result in the reduction of adjacent fishery yields. The question here is : Will the ponds be more efficient than the natural ecosystem ?

Unless pond production can surpass the natural system's production, it becomes an economic disbenefit with the possible added cost of upsetting the ecological balance of the adjacent mangrove areas.

The one important question here is how much can be converted (within a particular area) to aquaculture ponds without significantly affecting the mangrove ecosystem as well as its adjacent coastal fishery. There is a need for more research to answer this question.

This is an important question to answer as increasingly large areas are being earmarked for aquaculture use – particularly in fhe Asian region.

(e) Coastal Development

The destruction of the mangrove forest and the conversion of mangrove lands to domestic and industrial development is a major problem in high income countries and is beginning to become a problem of consequence in developing countries. The most common forms of conversion are to housing and residential development, coastal tourist facilities and industry, including small port development. Although many of the countries where this is a problem (e.g. Australia, New Zealand, United States) have laws protecting the mangrove environment for its high ecological value, such regulations only seem to have slowed the pace of conversion. Developers frequently search for legal loopholes and in one instance in the United States persuaded the federal court to rule in favour of private economic interests as opposed to the interest of the public in protecting the public domain.

The three major types of development (urban, industrial and airport) are very different in character but agree in being high–intensity land use with extensive peripheral impact. Industry and airports tend to be close to major population centres.

Urban development represents maximal density of human settlement and the focus of all regional development. Road concentration and land modification is maximal and the on–site mangrove will be totally reclaimed. On–site rainfall runoff is accentuated by steeper and redirected outflows and reduced penetration. Runoff carries high levels of automotive pollutants. Organic effluents would be treated but still constitute a high nutrient input downstream. Bulk garbage disposal is likely to affect nearby catchments through seepage and often includes high–toxicity pollutants. Water demand is high and likely to affect freshwater output in neighbouring catchments over a wide area. Nearby ecosystems will be subjected to interference by public health management. Recreational pressure and vandalism will be serious. At medium distance, therefore, mangroves will be under heavy pressure. In areas where they provide defence against natural disasters such as typhoons or flashflood, the high human density may give mangroves high priority for maintenance in selected areas and at some distance. It is important to remember the time dimension. Cities attract further development and tend to spread and establish satellites. Long–term maintenance and development of peripheral systems must anticipate such processes.

Industrial development may make higher demands on water supply than residential settlement, though for lower quality. Waste outputs tend to be higher and more toxic. Transport–related effects such as road and channel construction and automotive pollution will be of high intensity.

Airports are a special case. They are likely to encroach on mangrove because of requirement for flat terrain and approach routes avoiding areas sensitive to noise pollution. They have special sensitivity to wildlife concentrations such as risk from bird–strike. Because the human population is highly transient, with throughflow of non–immunes and carries of exotic pathogens, public health management is likely to affect a wide peripheral zone. An airport may involve little structural change to nearby systems but a very high intensity of interference.

City development in a catchment is likely to strongly modify the quantity and quality of freshwater throughflow in mangroves. This will be reduced and redirected under low–rainfall conditions, and under high rainfall the response time is likely to be shortened so that special provisions for flashflood control may be needed.

Marine inputs will be correspondingly modified in that saltwater penetration will be higher upstream and closer to the surface. This will inevitably provoke changes in zonation. Death or modification of significant areas of mangrove can provoke release of organic and inorganic materials from the large organic pool within the system, perhaps fouling areas both up– and downstream.

In–system process will be directly affected by pest control (e.g. birds and insects), by recreation pressure and by pollution. Of these, routine spraying for the control of mosquitoes and other nuisance insects is probably the most significant detrimental impact. Pesticides will affect the natural fauna and it is now well known that mangroves can take up and accumulate a variety of synthetic organic biocides.

Many established cities are directly associated with estuaries for reasons of transport, water supply, waste disposal and accessibility to interior land areas. In the past, impact on mangrove has probably been disproportionately high because of the failure by planners to appreciate that estuaries including mangroves, are economically valuable resource areas.

Land reclamation for settlement

Mangroves have always been closely related to the historical development of tropical countries. Major urban centres like Maracaibo, Recife, Freetown, Douala, Lagos, Bombay, Madras, Calcutta, Rangoon, Bangkok, Singapore and Jakarta – to name but a few – originated from small stilthouse settlements in mangrove swamplands. Now–a–days, however, it is almost impossible to find any mangroves in or next to these urban conglomerations, most having disappeared in the course of urbanisation, establishment of squatters or been converted into industrial areas and real estate.

The Indian subcontinent renders an imperative example of how mangrove areas have always suffered from conversion due to colonisation, migration and population explosion. When Bombay was founded in 1670, the entire mangrove vegetation on the seven islands was clearfelled to make way for a thriving trade metropolis. The history of the Sundarbans, too, has been characterised by continuous land reclamation since the beginning of British colonial times in 1770. In the largest delta region of the world, extending between India and Bangladesh some 150,000 ha of mangrove forests disappeared during the past 100 years, reclaimed for agriculture, settlement sites and road networks for an ever–increasing coastal population. As in most places in the Indian subcontinent, coastal roads and industrial estates were placed in reclaimed mangrove areas to minimise conflict with existing agricultural interests. With the expansion of paddy cultures and coconut plantations, greenfodder became a rare commodity and as a makeshift solution cattle are often driven into mangrove forests.

In India's southernmost state of Kerala, where the rate of population increase is the highest in the country, some 70,000 hectares of mangrove were cleared at the turn of the century. At that time Kerala had about 1.5 million inhabitants. Today, some 22 million people crowd the coastal belt of 600 km by 25 km, resulting in 1500 people per square kilometer. It is no surprise that only 3.5% of the former mangrove vegetation survives. The major part was cut for fuelwood – unquestionably the cheapest and most readily available source for the needy population. The after–effects of this extensive mangrove clearance were felt by people living around the Cochin Backwaters in Central Kerala. Formerly a highly productive lagoon system bordered by mangrove forests, the bay is now secluded. Water quality if poor with many parts of the lagoonal waters stagnant and acidic. Vast areas are covered by water hyacinths, causing the shading out of phytoplankton and consequently the decline of planktivores. The formerly thriving shrine production of the Cochin Backwaters have fallen almost to nil. Recently, probably too late, a Forest Regulation Act was passed to protect the remaining mangroves.

The Bay of Bengal coastal stretches frequently become victims of severe cyclones, flooding and storm surges. In the past, when unharmed mangrove belts functioned as firm wave and wind breaking barriers, many of those regions enjoyed some natural protection. Their significant value for coastal protection was first appreciated in areas where mangroves were gradually vanishing. To mitigate the effects of such natural events, hundreds of kilometers of flood protection dykes were constructed in Eastern India and Bangladesh in areas where the artificial dams provided far less protection than the former mangrove belts, and also incurred substantial costs for building and repairs. Moreover, when the dykes breached, results were often even more devastating. In such circumstances vegetation, livestock and humans was swept away by the sudden sediment loads which were funnelled through the broken dams. The conclusion drawn is that only a broad and intact natural mangrove belt could effectively act as a buffer against these recurring natural forces and give reasonable control against coastal erosion.

(f) Salt Pond Construction

Salt ponds or brine evaporation ponds are built on mudflats or more commonly in mangrove areas. The decision to grant concessions to salt pond operators is one involving the total, irreversible clearcutting of a certain mangrove area.

Salt ponds require a complete eradication of the trees and shrubs, levelling and diking of the land, construction of a flooding canal system, and intensive mechanical compaction of the soil surface, and their operation if facilitated by solar heat input. Under operation, the salt ponds are subjected to an inundation regime dependent on local evaporation rates and labour available for gathering the raw salt. The repeated inundation and drying of the soil surface increases the salt content of the soil and also alters the soil structure. Such ponds may later be abandoned for various reasons. However, as experience shows, the drastically changed physical and chemical soil conditions will not allow recolonisation from adjacent mangrove stands, as may occur in areas that have been clear-cut without further changes of the land.

Brine evaporation ponds in mangrove areas are widespread over the tropical coasts of Asia and Africa. They are, however, almost unknown in everwet areas such as Malaysia, or Sumatra and Kalimantan in Indonesia.

Under subhumid climates, as for example in the Cauvery Delta in south–eastern India, these ponds usually do not exceed a few hectares. However, under arid and semi–arid conditions the situation is entirely different and the damage caused to the mangroves by salt pond construction is much more extensive.

(g) Construction of Channels and Harbours

Since the early 1970s, there has been an increase in industries harbour / port construction for the transhipment of ores, fuels and other raw materials. This increase has been due largely to the development of new mines and oil and gas fields in remote areas, and to an increased tonnage of transporting vessels. Included in these new initiatives are liquified natural gas facilities such as the major installation proposed for Natura Island (Indonesia) in the South China Sea. Owing to the high costs of harbour development, coastal areas are sought which involve the least amount of dredging. In at least one example (Port Hedland, Australia) the preferred location was a pre–existing small harbour dominated by one of the most extensive mangrove forests in the region. In Latin America a Japanese company is engaged in the preliminary design of a new oceanic canal in Panama at the same time that others are discussing an oil pipeline through either Panama or Costa Rica, connecting with new ports on the Atlantic and Pacific coasts. Either of these projects will involve major coastal modifications in areas dominated in parts by mangrove forests.

Creation of harbours, ports, anchorages and loading/unloading facilities, and the dredging of deep channels has certain obvious impacts on the mangrove environment, and some that are less obvious. Dominant among the latter is the change in the local hydrology, particularly with respect to mean water levels and the mixing of freshwater and seawater in the nearshore environment. Such alterations have profound effects on the intertidal environment, resulting in chronic salinity stresses and eventual changes in species composition and dominance. It is interesting to note that the presence of mangroves is seldom considered in the early planning or initial feasibility studies, and is a topic of concern only to construction contractors who may have to remove them from a construction site in the intertidal zone and cope with mangrove sediments for foundations.

In addition to the effects of harbour and channel development and subsequent operation, loading/unloading facilities also become potential spill sites both for routine discharges and accidental spillages. Contingency plans and spill control equipment are often not incorporated.

(h) Solid Waste Disposal

Where human populations have been doubling in urban areas, the garbage and solid wastes generated have increased three to four times in most countries. Characteristically, this waste has been transported and dumped where it would not be highly visible to people. It has therefore been put in depressions or on flat lands. Since most major tropical and subtropical urban centres are located on coasts or estuaries and since mangrove areas have traditionally been regarded as wasteland, much solid waste and garbage refuse has been dumped into mangrove ecosystems.

Apart from the loss of the mangroves resulting from this activity, disposal of waste so close to waterways has produced many unforeseen problems. The leachate of materials, plus any toxic materials included in containers, will flow to adjacent areas unless bunds or dykes have been built to prevent surface flow away from the site. Even plastics, believed by many people to be biologically inert, slowly weather and break down, releasing toxic phthalate esters (a plasticizer) into the environment. In those countries where laws limit toxic waste disposal, there is an increasing tendency for illegal dumping of liquid toxic wastes in mangroves.

The near–urban–centre mangroves, which are frequently thc locate for solid waste disposal, are often important areas of open space and, as a forest, offer opportunities for education and nature study. They may also represent an aesthetic resource which is destroyed by society's refuse. Animal scavengers such as rats and seagulls replace natural fauna, and pathogen–bearing insects may present human and animal health hazards.

Though the area affected by this activity is very small in global terms, these areas are very important ones because of their proximity to urban population centres.

(i) Liquid Waste Disposal

Many agricultural, agro–industrial, chemo–industrial and domestic processes create varying quantities of basic organic material as unwanted waste, and discharge it – in contrast to the solid waste and garbage – as suspended or dissolved matter into nearby rivers and coastal waters. Though the major part of this material is naturally occurring organic substances and not toxic *per se*, the load in the aquatic environment can become too great for natural recycling. Excessive concentrations of livestock waste, food processing wastes and domestic sewage can deteriorate riverine environments to conditions (e.g. oxygen depletion) unfavourable to aquatic living resources.

Waste material which is potentially toxic to the living environment is generated in most of the industrial production processes, whether agricultural, forestry, chemical or mining industry. Also, urban and domestic human activities generate effluents containing substances with potential toxicity.

Virtually all industrial processes involve at various stages one or several organic or inorganic compounds necessary to clean, change or convert the basic material or finish and upgrade the end–product. The process additives range from inorganic and organic salts, often heavy metal salts, and petrochemicals of simple as well as complex structures, to an array of newly synthesized pharmaceutical and agro–chemical compounds unknown and undetectable to the environmental analyst.

In many cases, a certain percentage of the original material, intermediate process products, or even the finished product is discharged as wastes with the effluents. There are various technological and economic, justified or unjustified reasons why these compounds are discharged to the environment.

The potential toxicities of these numerous waste products do not necessarily make them toxic *per sė* to the environment, but their way of discharge into the water courses, their resulting concentrations and accumulation in the biota do.

The introduction into the food chain at very low levels and rates can give rise to bioaccumulation of these compounds or elements which then reach concentrations that become toxic or lethal at certain steps of the ladder.

The toxic effect may be visible in a single, particularly sensitive species, but its direct effect may also remain undetected and only be revealed by the disruption of a particular food chain, or break of a basic recycling or regeneration process.

Though many individual mangrove species are resistant to various degrees to these impacts, the equilibrium of the system may be upset, and result in a shift to another ecosystem structure.

(j) Oil and Other Hazardous Chemicals

The increasing ocean transport of bulk liquids, particularly crude and refined petroleum products, is associated with an increasing frequency of spills, Owing to the presence of mangrove-dominated shorelines close to major shipping corridors, mangroves are more frequently being impacted by accidental or wilful spillage. Although certain governments (e.g. Malaysia, Peurto Rico, United States) have recovered damages from responsible shipping interests, spill areas have not been restored to former levels of productivity.

The problem of oil spillages in mangrove zones is that, in contrast to rocky shores which have a greater self-cleaning capability through highly energetic wave action, mangrove forests are excellent traps for drifting oil slicks. The tidal water circulation in the mangroves is favourable to deposition of slicks on the mangrove aerial root systems and to accumulation by the soil.

Two classes of effects are observed following oil spills in mangrove forests. The first class of effects is acute, immediate and relates to the physical something effect of oil on the plant surfaces (e.g. sediment, bark, prop roots, pneumatophores) responsible for gas (carbon dioxide and oxygen). Under circumstances of heavy oiling, mangrove plants can be killed within 48–72 hours. The second class of effects relates to the long-term chronic poisoning of mangroves and associated fauna by the toxic components of the retained oil. Surviving mangroves exhibit numerous symptoms of chronic stress, reduced productivity and lowered rates of leaf litter production. In this state of chronic stress, mangroves are highly susceptible to any additional perturbation or stress. Years following the *Peak Slip* spill i Peurto Rico and the *Howard Star* in Florida, USA, mangroves died, presumably as a secondary effect of the original spill that was triggered by an unrelated perturbation.

Though the long-term consequences of a spill, particularly as they relate to the toxicity of weathered oil components and other hazardous substances in the bleedwater from mangrove soils, are still not well established, it is clear that the increased retention of oil and other substances subjects the system to a prolonged impact.

With respect to all of the destructive activities in mangrove areas, it is only in the area of oil spills that legal / political action has been taken to recover damages and / or prevent the wilful dumping or spillage of hazardous liquids. Predictably, this has led to adversary litigation in which government agents argue for recovery of damages and agents of shipping interests argue that any observed damages are minor and / or ecologically inconsequential.

In view of the expanding oil exploitation and shipping activities in a number of countries with mangrove forests, spillages can be expected to increase in the future.

(k) Mining and dredging near mangrove belts

Mangrove and estuarine sediments of Thailand are rich in metals of high industrial value, namely tin, chrome, aluminium and occasionally, titanium. Until the late 1980's 427000 hectares along Thailand's coasts were concessioned for tin mining, many of them in the immediate vicinity of mangrove belts or coral reefs.

Exploitation of tin deposits from sub–sedimentation in mangrove is done *in situ* by means of industrial plants mounted on large floating barges. The ore extraction is based on repeated rinsing of the sediments through special sieves. The tailings discharged have various direct and indirect impacts on surrounding marine and mangrove habitats. Refining and washing of sediments releases significant amounts of heavy metals which pollute the surrounding waters and are deposited in the surface layers of nearshore bottoms. Purportedly, the fine-grained and toxic tailings have been the cause not only for many mass mortalities among marine life but also for the dieback of tidal forests in contact with the washed sediments.

Until recently, when tin prices of the world marked dropped significantly, mining these resources off–shore or within deltaic areas was a highly profitable business. Annual records ranged from US $ 60,000 to 256,000 per hectare. Now–a–days floating tin mines have almost completely disappeared and conservation values are accorded somewhat more recognition than before. Thailand's government agencies appreciate the underlying problems and try to minimise environmental impacts by imposing technical directives on any kind of nearshore dredging operations. As well as that it has to be noted that mining or dredging companies are requested, whenever applicable, to restore and reforest sites following extraction.

Ecological Consequent of Mangroves Destruction

Mangrove forests tend to react sensitively towards disturbances from outside the system. Even short–termed interferences may affect species that are strictly adapted to this habitat, particularly when their life cycle is interrupted in one way or another. Each intervention mentioned above in specific cases has its intrinsic set of environmental problems. Generally speaking, mangrove destruction results in adverse effects at different levels.

Locally, an important habitat for plants and animals is lost and many species may have disappeared from an area, either through extinction or outmigration. This may involve declines in large waterbirds and mammals, mass fish mortality and decreasing numbers of spawning fish. The loss of even a few species from an ecosystem will have serious knock–on efforts for other organisms in the food chain. For instance, insect, bird and bat pollinators may disappear if trees stop flowering or are selectively removed.

Regionally, the micro–climate and soils are effected by large–scale clearance of trees. This may locally lead to lower water tables and impoverished soils which then might seriously effect ecological functioning in other coastal habitats and even in inland forests.

Lastly, on a global scale, the loss of mangrove wetlands means the loss of critically important wetlands for many migratory species. Many species of waterbird and fish for instance link their migration routes to this ecosystem. The World Conservation Union (IUCN) recognizes many mangrove associated species in their Red List (this list of mangrove-associated plants and animals at risk, includes, among other, 14 orchids, reptiles, 65 birds and 15 mammals, Saenger, Hegerl and Davie, 1983), i.e. species which are threatened or vulnerable to extinction. Many of the medicinal applications of mangroves – at least the traditional knowledge of how to use these plants – may be lost when the global destruction of mangrove forests continues its present scale.

The Scarlet Ibis is representative of the common threat to large birds associated with mangroves. Already in pre-colonial times the native Indians hunted the ibis throughout Caribbean coastlands to use its feathers for their emperor's ceremonial coats. Portuguese seafarers hunted the species for food. Today, this rare bird is still intensively pursed in Brazil and French Guyana where it is traded as a delicacy in gourner restaurants and delicatessens. In Cayenne, where hunting of this bird is still legal, a whole craft industry specialized in making ornamental blossoms out of ibis feathers. A French environmental organisation is currently supporting a vocational programme to retrain these floristic artists; the intention is to encourage them to replace ibis feathers with coloured hen feathers.

It is widely accepted that the bioproductivity of nearshore or estuarine fish populations is correlated with intact mangrove ecosystems. Only if the productivity of mangrove forests is safeguarded will sufficient food stocks develop to foster and maintain local fisheries.

The negative effects associated with the prevailing practices in large-scale brackish water fish and shrimp farming, particularly when sited in mangrove wetlands, are well documented and discussed in other publications in this volume. Typical environmental problems in this context are :

- intensified mangrove cutting to obtain construction materials used in maintenance work for the aquaculture farm ;
- locally reduced water exchange by blockage of tidal creeks, resulting in adverse effects on vegetation patterns and impairment of local navigation;
- lowering of ground-water table in landward zones, caused by drainage and ditch excavation;
- increments of local sedimentation rates, resulting from dam ruptures, repairs of bunds and pond draining;
- outwelling of toxic substances originating from prophylactic and medical treatment of ponds, affecting adjacent communities;
- concentration of toxic substances in the culture product; for example endosulfan (*Thiodan*), an organochlorine widely used throughout Southeast Asia to 'disinfect' shrimp ponds; the indiscriminate application of this insecticide, which is chemically similar to DDT, is hazardous to humans who consume shrimp cultured in this manner;
- selective exploitation of wild fry, leading to thinning out or extinction of natural stocks;
- decrease of mangrove crabs resulting from habitat losses and from over-exploitation caused by crab farmers;

- intrusion of insect vectors of waterborne diseases (e.g. *Anopheles sundaicus*) into abandoned and stagnant ponds.

The problems of direct interference and the destruction of mangrove habitats due to agricultural land reclamation is often compounded with long-term environmental impacts resulting from the intensive use of fertilizers, pesticides and herbicides. Such problems occur almost everywhere where rice fields adjoin mangrove wetlands. High dosages of various agro-chemicals are washed into creeks and rivers via drainage canals until they accumulate in mangrove waters, and eventually enter the aquatic trophic chain. Biocidal chemicals are the most likely cause of a striking diminution of *Pteroptyx* fireflies in Malaysian mangrove estuaries during the past 25 years.

The deleterious effects on mangrove habitats resulting from nearshore sediment extraction, mining or large construction projects (e.g. dam or harbour construction) are commonly described as :

- increase in water turbidity and sedimentation rates, caused by both dredging and tailings, which reduces the photosynthetic rate of local flora, inhibits growth of corals and impairs marine filter feeders;
- commercially useful organisms either die or emigrate to less polluted areas, causing local declines in local fisheries production (e.g. bivalve shells and small shrimp);
- dissolved heavy metals from refining processes accumulate in the trophic chain, leading eventually to maximum concentrations in commercially used fish and shellfish, which in turn become a health risk when consumed by humans;
- the metabolic intake of dissolved heavy metals reduces growth and resistance to diseases in various plants and animals;
- in the course of continuous sediment discharges large sand banks develop which obstruct tidal water flows into and from mangrove habitats, causing a gradual die-back of the vegetation;
- change of level and quality of groundwater due to dredging operations.

Economic and social disturbances following mangrove destruction

Inevitably most people living in or associated with mangroves suffer economic losses once this ecosystem is disturbed or eliminated. While healthy mangrove forests may provide the primary sustenance for large rural communities, barren and salty desert landscapes or waters void of fish do not.

Regional economic setbacks resulting from mangrove destruction can sometimes reach enormous proportions. If the basis for local fisheries is affected, economic and social problems follow. Both subsistence and industrial fisheries experience decline in nearshore fish and shrimp catches which are directly or indirectly linked to the status and spatial extent of tidal forests. In brackish water aquaculture too, a decrease in mangrove wetlands has detrimental effects on production since most cultures (e.g. mullet, milk-fish and prawn) depend almost exclusively on wild fry. Frequently fish and shrimp farms have to be abandoned because of insufficient fry supply; what remains is fallow, unproductive lands. Thus, the actual damage to regional economics is even higher than the immediate loss to the individual fish farmer as the abandoned ponds cannot be utilised for other purposes (Kapetsky, 1986; Mastaller, 1995).

Last but not least, a number of other economically significant services are lost when the mangrove forests are cut, disturbed or devastated. Their losses and associated costs will become more obvious in the context of the functional and monetary values of the mangal, as viewed by the local communities rather than by the external users. This dilemma becomes particularly obvious in the sector of industrial mangrove exploitation. Again, the Indonesian case of Bintuni Bay in Irian Jaya serves as an example. In these concessions predominantly Javanese immigrants are employed as loggers (Erftemeijer *et al.*, 1989). Such a situation is likely to disrupt village life and to interfere with traditional customs, which eventually leads to the loss of cultural identity of the local communities.

8.5 THE PERSPECTIVES

Mangroves have a weak lobby. As few people are aware of their functional values, next to no one cares about mangrove protection. It is no wonder that investors and entrepreneurs have few scruples in seizing areas of swampy forest that nobody else likes. Developers are quick to 'use' mangrove land in preference to other lands on the pretext that they are cheap to buy, development permission if easily obtained and there are few problems when negotiating with or compensating powerful landowners or appearing politicians. In many countries mangrove lands are still regarded as common land, largely neglected by the authorities.

The case studies bespeak the many different types of mangrove destruction. Short–sighted planning, short–term profiteering and, above all, social misery in the majority of coastal populations fuels the world–wide decline of mangrove forests. Coastal zones in the tropics are particularly affected by surging increases in population where, in contrast to the hinterland, birth rates are commonly high. As coastal regions absorb the influx of people the pressure on local natural resources are not protected by local authorities or a responsible public, they are inevitably at risk. Needless to say, mangroves are among the first to be exploited in such circumstances. The blame for mangrove loss should not be focused on the (mostly impoverished) communities in countries with mangrove forest resources. The most radical and extensive destruction of mangroves must be attributed to non–residential businessmen and big industries which reclaim and exploit these areas for their own quick profit.

The international community, too, bears certain responsibilities for the global destruction of mangrove forests. National or international development projects and technical aid programmes have not always aimed at an integrated management concept in tropical coastal zones. Even the Forestry Convention of Rio de Janerio, discussed at length during the Earth Summit in 1992, barely contains proposals to save mangrove forests in contrast to the support offered by signatories declaring their readiness to defend tropical rainforests, such as the Amazon. Mangrove wetlands have been internationally neglected since the policy priorities of most countries have been oriented towards supporting production and technical projects. Unmistakably, this strategy is justified when the main objective of development is restricted to immediate economic returns – especially when measured in terms of foreign exchange and securing future markets. As a result quite a few projects funded with international aid money have had extensive negative impacts on pristine mangrove areas. For a short time, such projects may yield some profit (at least to a few people), but in the long run they destroy more than they achieve.

The international scientific community can play an important role in the efforts to contain global mangrove destruction. When compared with other research sectors, brackish water wetlands receive only modest scientific interest, and it was only two decades ago that research into the ecological functions and food webs in the mangal began. If research had started earlier and the results been published globally, as with tropical rainforest issues, the destruction of mangrove wetlands would probably not have reached its present dimensions.

To summarize, stringent management plans are needed. Such plans, devised as integrated approaches and able to tackle the issues from different angles, are seen as the basis for re–establishing or sustainably using mangrove wetlands (Mastaller, 1996). To meet these ends, nine elements have been identified –

- Proper land–use planning
- Comprehensive inventory of the natural resources
- Innovative research programmes
- Public education (awareness building) and training
- Creation of an adequate legislative base and institutional arrangements
- Restoration of damaged mangrove forest areas
- Introduction of social forestry schemes
- Appropriate technical assistance (particularly in brackish water aquaculture)
- Creation of nature reserves.

Chapter - 9

MANGROVE CONSERVATION AND MANAGEMENT

9.1 CONSERVATION

The action of conserving; preservation from destructive influences, natural decay, or waste; preserving in being, life, health, perfection, etc.

Conserve

To keep in safety, or from harm, decay, or loss; to preserve with care; now usually, to preserve in its existing state from destruction or change.

As a word in the English language 'conservation' can be defined by the quotations from the Oxford English Dictionary. However, conservation of biological resources is not solely a problem of preservation and protection. It is not hoarding or leaving well alone as this definition may imply, since nature is not static and cannot be maintained in an unchanging state even if this were desirable. Thus, within a biological context any definition of conservation, or search for its meaning, must explore further than a concise form of word.

Any human intervention in nature, even presupposing good intentions, can rarely be reconciled with the idea of strict conservation. True, homeostasis in natural systems is always active, and although man tries again and again, he rarely produces truly catastrophic changes in the biosphere. But this is a credit to the efficient organization of ecosystems and not to the wisdom of man. Genuine conservation forbids any interference (Margalef, 1968).

The view that Margalef is advancing is essentially that man is not part of the natural environment, and that true conservation of the environment is not compatible with any human activity. Since species of the genus *Homo* have evolved as part of the biosphere, man must be a part of that environment and it is false to consider an environment without man. However, the characteristic of the species, the ability of environmental modification, has meant that man has had a greater effect on the environment than any other species. This alone does not justify the view that, wherever and however this effect is felt, there is no longer a natural environment. However, one should try and assess just how much environmental modification there can be before an ecosystem changes from being natural to unnatural or man-made. Hence, there must be a definition of conservation that includes man as part of the ecosystem.

Conservation is not to be regarded as a negative affair but as a positive, if rather long–term, attempt to set up ecological systems which are viable, stable and productive. (Macfadyen, 1963). This definition suffers, as did Margalef's, from being too restricted. It is essentially management oriented, being concerned with the setting up of ecosystems – does this imply that there is no ecosystem that is natural or unaffected by human intervention? The pre–requisites of this form of management are that the ecosystem should be stable, a concept discussed by Williamson (1972), and that it should be productive. But what is the purpose and aim of this stability and production? It is, for example, known that some of the ecosystems that are in most need of protection are those that are least stable. Should stability therefore be an aim of conservation management?

Conservation is the wise us of the country's resources of land and water and wildlife for every purpose including amenity and recreation (Countryside in 1970). Biological conservation .... is concerned with the maintenance of natural systems and, where possible, with their utilization – either directly or by way of information obtained from their study – for the long–term benefit of mankind (Cragg, 1968). These two definitions share in common the maintenance or wise use of a biological resource and imply management for the benefit of mankind. Both suffer from the subjectively of the first premise; what is wise use to one person may be unwise to another, what is maintenance to one may be considered as over–emphasis or despoliation to another. The statement of the utilization of the resources is also open to question. Should this utilization be for the benefit of man, or should there be more of a two–way flow of benefit between man and the remainder of the biosphere?

A conservationist is a person .... who is concerned that the relationship of Man and Nature evolves in directions that are more beneficial to Man and more acceptable to Nature.

In dealing with the present we must look forward to the future to find better ways of living in and with Nature, bending and shaping Nature to serve our purposes but never breaking her. Conservation must be for people as well as for the higher plants and mammals, birds, fish, soil, vegetation and water (Fisher, 1969).

Fisher's definition of a conservationist and his statements concerning conservation establish his belief in the two–way interaction.

[Man = Nature }

Fisher has therefore attempted a definition which fits into Black's (1970) category. Black argues that such a rationale is a more modern replacement of the older system of Stewardship, based on the interactions

[God = Man = Nature]

This system implied that Man was accountable to God for his actions and for his management of the biological resources. However, the definition of the type put forward by Fisher fails to explain why there should be any care for the environment that posterity will inherit or for their opinion of this generation. Black concludes – There is as yet no fully worked–out and satisfying philosophy of conservation, but only a collection of generalities and catch–phrases. These may certainly be useful if not overstretched, but may turn out to be dangerous in that they incorporate their own built–in mechanism for devaluation.

It is not the intention of this chapter to attempt to find such a philosophy of conservation, but rather to establish a framework within which the remaining chapters of this book discuss some aspects of conservation of mangrove.

Biological conservation is essentially concerned with the interaction between man and the environment. Fulfilment of conservation objectives for a biological resource requires the resource's management in perpetuity on the basis of a sustained production of the resource or biotic component of the environment and a sustained demand on the resource or environment by man.

9.2 MANAGEMENT OF MANGROVE RESOURCES

Because of heavy utilization of mangrove resources in many countries, the opportunity to protect pristine mangrove areas are rapidly disappearing. It is important that Government should act quickly and should establish reserves which may be representative of the untouched mangrove forests of the country.

As regards the managements of the Indian mangrove forests very little attention has been paid. This is because of age–long neglect or the view that mangrove forests are not important to the National economy. However, as the economic potential of the mangrove forests is being recognized on a worldwide basis, the Central Government and many state Governments have been recognized the importance of mangrove and various steps have been proposed to protect and develop this ecological zone.

Each country will need to assemble an expert group to develop its national plan. Some countries have already formed National Mangrove Committees and in many cases these committees will be most appropriate group to assume responsible for the preparation of National plan. Once Government have a National Mangrove plan, they will have the necessary framework to go and to develop effective coastal zone management plans.

In India, Worldwide fund for nature – India (WWF–India) is taking keen interest in mangrove awareness and management. In 1990 WWF–India Maharashtra and Goa states committee of Nehru Seminar on mangrove awareness. Nearly, 200 participants were taken part in this seminars from all over the country. It was unanimously decided to establish, a long awaited, mangrove society of India.

The objectives of this forum was to bring together all those who are interested in mangroves :

(i) To create awareness among various groups of people.
(ii) To initiate action oriented integrated programmes.
(iii) To take–up four prolonged programmes with research and training, conservation, afforestation and awareness.

The work on mangrove management in various parts of our country has been started by various workers with following aims –

(a) Protective function,
(b) Soil conservation and proper utilization of land,
(c) Economic utilization of natural resources,
(d) Fodder and forage supplies,
(e) Maintenance of the cottage industries supported by the forest resources, and

(f) Development of community forestry and selection of suitable plant species.

The mangroves are essentially a forest formations and like other forests are a renewable natural resource. It is thus possible to manage the mangrove ecosystems on a sustainable use basis, although unfortunately this often is not done. The management concept is for the two purposes :

(i) Either sustainable harvest or
(ii) Sustainable economic returns

While at the same time the system can be maintained in as natural or close to its original state as possible. This can be achieved almost solely through preservation. Preservation can be part of sustainable use basis often does not involve added costs–not on a long term basis.

Sustained yield management can be based on :

(i) Management for forestry,
(ii) Management for coastal fisheries,
(iii) Management for foliage and grazing.

Mangrove preservation areas can have many objective in addition to ensuring that essential ecological process and genetic diversity are maintained. The objectives and uses of preservation areas are :

(a) Preservation of essential ecological processes.
(b) Natural restocking of adjacent exploited areas.
(c) Preservation of rare and endangered species and habitats.
(d) Protection of commercial and recreational fisheries.
(e) Establishment of reserves for scientific study.
(f) Establishment of reserves for educational study.
(g) Maximization of recreactional opportunity and preservation of aesthetic qualities.
(h) Protection of coastlines and river banks from erosion and storm damage.
(i) Maintenance of navigation channels.

While, the full list of preservation objectives is not applicable in every situation, it should indicate that preservation areas are valuable economic and social assets and prevent the extinction of species that may be of tangible economic value in the future. The management plan for mangrove is given in Fig. 35.

Sustained yield management of mangroves for timber is neither a new nor a localized practice. The mature trees are clear felled in batches and the timber is removed. If no natural regeneration occurs, these areas are planted. The trees are allowed to grow unattached. Thinning of the trees is carried out once or twice. There are quite a number of advantages in sustained yield management. It is possible to harvest a large amount without disturbing the vital ecosystem processes.

In Andaman mangrove timber exploitation occurs on a very large scale on various parts of Islands, without sustained yield management practices. According to working plan of forest department mangal get to exploitable size in 30 years and fuel yield expected in 35 cards/ha, which means in 30 years the mangal of Andaman, can produce 15, 51, 550 cards. Demand for fuel for industries and public for making charcoal and fire comes to a total 12,000 cards/yr or 3,60,000 cards for 30 years, if there would be no change in socio-economic pattern of these islands than a much less demand that expected production. Thus mangrove of Andaman are in excess for this purpose and can be utilized under proper care and management and management must be based upon a careful blending of ecological and economic principle.

Fig. 35.

(a) Establishing National Plans

In the majority of the countries containing mangrove resources, both the types of plants and animals associated with mangrove forests and the distribution and extent of the mangrove plant communities are poorly known.

Decisions on future use of mangrove ecosystems which are based on inadequate knowledge of the resource may result in unanticipated and irrevocable loss of valuable mangrove resources is to develop a National Mangrove Plan.

The National Plan should –

(a) define the total national resource by means of maps and inventories;

(b) assess people's needs in relation to sustainable uses of the resource while ensuring adequate reserves for preservation purposes;

(c) assess the national and international significance of the resource in relation to:

(i) waterfowl migration,

(ii) genetic reservoirs,

(iii) regional sedimentary stability, and

(iv) marine species migration.

In countries where opportunities to protect pristine areas no longer exist, it will still be valuable to establish reserves in the better quality stands in disturbed areas. Despite the fact that areas are no longer in pristine condition, they may still be essential nursery areas for fish and prawns, stopover places for migratory birds, and important habitats which, if properly managed in the future, will protect rate plants and animals associated with the mangrove ecosystem.

(b) International Co–operation

Aside from considerations for the optimal use of the resources at a national level, development plans of two or more countries may have conflicting interests. Two types of problems arise :

– geographical conflicts between neighbouring countries; and

– conflicts of countries linked through economic interests.

The first type of problem is represented in all those cases where the destructive activity is separated from the affected mangroves by national borders. This applies, for example, to freshwater diversions, upstream mining and logging, or upstream waste discharge.

The second type includes the large scale exploitation of a mangrove forest by a foreign company, or a transnational mining or oil company exploiting the mineral resources in or near mangrove forests. Oil transportation through mangrove–lined shipping lanes also poses a risk for large–scale destruction which may affect the mangrove forests of more than one country.

It is clear that in the first case a national management plan alone cannot provide guarantees for the optimal use of the mangrove resources in question. Here, dialogue between the respective countries is called for, which it is hoped this report will help to initiate.

In the case of destructive resource exploitation through transnational interests, the principles for optimal management on a national basis may also be subject to international compromises. To optimize management in this situation co–operation may be required between national governments and a number of international bodies.

However, such international co–operation requires to willingness on all sides to accept long–term management principles. Some examples are already apparent in various regions through the stimulating assistance of the UN system, e.g. the Regional Seas Programme of the United National Environmental Programme which helps to develop national and co-operative international management schemes for protection of coastal zones, including, where appropriate, mangrove ecosystems. One example of international arrangements for protection of mangroves is the regional oil spill contingency plans.

Summing up, the need for international agreements on national management plans for mangrove resources is becoming increasingly urgent as the influences of widespread development and exploitation activities encroach alarmingly on neighbouring, as well as distant nations.

(c) Policies for Sustainable Uses

The mangroves are essentially a forest formation and like other forests are a renewable natural resource. It is thus possible to manage the mangrove ecosystem on a sustainable use basis although, unfortunately, this is often not done. The concept of sustainable use involves either sustainable harvest or sustainable economic returns while at the same time the system can be maintained in as natural or close to its original state as possible. The latter part tends to be much more difficult to attain except for a few cases (e.g. the use for tourism). Thus, sustainable use often does not mean sustaining the original natural system : this can be achieved almost solely through preservation. Preservation can be part of a sustainable use management plan where it has the added advantage of being placed in a buffered system.

Management on a sustainable use basis often does not involve added costs – not on a long term basis. There is no rational basis for governments not to insist on such a management regime when it has been demonstrated to be possible.

Sustained yield management for coastal fisheries

It is well known that the mangrove forests support the fishery resources within the ecosystem. There is evidence that the mangrove swamps are used as sheltering nursery grounds, as permanent habitat for some species, and as breeding grounds for some coastal species – including commercially important ones. It has been shown that in some mangroves, most fishes feed at least partially on mangrove detrital material. Other groups of aquatic animals depend on planktonic or benthic organisms – a food supply generated within the ecosystem. The species that are fished, trapped or collected in the mangrove creeks, rivers and estuaries comprise finfish, crustaceans (prawns, crabs) and molluscs (e.g. oysters growing on the mangrove prop roots).

The influence of the mangrove ecosystem, however, extends beyond the mangrove forest limits into the coastal waters. Organic detritus produced in the mangroves is exported at high rates into the coastal zone and supports the productivity of these waters. The link and significance between mangroves and its adjacent fisheries is still being investigated. At a number of localities a positive relationship has been found between the area of intertidal vegetation and commercial yields of penaeid shimps. A similar relationship appears to exist with certain finfish species.

If a world-wide link does exist between mangroves and its fisheries, internal and coastal, then this link is mediated through the litter production and decomposition. Under this aspect a sustained production of mangrove litter and nutrient cycling is vital. The influence of this productivity factor on fish populations cannot be denied, and should be given full consideration when allocating the mangrove resource.

Fisheries in the mangroves, i.e. in the rivers and estuaries, are commonly subsistence fisheries for local, small-scale fishermen, fishing below the optimum fishing level.

A management plan for maximum sustained yield fishing should consider the parameters for productivity of fisheries, in addition to fishing effort and stock assessment. The input of nutrients from the mangroves is one of the factors responsible for productivity of fisheries, and it is readily influenced by the manipulation of mangrove ecosystems. Destruction of mangrove forest, direct or secondarily through other activities, will result in lower litter production and consequently influence the productivity of the coastal waters. Even at maintained levels of fishing effort, the diminished fish stocks would then become over-fished. Overfished coastal stocks can, in turn, also affect the mangrove fisheries as breeding or juvenile populations of the transitory species may decrease in size.

Management plans should also recognize the need to attempt to protect any adjacent seagrass beds, tidal flats and tidal marshes as they are important not only to the mangrove ecosystem, but also to the fisheries they help support.

9.3. RESTORATION ESTABLISHMENT OF MANGROVE ECOSYSTEM

The restoration or establishment of mangroves is a management option that can be applied to stabilize unvegetated areas (such as river banks, shorelines, canals, or spoil heaps) or to enhance or regulate natural regeneration.

In general, planting of large tracts for non-silvicultural objectives is best left to nature because of the expense involved. It is possible, however to prepare the terrain and accelerate the recovery of an area by planting in naturally 'bald' spots or thinning dense clumps of arrivals to avoid 'stagnation'. If an area is subjected to stress (e.g., chemical pollution), then for restoration to be successful the source of stress must be largely removed and only tolerance amounts left.

Rhizophora is one of the easiest and most economical genera to use for planting. Propagules may be collected and stored for more than twenty days in a moist container before use, and they can be allowed to develop primary roots without impairing their vigor. Propagules also have been stored in dry rooms for up to two months in the Philippines at the Mangrove Research Centre. Mortality of these seedlings is site-specific and may range from 10 to 70 percent. Although recently collected mature hypocotyls are usually used for planting, in some circumstances there is a better survival rate and more rapid growth if they are grown in a nursery before planting. This allows the seedlings to develop a healthy root system before implantation. The higher nursery costs and increased difficult of planting, however, may offset this advantage.

Seeds of other species (e.g. *Avicennia, Heritiera, Laguncularia, Cynometra*) may be sown by hand. They should be left on the surface and not buried, although gentle pressure may be used to push them into the soft sediments. They must be seeded at a time when the area is not expected to be flooded for at least a week so that the seeds are not dispersed before they become established. It is to be expected that less than 50 percent of the seedlings will become established so it is important to ensure that distribution of seed to the site is repeated as necessary. Use of nursery grown seedlings might be considered for these species.

Small trees of *Rhizophora mangle, Aegiceras corniculatum, Avicennia germinans, A. marina* and *Laguncularia racemosa* (five or more years old, 0.5 – 1.5 m high) have been successfully replanted. Planting trees of this size may be desirable where seedlings may be uprooted and washed away. The larger size of the trees and their more extensive root systems offer a promise of greater and faster shore protection than can be achieved using seedlings. This technique also may be used to salvage individual trees in areas to be used for other purposes, and the trees can serve as a source for restoration purposes in other areas.

Small trees of other species (e.g. *Avicennia marina, Aegiceras corniculatum* and *Rhizophora stylosa*) also can be transplanted. For plants less than 1m high, success rates of approximately 80 percent have been achieved for the *Avicennia* and *Aegiceras* in mixed plantings of more than 50,000 plants.

The survival of larger trees following transplanting appears to be poor. For example, in Peurto Rico, the transplantation of thirteen *Avicennia germinans* and *Laguncularia racemosa* mangrove trees 5 to 13 cm in diameter and upto 6m high resulted in the death of all trees within six months. These trees had been root pruned five to six months before transplantation and were top pruned to 1/4 to 1/3 of their height at the time of transplanting. Moving the trees rather than pruning caused the mortality.

When root–ball and soil–free transplanting techniques for planting small trees are compared, the root–ball method yields higher survival rates (85 percent *vs.* 65 percent after three months) when the trees are planted at natural depths with leaves intact. Leaf stripping resulted in an initial high rate of leaf production but after four months the normal (unstripped) trees had more leaves. Leaf stripping did not seem to affect survival.

Twig and branch pruning before replanting apparently enhances recovery; pruned trees grow better and suffer less defoliation than unpruned ones. Reduced foliage also decreases water loss during the adaptation period following transplantation. Trees or bushes to be replanted should be pruned to two–thirds of their original height.

Unpruned and some pruned individuals of *R. mangle* and *L. racemosa* have exhibited defoliation following transplantation. In *L. racemosa* and *R. mangle* defoliation has occurred within thirty six days after replanting but in *A. germinans* this occurred some eighty–four days after replanting. *R. mangle* suffers little defoliation even if unpruned, and pruning resulted in a 23–percent decrease in growth after transplanting. Pruned and unpruned *L. racemosa* have had difficulty in becoming established, but pruning does increase growth rates of this species following transplanting. Pruned and unpruned *Avicennia* respond well to transplantation although some trees will show partial defoliation.

Care must be taken in pruning *Rhizophora*. If branches greater than 2.0 cm. in diameter are pruned back there will be no sprouting, and the branch will continue to die back. In contrast, *Avicennia* and *Laguncularia* are extremely tolerant of pruning, and this can include major branches as well.

Where individuals are being pruned for transplantation the root ball diameter should be as large as possible, no less than one–half the original (unpruned) tree height. The root ball must be dug from a depth of at least 20–25 cm to reduce damage to the root system, and the root ball should remain intact. In the case of *Rhizophora*, prop roots should be included in the root ball, but if severed they should be left on the tree for possible regrowth.

Replanting should be made at approximately the same tidal level as the original habitat. The root ball should be placed in the prepared hole, watered, and stamped on while replacing soil to seal the space between the root ball and the sides of the hole. During this stage, care must be taken not to burry or damage prop roots or pneumatophores. Slow–release fertilizers may be placed in the hole before planting to aid in the recovery process, although no data are available to demonstrate the value of this practice. The determination of proper spacing intervals should be based on the distance that will minimize early competition. Thinning will occur naturally without human intervention; however, a stand can be managed to avoid periods of slow growth due to competition. By selecting an appropriate spacing, artificial thinning is postponed to later stages of development of the stand.

A planting distance of 0.4 – 0.6 m has been used in plantations of *Rhizophora mucronata* in the Philippines. For mixed stands of *R. apiculata* and *R.mucronata* a spacing of 1.8 x 1.8 m has been recommended. For mixed stands of *Avicennia marina* and *Aegiceras corniculatum* a spacing of 1.5 m has been used in Australia. In general, at a mean tree spacing of more than 1m, self–pruning does not start until the seventh year.

(a) Habitat Selection

Each species should be planted within its tidal and flooding tolerance range and in suitably sheltered locations. In most cases it will be possible to ascertain these site characteristics by observing the growth of mangroves nearby. If this is not possible, a manager must ascertain the elevation of site above the tidal datum and predict flooding depths. Species tolerant of submersion should be planted at the seaward edge (e.g. *Rhizophora mangle, R. stylosa, Avicennia marina, Sonneratia* spp.).

Ideally planting should be behind a protective mangrove fringe or a protective barrier of transplanted small mangrove trees or a native nurse plant species to provide initial wave and current protection (i.e. *Spartina*).

At the landward margins, species tolerant of drier conditions should be used (e.g. *Lumnitzera racemosa, Ceriops* spp., *Excoecaria agallocha, Laguncularia racemosa*).

Air layering has been suggested as a promising technique to provide stock plants for transplanting without removing mangroves from a source area. It is a commonly used horticultural technique in which short segments of bark and phloem are stripped to the cambium and the exposed area is wrapped with sphagnum moss and aluminium foil or plastic to retain moisture. This promotes root development at the griddled area and allows the branch (the layer) to be removed from the parent plant and planted. It has been suggested that this technique should be attempted at the beginning of the wet season to assure exposure to moisture (and in Florida to warm weather). Moisture appears to be a crucial factor determining its success. Drying impairs root formation, whereas too much water may promote fungal and bacterial infections and cause death of the branch above the wound. No branches greater than 2.5 cm in *Rhizophora mangle* have been used in experiments since there appears to be no regrowth on branches greater than that size.

In summary, these investigations found that the girdling was more easily done in *Laguncularia racemosa*, followed by *Rhizophora mangle* and *Avicennia germinans*. However, the technique was most successful with *R. mangle* (rooting success 39 percent) followed by *L. racemosa* (35 percent) and *A. germinans* (6 percent). *R. mangle* produced the longest roots. These roots were firm, had good colour, and often has short secondary roots. *L. racemosa* and *A. germinans* produced smaller roots. *A. germinans* often produced roots off the stem above the layer site. allus formation occurred in 45 percent of the *R.mangle* airlayers, 20 percent of the *A. germinans*, and 15 percent of the *L.racemosa* that had not produced roots.

Additional work is needed to ascertain the practicability of this technique. Among the variables that still need further experimentation are type of wrapping material (aluminium foil, plastic, or other), effect of application of rooting hormones, and whether the plants can be planted directly or need to be stored under mist conditions for some time.

(b) Aesthetic Enhancement of Mangrove Trees and Forests

If restoration has been done near residences, there are many situations in which horticultural practices can be employed to enhance the aesthetic appearance of mangroves and create a panoramic view from inland sites. The practices can be divided into two types based on the ultimately objective. First, individual trees or groups of trees can be pruned, trimmed, or shaped for aesthetic purposes in residential and waterfront settings in urban areas. In general, standard horticultural techniques can be employed without any modification or concern about the foliar response of the trees. The only qualification is that some species of mangroves have regenerative limitations which must be taken into account during the pruning or trimming.

In the second type of horticultural management, the objective is to reduce the height of a afforested area for a panoramic view of adjacent areas. In general, this objective is met through the topping or removal of selected trees to stimulate coppice regrowth or rapid growth and development of seedlings and small saplings. When the coppice and new sprout growth form a full but lower–level foliage, the remaining trees can be topped or removed. Annual pruning and trimming of the new growth maintains the vegetation at a desired height. Based on the regeneration characteristics of the species, different strategies are employed to achieve the final result. For example, *Rhizophora* can not survive a complete removal of branches and foliage. However, when an abundant propagule source is available, total removal of selected trees (approximately 50 percent of population on site) is recommended to stimulate seedling or sapling growth. In the event of an inadequate propagule source (unlikely in most areas) or where there is a high probability of competitive domination by other mangrove species, the *Rhizophora* should be pruned or cut back in stages over several years.

An advantage afforded by the ability of mangroves to be managed using standard horticultural practices is that certain objectives, albeit limited, can be attained without the destruction or total removal of the mangrove forest. The practice has been clearly demonstrated and may be considered and alternative to the complete removal of mangrove forests for competing aesthetic reasons.

(c) Suggestions for Mangrove Restoration

Before attempting to restore a particular area, a manager should consider the following :

- Select target sites carefully. If the area has contained mangroves before, determine the cause of loss and make sure the stressor has been removed. Naturally blank areas may be hypersaline or too elevated for successful establishment. Natural succession and accretion may have rendered the site unsuitable for the previous mangrove species.
- Select the species to plant on the basis of the dominant species at nearby locations having similar elevations, frequency of inundation, and exposure. Because of their ease of handling and planting species of the Rhizophoraceae (*Rhizophora, Bruguiera, Ceriops* and *Kandelia*) are the most suitable species for restoration projects, but they may not always be appropriate.
- Be sure, soil elevation is low enough so the area if flooded frequently by tidal water and there is adequate circulation. Stagnant areas tend to overheat.
- Measure interstitial (soil) salinities. Interstitial salinities will determine the most suitable species. For example, areas above 55 percent are not suitable for *Rhizophora mangle.* Salinity tolerance for some mangrove species may be found in Snedaker and Brown (1981).
- Shelter areas from wave action and current scouring.
- Remove dead standing wood and clean-up the area. Floating or falling debris will uproot and damage planted seedlings.
- Control the spacing of seedlings. Dense plantings result in slower growth due to competition. For *Rhizophora* spp. the recommended spacing for seedlings is 0.6 – 1.2 m, and for mixed *Avicennia marina* and *Aegiceras corniculatum* the spacing recommended is 1.5 m.
- Collect recently fallen propagules from nearby mangrove areas or from trees. Use only ripe propagules. For red mangrove (*R. mangle*), use those propagules in which the abscission layer has developed. (These are easily detached from the parent tree). Select healthy propagules and discard malformed or damaged ones. Maintain the collected material wet and avoid exposing it to overheating during transportation and before planting. Propagules can be carried in a foam ice box or a wet burlap (hessian) sack.

- Do not insert propagules too deeply into the substrate. For example, *Rhizophora* should be planted only deep enough that it will not fall over. For a 20–30 cm seedling the planting depth should be 4–7 cm.
- Use natural primary and secondary succession processes to improve planting success (e.g. use of nurse plant cover).

9.4 ECONOMIC CONSIDERATIONS IN MANGROVE MANAGEMENT

There are basically three options for the management or development of mangroves. The first option is *preservation* of the ecosystem in its natural state. The second option is *utilization* of the system to extract various goods and services on a sustainable basis. The last option is the *conversion* (or destruction) of the natural ecosystem, usually for single replacement use. Since these three management options are mutually exclusive, a framework is needed for evaluating the alternatives; economic analysis has a role to play in this process.

In practice, ecological and economic considerations cannot be separated in evaluating management alternatives for mangroves. This statement reflects the growing appreciation of the social and economic importance of mangrove ecosystems and reverses the once–common view that mangroves were worthless unless developed for major single purpose activities or converted to alternative uses. To measure the value of mangroves accurately, the value of the goods and services produced by these systems needs to be considered and incorporated into the assessment of the relative merits of development alternatives.

This section explores the economic characteristics of mangrove resource systems and discusses the use of economic analysis in weighing the costs and benefits of alternative forms of mangrove development. Although it may not be possible to put a monetary value on all relevant factors, they must be explicitly recognized and be incorporated into an economic valuation method to the extent possible. If planners and mangers are aware of the uses and limitations of economic assessments of complex biophysical resource systems, they can help improve the expression of the social and economic significance of these systems to society. In the actual decision–making process the results of the economic analysis are normally evaluated and weighed along with political, social, cultural, environmental, and bureaucratic, factors.

Mangrove ecosystems are difficult to value adequately using conventional economic analysis techniques. The difficulties arise because of two separate phenomena : first, many of the goods and services produced by a mangrove ecosystem are not easily monetized, and second, many of these goods and services occur off–site, that is, they are external to the physical system and thus become economic externalities.

If the full range of goods and services produced by mangrove ecosystems are included in an economic analysis of various management alternatives, quite different decisions may be made about how best to manage these unique ecosystems. The first task is to identify the important parameters involved and then discuss how they can be explicitly incorporated into an economic analysis.

Economics, the science of the allocation of scare resources, relies heavily on observed market prices to place values on various goods and services. This has been the conventional way that agricultural or natural resource development questions are analyzed. The values of inputs required and the resulting goods and services produced are compared to determine if a proposed activity is 'profitable'. When several options exist for use of the same basic resource (a field, a forest, a mangrove swamp) the net benefits of the various options are compared to determine the most profitable way to use the resource.

There are several problems with this approach when mangroves are the resource under consideration. The first, and a point that will be the focus of this section, is that only a few of the goods and services produced by the ecosystem are usually included in the analysis. For example, a decision on whether to fill in a mangrove area for housing development that is based only on the value of lost charcoal production may be very different than if the value of fish caught in adjacent waters but dependent on the mangrove for part of their food chain is included. Excluding the fish or shellfish in the analysis will understate the true cost of destroying the mangrove area.

Another major aspect of this problem is that an analysis done from a *private* point of view (a *financial* analysis) may well give different results from similar analyses conducted from *society's* perspective (*social* or *economic* analysis). Since the wide range of goods and services produced by mangroves should be included in an analysis of development alternatives, the social or economic analysis is often the appropriate framework.

In some cases, mangroves are privately owned. Private owners are usually concerned only with the commercial profitability of the use of the mangroves they control. As such, in assessing the relative metris of different forms of mangrove development, they are primarily concerned with the financial aspects of the funds, material, and labour required and the resulting goods produced and the prices these goods receive. This narrow *financial analysis* ignores many of the off–site or nonmarketed goods and services of a mangrove ecosystem.

Public bodies are also concerned with the financial aspects of development projects; however, their perspectives on the relative costs and benefits of alternative projects is more broadly related to the economic impact upon society as a whole. Society generally seeks to allocate its resources in a manner that makes the greatest possible contribution towards satisfying its needs and wants for goods and services. Therefore, factors that often are excluded from financial analyses by private owners–such as the off–site environmental effects of the pollution they may create – should be considered by public bodies in an *economic analysis* as described in this section.

There are, of course, direct parallels between financial analyses and economic analyses, but there are significant differences in the manner in which costs (inputs) and benefits (outputs) are identified and incorporated into the respective analyses. There are also may be differences in the way the costs or benefits are valued. In financial analyses, inputs (costs) are measured in monetary terms expressed as the price of goods and services purchased in a market. The outputs are also valued in terms of market prices. An economic analysis uses consumer willingness to pay as a measure of value; in reality such an analysis frequently uses market prices of inputs and outputs. However, additional concern is given to weighing what society gives up and what society receives in return from a specific form of resource development This concern may be reflected by the use of "shadow prices" (adjustments to market prices to correct for distortions caused by government intervention or regulations) or by quantifying public opportunity costs.

The relationship between the steps in a financial and an economic analysis are illustrated in Table 56. Although developed for economic analysis of forestry projects (Gregersen and Contreras, 1979), this comparison is valid for the analysis of mangroves or other natural systems.

Table 56. Relationship between steps in a financial and an economic analysis

Financial Analysis	Economic Analysis
Develop Physical Flow Tables (Inputs and Outputs)	
Direct inputs provided by the financial entity and outputs for which the entity is paid are included.	In addition to direct inputs and outputs, indirect effects are included, i.e., effects not included in the financial analysis since they are not directly traded in a market. These are effects on others in society.
Develop Unit Value Tables	
Market prices are used for inputs and outputs that occur in the future. Future market prices are estimated.	Consumer willingness to pay (w.t.p.) is used as the basic measure of value. In cases where market prices adequately reflect w.t.p. such prices are used. In other cases, 'shadow prices' are estimated to provide the best measure of w.t.p.
Develop Cash Flow/Economic Value Flow Tables	
Inputs and outputs are multiplied by market prices to arrive at total costs and returns, which are then entered in the cash flow table. Transfer payments (such as taxes, subsidies, or loan transactions) are added to the cash flow table.	Inputs and outputs are multiplied by unit economic values to arrive at total economic costs and benefits, which are then entered in a total value flow table. Transfer payments are not treated separately but are included only as part of the economic costs or benefits as appropriate.
Calculate Measures of Project Worth	
Using cash flow table, calculate chosen measures of project value or commercial profitability. Test results for uncertainty by varying values of key parameters in a sensitivity analysis.	Calculate chosen measure or economic efficiency or economic value using the information in the total value flow table. Test results for uncertainty by varying values of key relationships/parameters in a sensitivity analysis.

Although this section concentrates on the construction of a broadly based economic appraisal, it is important not to lose sight of financial considerations that may govern the decisions of private owners. A mangrove development programme that is considered economically efficient from a national viewpoint may not be attractive from a private viewpoint. Or more likely, a development that is seen as attractive and profitable from a private perspective may have major negative externalities and therefore be undesirable from a social point of view. In this case, controls, taxes, or subsidies would be required to achieve the socially efficient and desirable outcome.

This section presents a framework for an economic analysis of a mangrove ecosystem. The details of how shadow prices are derived or the technicalities of what to include or exclude in an economic analysis are not covered here. A trained economist should be able to carry out such an exercise, and various texts on project evaluation and natural resource economics provide detailed information on doing an economic analysis (for example, see Gittinger, 1982, Hufschmidt *et al.*, 1983, Squire and Van der Tak, 1975, or Randall, 1981).

Economic Analysis of a Natural System

Mangrove are just one of many natural systems that are utilized or exploited for the benefit of society. As they would for any of these natural systems, the decision makers want to know what will be gained and what will be lost as the result of any decision. Because of the unique nature of natural systems (when compared with a factory, for example), new approaches have been developed to value the benefits and costs properly. Much of the recent work in environmental and resource economics has been directed to these problems.

Chapter - 10

MANGROVE UTILIZATION

10.1 INTRODUCTION

In present industrializing world, there is a tendency to presume that any services can be provided if funds are available to pay capital and maintenance cost, it has become common practice to use engineering services to provide basic need required by a coastal urban community such as storm protection, erosion control, waste water clean-up and a variety of activities. In coastal areas, mangrove ecosystems provide all these services required by the coastal communities. Number of examples are there where mangrove ecosystems saved the areas from storms and cyclones. In the Bay of Bengal cyclones pose a major threat to human beings and property in India and Bangladesh. However, the loss of life and property is consistently lower in the areas of the Ganga and Brahmaputra Deltas (heavily forested by mangroves) than in similar areas to the east or west. In southern Florida, U.S.A. and some other countries the mangrove are purposely planted to minimise the shore–line erosion and provide fisheries, habitat for local sports. In addition to the multiple and sustained free services provided by mangrove ecosystems a range of direct and indirect useful products from mangrove are listed.

While many people are concerned about the accelerating pace of mangrove destruction, many others regard this resource as wasteland which ought to be converted to other forms of land use as rapidly as possible. This divergence of opinion arises because the uses and values of the mangrove system are many and extremely important, but are relatively little appreciated by decision makers, developers and many citizens, especially urban dwellers.

The importance of the resource derives both from the products taken directly from the mangrove systems, and from the amenities provided by the resource from within and beyond its boundaries.

Products taken from the mangrove ecosystem range from construction materials to reptile skins and honey. A partial list of products in their processed form is given in Table 57. Currently, mangroves in developing countries are being harvested and chipped for pulp and particle board on a very large scale by companies from developed countries. In addition, mangroves are used as an important and potentially sustainable source of fuelwood and charcoal to meet the increasing needs of developing countries for domestic fuel. Interest also has arisen in species such as *Nypa* palm, which produces alcohol that can be turned into transport fuel. Even though incomplete, Table 58 shows the substantial number of direct tree and shrub products.

Some grazing occurs in mangrove areas. For instance, cattle, camels and goats in India graze in the mangroves, camels are put onto mangrove islands in Pakistan, and caged sheep are grazed in Indonesian mangroves. Feral water buffalo in substantial numbers graze mangroves in northern Australia.

Other natural products are harvested from mangroves as indicated in the Table 58. Crabs are very common on the mud flats in most mangrove ecosystems and are often a very important subsistence of even commercial food source. In some systems, edible shellfish are supported on the roots and trunks of mangroves.

These direct uses often sustain communities whose economy is based on harvesting the fish, shellfish, crustaceans, wood and other minor products which may be gathered. The continued viability of these ecosystems and the well-being of these people depend on managing the resource in a sustainable fashion. Locally important industries, providing rural employment, are also based on the mangrove resource.

Mangroves reduce coastal erosion. They serve to dampen storm surges and to a minor extent high winds, both of which are associated with many tropical and subtropical storms. While the mangrove coastal barrier may be battered and damaged in serve storms, it will grow back naturally, without cost to man. No man-made coastal protection barrier is capable of self-repair.

The mangrove resource, where it occupies estuarine flood plains, performs a flood reduction function which may be lost if the area is filled and converted to other uses. Mangroves lining the banks of rivers also help prevent erosion of the river banks, which in turn helps protect adjacent property.

The mangrove area is a spawning and nursery area for many marine species of fish. Moreover, the particles of vegetation (detritus) and nutrients exported out of the mangrove ecosystem form the food base of a complex of marine organisms which, in turn, support valuable estuarine and near-shore fisheries (finfish, shellfish and crustaceans). Those whose livelihood depends on fishing have long recognized the interconnection between mangroves and fisheries, but these values have only slowly been considered in planning processes where slowly been considered in planning processes where decisions on allocations of intertidal land are being made.

The interesting and unusual fauna and flora of the mangrove community, particularly the bird life which feeds or shelters there, provides valuable opportunities for education, scientific study and tourism. For instance, thousands of visitors per year go to Trinidad's Caroni Swamp mangrove area specifically to see the very great numbers (as many as 20,000 birds) of scarlet ibis (*Eudocimus ruber*), and other birds which are rare or endangered. The endangered Bengal tiger (*Panthera tigris*) is dependent on the Sundarbans mangrove area in India and Bangladesh. In Borneo, the proboscis monkey (*Nasalis larvatus*) is a mangrove inhabitant. These are but a few examples. While it is difficult to put monetary value on these wildlife based activities, they are nonetheless significant uses which add to the importance of mangroves.

Table 57. Direct Products from Mangrove Forests

Uses	Products
Fuel	Firewood for cooking, heating, firewood for smoking fish, firewood for smoking sheet rubber, firewood for burning brick charcoal, alcohol.
Construction	Timber for scaffold, timber for heavy construction (e.g. bridges) railway ties, mining pit props, dock pillings, beams and poles for buildings, flooring panelling boat-building materials, fence posts, water pipes, chip boards, glues.
Fishing	Poles for fish traps, fishing floats, fish poison, tannins for net preservation fish attracting, shelters.
Agriculture	Fodder, green manure
Paper Production	Paper of various kinds
Food, Drugs and Beverages	Sugar, Alcohol, cooking oil, vinegar substitutes, fermented drinks; dessert topping, condiments from bark, sweetment from propagules, vegetables from propagule fruits or leaves for cigarette wrappers, medicines from bark, leaves and fruits.
Household items	Furniture, glue, hair dressing oil, tool handles, rice mortar, toys, match sticks, incense.
Textile and Leather Production	Synthetic fibre, dye for cloth, tannin for leather preservation.
Others	Packing boxes

Mangrove systems are complex and diverse and are important to many human populations. The complexity exists not only in the conventionally defined biosphere, but in the broader sphere of human–mangrove interaction.

The mangrove system serves a multitude of functions and there are many other competing demands for these resources. Foresters believe that if mangrove from mangroves can be maintained. Fish farmers convert them into fishponds and environmentalists want to conserve the area for the protection of the many dependent organisms. The increasing pressures of population growth will force further the direct and indirect use of mangrove products, and the encroachment of man into mangrove areas.

Mangrove forests have been depleted and subjected to various stresses originating from human activities. Man influences the growth and development as well as death and extinction of mangroves. With such a dynamic and useful resource as the mangrove, conflicts among the forestry and fishery sectors, and the mangrove dwellers are almost inevitable.

Decisions on the development, management and control of mangrove systems, require an understanding of the perceptions of the many users and managers of the resource.

For example — in many mangrove areas, particularly in South–East Asia and India, many of the people inhabitating them, or atleast living ner them, and collecting mangrove products are those who may be called 'living below the poverty threshold'. For these people, the basic thing is survival and they would do anything just to survive. Most of these people do not have alternative sources of income other than the direct monetary gain derived from the mangroves. Many of them are ignorant of the other values of mangrove ecosystems and of the consequences of increased exploitation.

Table 58. Mangrove products from woody formations and associated fauna

FUELS
- Firewood
  - for cooking
  - for smoking fish
  - for smoking rubber sheets
  - for burning bricks
- charcoal for cooking
  - for grilling
- lighting oils
- alcohol

BUILDING MATERIALS
- timber for construction
- scaffolding
- beams and poles for buildings
- rail sleepers
- timber for pit props
- timber for bridge building
- timber for boat building
- timber for quay construction
- timber for water pipes
- floorboards
- timber for wall filling
- shingles
- woven matting
- fence stakes
- chipboard

MATERIAL FOR FISHING
- poles for fish traps
- poles and beams for new cages
- timber for rafts and floats
- piscicides
- tanning material for impregnating
- nets, lines and ropes

MATERIALS FOR TEXTILE & LEATHER PRODUCTION
- fibres for weaving
- dyes for textiles
- tannins for leather production
- hides
- furs
- synthetic silk (Rayon)

PRODUCTS FOR AGRICULTURE
- fodder
- green manure

PAPER & PAPER RAW MATERIALS
- lignocellulose
- wood chips
- fibres for kraft paper
- miscellaneous papers

HOUSEHOLD GOODS
- furniture
- mortars and pestles
- handles for household items
- matches

FOOD & LUXURY FOODS
- sugar
- edible oils
- alcohol
- vinegar
- alcoholic beverages
- tea substitute
- spices
- vegetables from seeds, fruits and leaves
- confectionary from seeds
- cigar substitutes
- cigarette paper
- honey
- fish
- crustaceans
- mussels
- meat from mammals & reptiles
- eggs and meat from birds

CHEMIST'S GOODS
- wax
- aromatic substances
- skin care products
- hairdressing oil
- glue
- mosquito repellents
- substances against hair loss

OTHER PRODUCTS
- packing cases & baskets
- tar
- polished respiratory roots as souvenirs
- pharmaceuticals

Fishpond operators, on the other hand, look at the profits derived from converting the mangrove into fishponds.

Policy makers contribute to the exploitation as well as conservation of mangroves; for example, in line with the need and the government's desire to increase fish production, some governments have offered financial assistance to clear mangrove swamps for fishpond culture. At the same time, however, they may promulgate policies that can prevent unwarranted destruction. While extremely little quantitative information exists on the distribution, intensity and magnitude of uses of mangrove ecosystems. Unfortunately, this is not complete for each country having mangroves. It proved very difficult to obtain even this level of information. It is thought that the greatest amount of use is occurring in South–east Asia, where perhaps half the total mangrove area is presently being commercially exploited. Large forest clearfelling operations are being conducted in Malaysia (Sabah and Sarawak), Indonesia (Sumatra, Kalimantan, Sulawesi), and Bangladesh for either paper pulp or cellulose derivatives such as rayon.

10.2 SPECIES DIVERSITY AS A BASIS FOR DIVERSITY OF UTILIZATION

The basis for the traditional diversity of utilization is the horizontal and vertical differentiation of the mangrove forests. The horizontal differentiation is expressed in clear zonation as the result of different duration of water submersion and corresponds to the transition from the marine to the terrestrial habitat, in which more and more marine elements decline and more and more terrestrial elements increase from the sea further inland. Vertical differentiation is expressed in a clear layered structure, in which seven faunistic habitats are arranged one above the other:

- the canopy of the trees as a habitat for birds and insects,
- the trunk and branch area, where rainwater collects in the branch holes and axils, making a fresh–water supply possible above the salt and brackish water, as a habitat for mammals, reptiles and birds,
- the subaerial soil surface as a habitat for mammals, reptiles and snails,
- the permanently or periodically filled ponds as a habitat for crabs, frogs and shrimps.
- the mangrove channels and tidal canals as a habitat for pelagic fishes and reptiles.
- the submarine soil surface as a habitat for benthic organisms like demersal fishes, shrimps, crabs and molluscs,
- the submarine soils as a habitat for mud–burrowing molluscs and worms.

With the changing tides the levels shift periodically. The multistorey structure and the zonation make the high species diversity possible.

The species diversity has its counterpart in the diversity of utilization. The demonstrable uses are so highly differentiated and versatile that they can only be shown in a table. For the wood formation alone there is an enormously wide range of products from traditional and current exploitation.

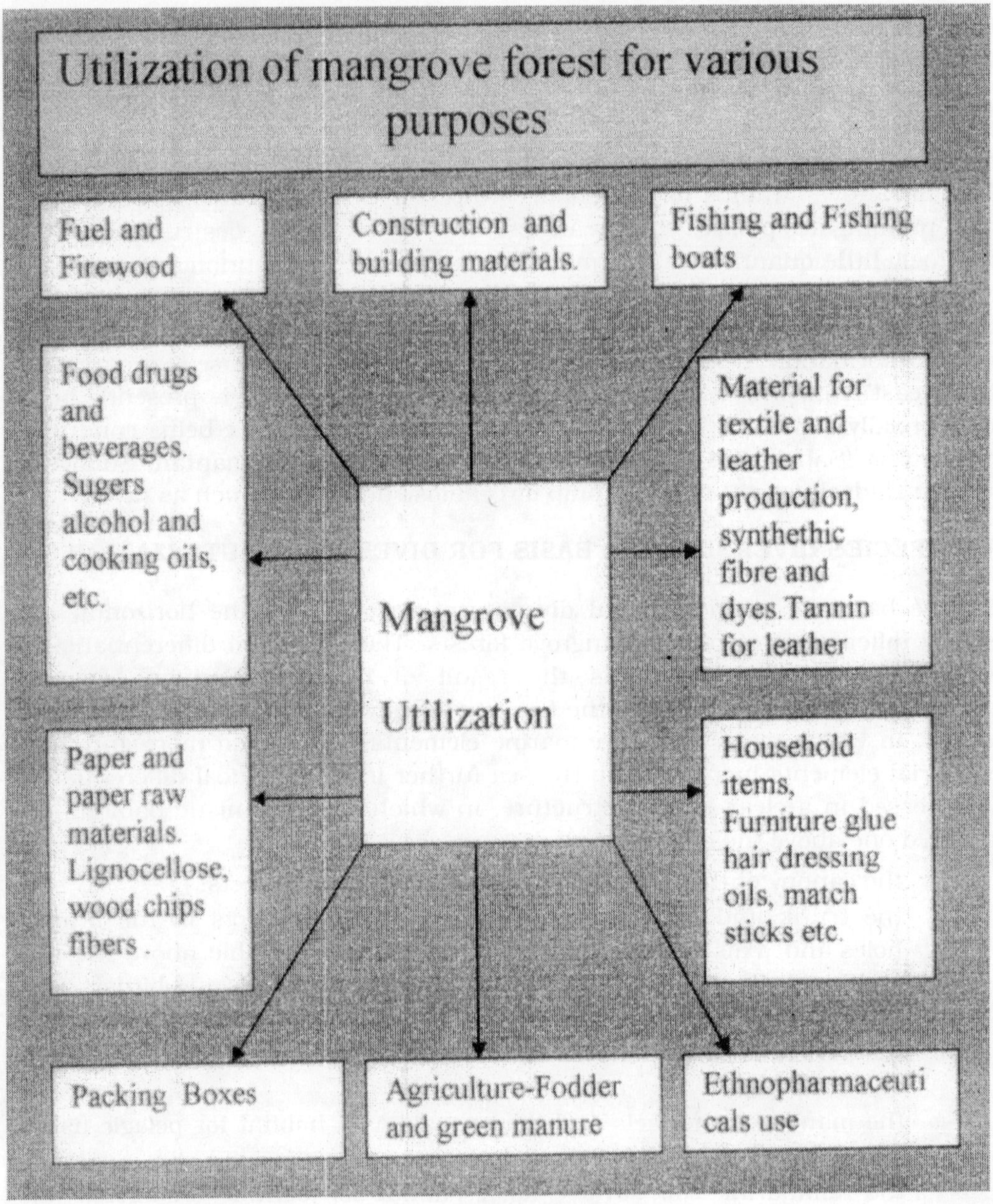

Fig. 36.

The substances contained in the mangrove woods are also used as ethno–pharmaceuticals in traditional medicine. Pounded or pulverized, as extracts, infusions, juices, baths and inhalers, tree bark, roots, leaves, seed resin and heartwood of individual mangrove species, they are used as remedies especially for treating wounds, because of their anti–inflammatory, haemostatic, wound–cleansing and antiseptic properties, and also against chronic illness. Occasionally seeds (*Xylocarpus granatum*) and leaves (*Excoecaria agallocha*) are taken directly. Pharmaceutical analyses of the effects have so far been largely lacking. All information is based on interviews with dwellers in the mangrove areas and thus on traditional practical experience. Some of the ligneous plants occurring in the mangroves can be classed with certainty as medicinal plants.

Table 59. Selected ethnopharmaceuticals from mangrove forests

Mangrove species	Parts used	Preparation	Treatment
Heritiera littoralis	seeds		diarrhoea
Rhizophora mucronata	bark	infusion	diarrhoea, dysentery
Ceriops decandra	bark	infusion	diarrhoea, dysentery anti-emetic
Xylocarpus granatum	seeds		diarrhoea, cholera
Bruguiera parviflora	heartwood	extract	constipation
Sonneratia caseolaris	fruit	sap	skin protection
Rhizophora mucronata	fruit	sap	insect repellent
Rhizophora mucronata	root cortex	extract	insect repellent
Xylocarpus moluccensis	seeds	extract	insect bites
Avicennia spp.	bark	extract	skin parasites
Clerodendron inerme	leaves	extract	skin parasites
Acanthus ilicifolius	bark, root	bath	skin diseases & allergies
Thespesia populnea	fruit & leaves	ointment	scabies
Acanthus ilicifolius	bark	extract	small pox
Avicennia alba	seeds	ointment	small pox
Avicennia spp.	resin		tumors and ulcers
Ceriops tagal	bark	extract	wound cleansing
Xylocarpus moluccensis	seeds	extract	wound cleansing
Acanthus ilicifolius	fruit pulp		snake bites, blood cleansing
Excoecaria agallocha	wood	smoke inhaling	leprosy
Rhizophora mucronata	bark	infusion	leprosy
Excoecaria agallocha	leaves		epilepsy
Acanthus spp.			kidney stones
Ceriops tagal	bark		haemorrhage
Acanthus ilicifolius	leaves	ointment	rheumatism
Cerbera manghas	fruits	sap	reheumatism
Bruguiera parviflora	heartwood	extract	stomatitis
Clerodendron inerme	roots	extract	hepatitis, chills
Bruguiera gymnorrhiza	fruits	sap	eye diseases
Hibiscus tiliaceus	blossoms	infusion	infection in ear canals
Avicennia officinalis	roots, heartwood	extract	stimulants and tonics
Avicennia alba	resin		contraceptive, birth-control

The products range of the mangrove ecosystem becomes even broader when the animal products are also included. Some of them individually attain high market values. Among the crustaceans they include the giant tiger prawn (*Penaeus monodon*), the somewhat smaller banana shrimp (*Penaeus merguiensis*) and the mud crab (*Scylla serrata*). Fish use the mangroves as spawning grounds, nurseries, feeding areas, shelter or permanent habitats, depending on species, and accordingly inhabit the tidal forests temporarily, permanently or in particular phases of their life cycle. Economically important species are the milkfish (*Chanos chanos*), mullets (*Mugilidae*) and various tropical perch species (*Epinephelus* spp., *Lates calcarifer*). Among the molluscs the oyster (*Crassostrea commercialis*) has the highest market value. It lives wild on the tidally flooded stems and roots. Marine snails of the family *Littorinidae* are used commercially too, as are the cockle (*Anadara* spp.), which inhabits the mud flats lying off the mangroves.

Because of its wide variety of uses the Nipa palm (*Nypa fruticans*) occupies a special position within the mangrove ecosystem. Nipa occurs especially on the landward edge of the mangrove areas where there is little periodic water covering. It is only to be found in the tidal forests of South and South–East Asia. Nipa is often described as the most valuable plant within the mangrove and is occasionally cultivated in plantations.

The fronds are processed into shingles used in the countrywide to thatch houses and as outer wall elements. One ha of nipa plantation provides about 15300 of these palm–leaf elements each year (FAO, 1982). But mats, baskets, hats and rain caps are also woven from nipa fronds. The leaf shoots serve as a substitute for cigarette paper. Young leaves are used to wrap food, while the ribs are used as fuel. The sap of young inflorescence is tapped for sugar production, alcohol distillation and vinegar production. The 'arak' of Indonesia and the local 'whisky' of Sarawak are nipa distillates. There are comparable drinks in the Philippines and New Guinea. One ha of nipa plantation provides 3000 l of pure alcohol per year (Fong, 1980). The fruits too are used for the extraction of alcohol (Aksornkoae, 1993). The soft endosperm of the fruits is edible and is highly prized in Thailand, Indonesia and the Philippines. The hard shells of the ripe fruits are used to make buttons.

10.3 SELECTIVE MANGROVE UTILIZATION AS A CAUSE OF QUALITATIVE DEGRADATION

What was characteristic of the traditional utilization of mangroves was its variety. The tidal forests have wide–ranging potential which a few decades ago was fully used, not quantitatively but qualitatively. It allows largely subsistent management. One deficiency was the seasonal absence of freshwater, especially in the seasonally dry locations in Southeast Asia, which was successfully overcome by storing rainwater in large–capacity vessels for a short while. The flora and fauna, given a low human population density, provided an adequate nutritional basis, guaranteed the energy supply and provided the raw materials needed for building houses and producing household equipment and for making boats, fishing and catching gear.

Traditional dwelling houses are almost exclusively constructed of mangrove. The various building elements are made out of the various species depending on the specific requirements. *Bruguiera, Excoecaria* and *Xylocarpus* provide the posts for the working platform and the building. Hardness and resistance to salt water are particularly sought–after properties of the timbers used for the base of pile dwellings. *Avicennia* and *Rhizophora* provide the constructional beams for the walls and roofs. The floor boards and the platform are made from *Sonneratia* and *Bruguiera*. The roof and outer walls consists of *Nypa*, the fronds of which are woven into shingle like elements. The internal walls are frequently made from *Avicennia*. Because of its attractive textures the timber from *Xylocarpus* and *Avicennia* are chosen for furniture production. Stand diversity is the basis for the traditional pile building in the mangroves.

The high degree of self–sufficiency of the population in the mangrove forests of Southeast Asia was made possible by the great range of uses with a high diversity of species. Goods which were lacking, for example river, were obtained by exchange for valuable products from the system (fish, shrimps, crabs or firewood). Exploitation was well adjusted, with extraction of small amounts, spread over many products. Interventions were confined to the local level. The regenerability of the ecosystem was never threatened by this kind of extraction.

In the course of time until recently, many of the kinds of utilization geared only to self–sufficiency have been dropped. Market integration is displacing subsistence. Products once extracted from the mangroves are being replaced by purchaseable goods. Conversely, within the mangroves the extraction and increasingly also the production of individual marketable goods is increasing. Utilization diversity is becoming poorer and is being replaced by selective use. The product range is being reduced to fuel wood, charcoal, wooden piles and valuable round timber, sugar and alcohol and to marketable living resources such as valuable fish, prawns, crabs and oysters.

In order to guarantee production in the long term, regulatory interventions are made in the stands. The species providing valuable products are encouraged while others are thus being suppressed. In the whole of Southeast Asia it is the *Rhizophoraceae* that are preferred on account of their versatility of application and their outstanding burning properties. It is species of this family that are mainly used as reserved trees for the natural regeneration of clearings and also for afforestation measures as far as zoning allows. Differentiated wood formation is being replaced by *Rhizophora* monoculture or, where sugar and alcohol are to be produced, nipa cultivation. The range of species is thus enormously restricted. The ecosystematic connection between the site–specific woody plants and the marine and terrestrial fauna is interrupted. The stage of more or less advanced degradation has been reached.

This development has taken place gradually. It began with the sale of coveted market products on neighbouring markets or to middlemen. Interventions in the mangrove forests were limited to the neighbourhood of ready markets. Areas distant from the market were initially unaffected and as a rule remained largely untouched. The results of this partially market–oriented mangrove management are the degradation of stands locally and partial overexploitation.

The following phase of complete market integration was initiated by the granting of land use and ownership rights of large areas to outside users and the extended state use by the forestry economy. Usually the only economic goal is the extraction of wood (for charcoal production, poles and chippings). There are wide regional differences in the exploitation systems. They range from pure extraction to sustainable cultivation. The regeneration of the forests takes place, depending on intensity of exploitation, by natural regeneration (hold–over trees, strip felling) or by afforestation. Independently of the utilization system what emerges is tidal forests with little species diversity. The final stage is *Rhizophora* monocultures. The traditional forms of exploitation are displaced. The mangrove settlers, only very few of whom have legal possessory titles, become dependent wage workers. With the reduction to few industrially exploitable intertidal commercial plants the mangrove ecosystem is partially destroyed by degradation. Many of the system–immanent functions are lost. Intertidal commercial forests remain. However, with exceptions, these are also still spawning grounds, growing areas and permanent habitats for fish, crustaceans and molluscs, and above all feeding grounds for commercially important fish species in their early stages of development. Thus within and especially on the seaward side of the Matang Mangrove, which was taken under management as early as 1908 (Larut–Matang District on the western coast of West Malaysia in Perak State), and is regarded as the oldest commercial mangrove forest in Southeast Asia, high yield coastal fishing is still possible. It has been supplemented recently by the development of aquacultural enterprises for cockles (*Anandara granosa*), sea perch and prawns (*Panaeus monodon*). The close relationship between the fishing exploitation and the functions of the mangroves have been repeatedly documented, although they have been greatly degraded in their species diversity by sustained exploitation (Sasekumar, 1993; Choy, 1993; Sasekumar *et al.*, 1993). Also the function of the mangrove as a biogenic coast and bank protection forest is largely preserved after its transformation into intertidal commercial forest, but its effectiveness is reduced through clearings and areas of young growth.

The destruction of original mangrove ecosystems by degradation, where there is complete market integration, is no longer confined to areas close to the market. With the development of the transport infrastructure and the improvement of transport facilities by land and sea, even peripheral mangrove forests are no longer safe from degradation. Especially felling for the production of chips has recently also integrated the peripheral outer islands of Southeast Asia, Borneo, Sumatra and New Guinea into an export oriented mangrove exploitation.

The degradation of the mangrove forests can be traced especially to the timber industry and clearly to a far lesser extent to the setting up of nipa plantations. The most important products in Southeast Asia are firewood, wooden posts, charcoal and wood chips. The goals of production vary regionally.

Fuel wood from mangrove species is still used in all countries in Southeast Asia, mostly only locally, and in the Philippines, in Vietnam and in Java in rather large quantities, especially near the mangrove areas. The often indiscriminate cutting down of young trunks and branches provides the mangrove settlers with an important source of income. The once important export from Sabah has stopped because of the high costs of transport and competition with the fossil fuels. The wood of all mangrove species is used. *Rhizophora* wood is particularly highly prized for cooking because its wood produces little smoke, has a high calorific value and burns very evenly. The light wood of *Avicennia* is particularly popular for smoking fish because of its low heat value and its pure, white smoke. Firewood from mangrove is used industrially in some countries for the production of sugar from the sap of the nipa inflorescence and from coconut palms, for distilling alcohol and burning lime. This type of exploitation, however, remains confirmed to an extremely small geographical radius of exploitable tidal forest.

Wooden posts of very hard mangrove woods (*Excoecaria, Bruguiera, Xylocarpus, Rhizophora*) are used for pile foundations, the construction of standing and floating cages and standing traps and for scaffolding. The hardness and the resistance to salt water and parasites are much sought–after properties. The production of wooden posts takes place mostly on the basis of cultivation schemes. In the 407 km^2 large Matang Mangrove of West Malaysia, where besides supplying fuel and charcoal there is concentration on post production, in a management cycle of 30 years the stands are thinned out three times. On 2000 ha in Larut Matang District 3 to 4 million posts are produced each year (FAO, 1985; GAN, 1993).

Charcoal is the most important product of the mangrove forests in Thailand. About 90% of the felled timber is used for charcoal production (Aksornkoae, 1978b). The raw material comes from stands that are meanwhile commercially used and qualitatively degraded. In West Malaysia too, and in Vietnam and Indonesia, specifically in the Riau Archipelago and in Sumatra, charcoal is one of the most important products from the mangrove forests regarding volume and gains. Charcoal production began there as early as 1887 during the Dutch colonial time. The first cultivation schemes were worked out in Central Java in 1930 (FAO, 1985), but forestry management of the tidal forests did not begin to really take effect until the 1970s. One exception is the Matang Mangrove (Perak), which has been completely under forestry management since 1908. There, however, charcoal is only a by–product resulting from the thinning of the stands. The primary goal of production is posts.

Because of their good combustion properties and the high calorific value, it is largely *Rhizophora apiculata* and *Rhizophora mucronata* that are used for charcoal production. For this purpose, timber of at least 5 cm diameter is felled. The charcoal burning process takes place almost exclusively in traditional brick kilns which take from 50 to 200 m^3 wood. In Thailand about 1300 charcoal kilns in or near mangrove areas were counted (FAO, 1985), in Riau 836 kilns existed in 1984 (Soemodihardjo, 1987). Originally charcoal was used by the coastal population for cooking. Accordingly, there were many small kilns, mostly operated by families, with capacities of upto 10 m^3. Today mangrove charcoal is a coveted product on the national markets, and in Thailand and Indonesia it is also an export product. The customer countries include above all Singapore, Malaysia and Hong Kong and other Asian countries. Production is shifting more and more to larger enterprises with batteries of large–capacity kilns. In 1990 in Thailand alone 790,000 m^3 of mangrove wood was felled for charcoal production. They produced 359,000 m^3 or 269,000 t of charcoal. The target for the year 2000 is 530,000 t of charcoal (Aksornkoae, 1987b).

Charcoal production is taking place increasingly in medium–sized and large–scale commercial processing plants in the middle of *Rhizophora* monocultures. These enterprises have their own docks and terminals, fleets of open boats with outboard motors, large warehouses, packing halls, workshops, canteens, shops, dwelling houses (mostly simple row settlements or "lines") for permanent employees, administrative buildings and managerial bungalows. For the wood production areas under management there are long–term concessions, which guarantee regeneration. Individual enterprises also work on their own properties.. As wood charcoal has now become an export product, such plants fulfil all the criteria of a plantation. Thus one may speak of *Rhizophora* plantations.

Wood chips are the latest product of the mangrove forests. They are produced exclusively in Indonesia and East Malaysia, mainly for export. There is further processing into cellulose, paper and artificial silk (rayon). Felling is carried out over large areas in hitherto untouched mangrove areas. Regeneration is reported to take place only very slowly (FAO, 1985). The beginning of this production line in the 1970s started a boost in felling operations. Fortes (1988) reports that in Sabah and Indonesia more than 3200 km^2 of mangrove area have been released for felling since 1970. According to Soemodihardjo (1987) 4010 km^2 of area were being used in Indonesia under licences granted. A further 2732 km^2 were available for licensing. FAO (1985) puts the felling area for Sabah at 40 km^2 per year. In Sarawak 6 km^2 per year are said to have been felled. In areas covered by licence at Bintuni Bay in Irian Jaya alone, a Japanese consortium will fell 1350 cm^2 of mangrove forest over a period of 10 years and in the final phase produce 400,000 t of chips per year (Hamilton *et al.*, 1990). So far the literature contains no reference for this area to any measures to regenerate the stands or to protect them in the long term.

Indonesia is today the main producer of wood chips from mangroves. Mills built for chip production are located in Kalimantan, Irian Jaya and Riau Province. The major customer country is Japan, followed by Taiwan, whose cellulose industry is heavily dependent on supplies from Indonesian tidal forests. It is therefore mainly Japanese–Indonesian joint ventures that are involved in the extraction of the raw material and they have also set up the processing plants for chip production. The value–increasing further processing of the wood chips is all carried out abroad, apart from a few exceptions (paper production in Sulawesi and cellulose production in Sumatra). A transition to sustainable utilization of the mangrove areas so far used for chip production, coupled with the achievement of stronger regional effects, is a matter of urgency.

Selective utilization of the mangrove by intensifying the production of individual species began in those mangrove forests which have larger neighbouring population concentrations. Demand from outside the mangrove areas made individual products from the range of self–sufficiency marketable, and they were then sold in neighbouring markets. Most mangrove products are highly perishable, and likely to crate high transport costs, so that production in tidal forests could only be carried out for the neighbouring areas. For this reason degradation or destruction of the mangroves in Indonesia, Java and the Riau Archipelago is the furtherest advanced, then at the southern tip of the Malaysian Peninsula in Singapore and in the Malaysian state of Johor, in the Philippines in Luzon and in the Visayas and in Thailand in the upper Gulf.

It was only the improvement of the transport infrastructure, the increase of traffic speed, the reduction of bulk transport costs, the introduction of cold–storage transport and the setting up of frozen food chains that ended the bond of mangrove production to neighbouring markets. Market production thus spread to further mangrove sites and with it the degradation and destruction of the tidal forests. Individual mangrove products, especially charcoal and shrimps, have meanwhile become profitable exports. Extreme market distance and insufficient transport links are the only really effective protection for the remaining natural mangrove ecosystems in Southeast Asia today.

Summary of Current Utilization

As major competitors for the natural mangrove sites the following must be listed :

- wood production and wood exploitation,
- rice, coconut and oil palm growing areas,
- marine aquaculture of shrimps and milkfish,
- transport facilities such as roads, docks and airfields,
- settlement area development,
- development of coastal industrial areas,
- raw material extraction in the intertidal zone and
- areas used for tourist demand.

Together these utilization have led to the degradation or destruction of about 50% of the mangrove stands of Southeast Asia. In the Philippines well over two thirds of the mangrove areas have been cleared. In isolated cases, such as in North Java, mangrove destruction has been almost total.

Protection of the natural stands, however, has not so far proved itself to be a competitive use of the land. For example, only four of the total of 52 national parks in Thailand include large mangrove stands. And these four protected coastal areas do not even cover 4% of the total national park areas. The mangroves do not so far have any effective lobby.

Even existing protection regulations are being withdrawn under increasing pressures of population and exploitation. As early as 1884 the colonial government in Singapore, alarmed by the rapidly advancing forest destruction, acted in an exemplary fashion by setting aside the Pandan Nature Reserve with an area of 1012 ha of mangrove. Half a century later it had been reduced to 556 ha and in 1962 the remains were devoted to the development of industrial areas. The second mangrove protection area near Kranji was progressively reduced from 306 ha to 21 ha by 1937. In 1955 it only measured less than one hectare until, in 1973, it was given up in the framework of setting up a water storage area (Wee, 1993). Some 1319 ha of mangrove protection areas have shrunk to nothing within one century. Today Singapore possesses altogether 4.5 ha of mangroves protected as a national park within the much larger Pasir Ris Park established in 1988. The protected area corresponds approximately to the average size of a shrimp farm. If all the islands are included, Singapore today has less than 500 ha of – mostly degraded – tidal forest areas left of the previous 7500 ha in 1819 (Corlett, 1987). Even of this small relict area only one hundredth enjoys protected status.

The lack of any lobby for mangroves is also shown by the current discussion of the destruction of the tropical rain forests. Although, proportionally the destruction of the tidal forests is much further advanced than that of the tropical rain forests and is taking place more rapidly too, the mangrove stands hardly receive any attention in the tropical forest protection programmes being started worldwide – and this, although, the tidal forests have at least the same effect for the global climate that the rain forests have, though on a reduced area. Through the interlocking of terrestrial and marine ecosystems in the mangroves they produce more benefits for nature and mankind than the rain forests do.

Chapter - 11

ECOLOGICAL NOTES ON MANGROVES FORESTS OF JAPAN, CONSERVATION AND MANAGEMENT — A CASE STUDY WITH JAPAN (A Special Contribution by Odaki)

1. INTRODUCTION

Mangrove plants form forests at the interface of land and sea. Constituents of this forest though considerably contribute for the formation of the evergreen vegetation in the sub–tropical belt of Japan, it could not draw adequate attention of the scientists and naturalists of this country. As a result only a limited few informations on this subject have so far been published by the scientists or the foresters of Japan. However, mangrove parks were created at national level and many mangrove areas are now protected. My idea is that, public awareness about the mangroves is still has not that much been created for the involvement of the local people for protecting this rare ecosystem in Japan. Mangroves attracted this author for last few decades. Several visits were made in different mangrove growing areas in Japan, certain parts of the western Pacific and many other places of the world including those in the Sunderbans on the Bay of Bengal, the largest mangrove forest in a single block. Beautiful landscape and the complexities of the ecosystem inspired me to write this book to introduce the Japan's mangroves to the readers in rest of the world. That mangroves grow in sub-tropical Japan is not known to many persons. This chapter expresses the personal experience of the author which he gained during working with the mangroves in field condition. Many scientific data were collected. Attempts were made to present some of those important data in this chapter is in simplified form with a hope that those will be the sources of information to the students of ecology and induce them to undertake research programs on the mangroves. Hope, the readers will find interest in this book. Mangroves need conservation. Basic principles of conservation have been discussed here in a nutshell in the context of Japan's mangroves. I expect that it would draw attention of the government administration, policy makers and the private organizations.

Major inspirations for writing book on 'Mangroves' was obtained from Late Prof. Makoto Numata, Director, Chiba Natural History Museum. I acknowledge him with thanks for his keen interest on the subject. I also thank my friends and well wishers for their constant inspiration. Last but not least, I must thank my wife Mrs. Kimoto Odaki. She was a constant source of encouragement and inspiration during my field work and writing this manuscript. Her help was essential in some critical phases of this work.

Mangroves are the very specialized communities of organisms adapted to the saline environment of tropics and occurring in areas which receive the influence of both the fresh water and the sea water. Minimum winter temperature in such areas does not exceed 5°C. Mangrove growing in some islands in Japan also, are not exception. However, the number of species of mangrove plants in Japan are only a few, varying from one to seven, depending on the temperature condition. Because, mangrove plants and its environment are not much common in Japan, author felt it necessary to write this chapter considering some basic scientific concepts. The natural habitat of the mangrove is amphibian as they require both land and the brackish water for their survival. Water is a rich source of nutrient. Nutrients are recharged in the soil when the land mass is flooded two times a day during the high tide. When water recedes during the low tide it carries back the litters of plants or other detritus from the land. Litters are decomposed by bacterial actions and nutrients are released in water (streams or rivers) around the land. Rate of growth and primary productivity of mangrove plants and the associated secondary or tertiary productivity (mainly the fishes, and other animals) in some mangrove forests are very high, not in all, in comparison to any other terrestrial plant communities. As a result, a large number of people living in the tropical coastal belts are economically dependent on the mangrove ecosystem. Mangrove plants are different from other plant communities as they develop certain structural peculiarities suitable for their adaptation in the saline environment. Many mangrove plants have additional root system which grow-up instead of going down in the soil. These roots are popularly known as the breathing roots, or the pneumatophores. In water logged soil respiration of root are not adequate. Aerial roots breath from air. In many mangrove plants the seeds germinate in the mother plant before those are dispersed. This type of germination is known as viviparous germination. Germinated saplings easily float or stick to the soft soil after they fall and starts growing there. Mangrove plants get enough water in the environment but they cannot use saline water for metabolic functions. Soil is physiologically dry, so, these plants need to conserve water. Plants develop thick cuticle in leaf with sunken stomata and check water loss by reducing transpiration. Some of the plants have salt glands in their leaves for elimination of the excess salt that they absorb as solutes from the soil. Normal roots are very elaborate and those are good soil binders.

Mangroves protect the shore line from the wind and surges. Organisms, that can grow in saline habitat are supposed to have unique genetic make-up. These plants and this special environment needs conservation from their over-exploitation and degradation which occurred in many poor countries due to human activities.

One of the objectives of the book (Japanese Language) is to make the common people of Japan aware of this very special type of plants and the environments, to help their conservation. Most of the people of this country might not have seen these mangroves. Although, the places where mangroves grow are famous for natural scenic beauty. These plants can be observed during sight seeing. Author (K. Odaki) of this book has the experience of doing research work with the mangrove plants and animals for a long time and have seen them from various angles. Different types of mangrove plants and also some animals – their habitat, their growth forms, association patterns in natural conditions and also, consequences of the introduction of mangroves in certain areas of central Japan will be discussed in this chapter.

11.1 MANGROVE FORESTS RESERVED AS NATURAL MONUMENTS

(a) Kyushu Island : (Fig. 37)

This mangrove forest was possibly planted in 1909, by the local king of Kiire region of Kyushu Island. It was the time of Ryukyu of Kingdom– the real ruler of the federal Japan. This mangrove forest consisting of Ryukyu–Kogai (*Kandelia candel*), is a natural monument and it exists in Kiire town of Kagoshima prefecture, in the southern part of Kyushu Island. It is a small forest and is being conserved since 1921 by the Agency of Cultural Affairs of Japan. It is important to note that this is the northern limit for the mangroves and is an as unique example for the students of Biogeography, Ecology and Plant Embryology. This author visited this *Kandelia candel* forest at March 1962. From the records of the old days, it is known that this small forest at the back marsh of sand dune along the road. Trees of *Kandelia* were about 1~3 tall, and plant density was high. Those plant communities grew in various places with *Phragmites communis* Trin. At present this forest has become more small and narrow growing along the road side only. Sand dunes are approaching towards the area where the mangrove forest grow, and coastal sand is gradually invading the community. As the result, some individual trees have shown die back phenomenon. In order to preserve this valuable *Kandelia* forest, fencing was made to check invasion of the sand into the forest. Also, water ways were dug into the inner portion of the forest, to keep the soil wet. It has became difficult to preserve this forest as it became narrow. Possibly the sea current has been induced to change by the industrial growth around the coast line in the Kogoshima Bay.

(b) Miinami-daitoh-jima Island (Fig. 37)

Miinami–Daitoh–Jima is relatively a small island of about 31 square km area. It is located at 370 km east of Okinawa–Jima on the Pacific ocean (Fig. 37). This island is circled by coral reef and is made-up of corals. The centre of the island is characterized by the occurrence of series of depressed land mass filled with saline water and appear like ponds around which mangrove grow.

Average water salinity is 1%. Sea water apparently cannot enter this forest. Salinity of underground water in the forest floor is about 4%. *Bruguiera gymnorrhiza* forest occurs at the edge of the one such pond (Ohike in Japanese).

This *Bruguiera* forest is not more than 10m in width and forms a narrow forest land along the pond. In a transect study, it was found that a few trees occurring in the middle of the forest are tall while trees in the side of the ponds or those at the end of the forest are of medium height. Forest floor remains full of young saplings of different heights. On the basis of tree height, plants can be arranged into four stratification.

Since the year 1900 this forest was exploited by the human being. Cut trees were mainly used for house construction. The emerging seedling during vivipery germination, before those were dropped in the forest floor, were collected for doll making. Now, there is no more exploitation, and the forests are growing normally.

(c) Okinawa Island (Fig. 37)

Gesashi is a village in the north–east of the Okinawa island. Okinawa is situated in the southern most of the Nansei–chain islands spreading south east of Kyushu at latitude of 24~27 degree north. Mangrove growing in this area is composed of *Rhizophora stylosa, Bruguiera gymnorrhiza* and *Kandelia candel.* Okinawa is the strategic island for the mangroves in the northern hemisphere because *Rhizophora* does not grow in further north of this island; *Kandelia candel* that are tall and plenty in northern islands become small and few in distribution in further south of Okinawa. Agency for the Cultural Affairs of Japan decided to consider this forest as the Natural Monument in 1972 and constructed a hanging bridge on the river at the Gesashi Bay. Standing on this bridge one can enjoy the beauty of the normal mangrove landscape.

Rhizophora stylosa forms a zone in frontline to the river or the bay with pure strand and remain in direct contact with salinity even in low tide. *Kandelia* when grow directly facing the bay water become low in height. The middle zone is a mixed forest of *Bruguiera* and *Kandelia* the former species being dominant. In the last zone pure strand is formed by tall *Kandelia* candel. Salinity is minimum in *Kandelia* forest floor. At the edge of the forest there are some associates (not true mangrove) like *Pandanus, Hibiscus* and *Clerodendrum.* Inside the forest there are many soil mounds (made by *Thossina anomala* a crustacean animal).

This mangrove forest is also affected by human activities and reclamation. Recent reclamation around the river side caused even death of some mangrove trees occurring at the edge of forest. Human activities have impact on this forest and death of edge communities.

(d) Ishigaki Island (Fig. 37)

Miyara river mouth is located at the east of the Ishigaki island on the sea. The Miyara is the biggest river in this island. The mangrove plants grow along this river at a distance of about 1.5 km from the river mouth. *Rhizophora stylosa, Bruguiera gymnorrhiza* and *Kandelia candel* constitute this forest and it was designed as the Natural Monument by the Agency for Cultural Affairs of Japan in 1972.

Present author visited this forest in 1959 for the first time. Since then the mangrove habitat has much been changed. At that time there were many sand burrs – small and large near the river mouth. Mangrove trees were smaller in size (like shrubs) and were scattered few on the sand bars, but dense in mouth of the river. Also, many rock pieces, broken coral reefs, were dispersed on the river bed. Dispersion pattern of the young trees on the sand bars were studied. It was observed that *Bruguiera stylosa* plants occurred at the outskirts of it. *Kandelia candel* were very few in number. Then in 1984, during the next visit, the author examined those sand var. vegetation again and it was observed that a large forest of *Bruguiera gymnorrhiza* was developed on the biggest sand bar lying there, but those occurring in the many small sand bars disappeared. A *Sonneratia alba* tree that grew near the coast line was found there after 19 years, but this tree was seen to have developed die back phenomenon and many branch were died and others were with only a few leaves. It seems that the tree has some environmental problem. An examination of the *Rhizophora stylosa* community indicated that the prop roots in those plants developed from one side of the trunk and this condition of root formation is known as "Shajohgata" in Japanese language. In recent years there were constructions of a Dam for farming in this area. Also, effluents discharged from factories contaminates this area. There are additional problems of cattel grazing, and all these factors contributed for the change in the mangrove environment there.

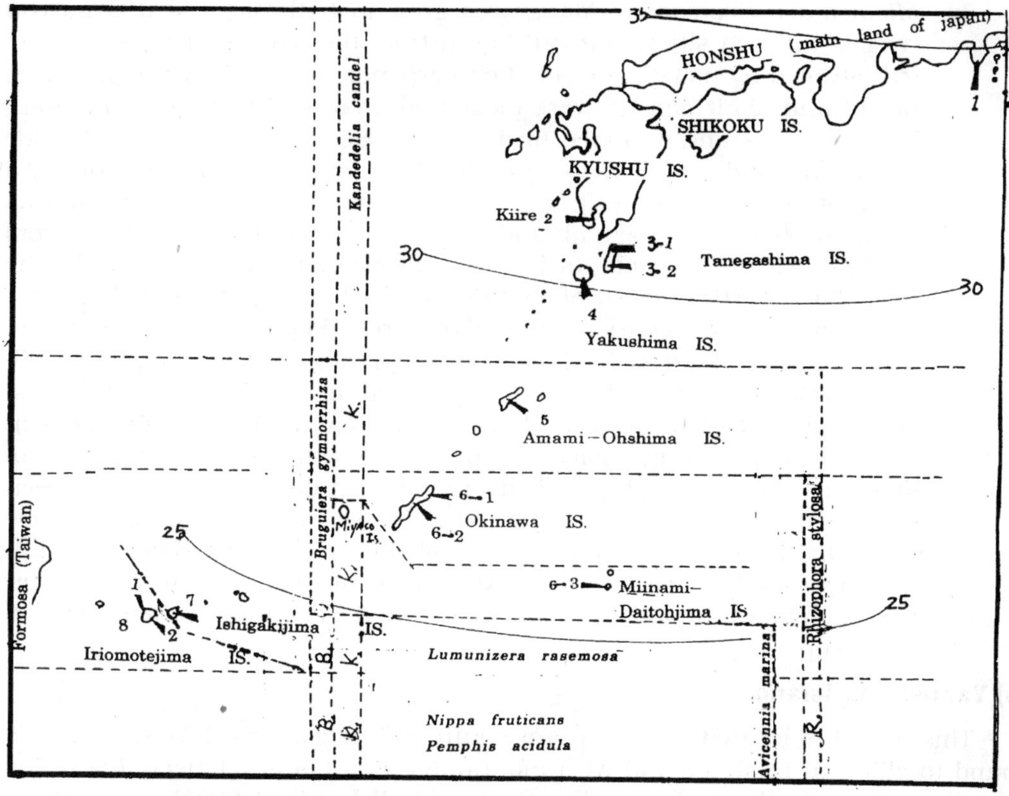

Fig. 37. Distribution pattern of mangroves of Japan

11.2 NATURAL MANGROVE FORESTS IN NANSEI–CHAIN ISLANDS

(a) Tanegashima Island

The sites of investigation as shown in Fig. 37 are : (i) Kumano–ura area and (ii) Minatogawa river side. This island is situated about 40 km south of Sata cape (south of Kyushu).

(i) *Kumano–ura area* : At the centre of Kumano there is a lagoon. Mangroves grow around this lagoon and is represented by only one species – *Kandelia candel.* Depending on the ecological condition this plant may be very small or dwarf trees or medium tall trees. Very small trees grow more like those in farm condition mainly in the western side of the lagoon. Dwarf trees grow in small patches near the stream that connect the lagoon with the sea. Those medium tall trees grow around the southern edge of the lagoon. Difference in tree height and growth of *Kandelia candel* in different situation seems to be due to differences in soil salinity. Places exposed to surface or the sea water are more saline. In high saline area plants become either small of dwarf trees. Those growing in low saline area like those in the southern edge of the lagoon are about 2m tall.

(ii) *Minatogawa river area* : Mangroves grow in both sides of this narrow river in an area situated about 700m from the river mouth. This forest was studied by using transects laid perpendicular to the river line. Tree height and plant density were measured. Plants were of three different height classes and occurred in different zones. Plants on the river side are young and smallest in size. In the next zone towards the land medium tall plants grow. Tallest plants occurs as the last zone towards the land. Zones (grouped a,b and c) occurring in different distance from the river side. These plants have been designated as Stand B. In the river bank, further south of the area with plants of stand B there is a small forest and occurs in a narrow area. Plants of this forest are designated as stand A. In this stand all plants are tall or medium tall. Tall plants may be adequately branched or with a few branches. Difference in tree height and branching is mainly due to differences in the duration and intensity of sun shine coming over the hills of different elevation. Light intensity radiated in different areas has been graphed and compared with normal radiation. Tree few branches grow in low light condition. The floor of a tall *Kandelia candel* forest (stand A) was clear in 1988. During my next visit in the same spot in 1996 (after 8 years), it was found that the clear forest floor was occupied by tall grass spp.

(b) Yakushima Island

This island is located at the 30 km south of Tanegashima island. Island is round to elliptical in shape and with mountains. Rainfall is relatively high. The river Kurio flows in the east west direction. A small *Kandelia* forest occur in the bank of this riverside. In south of this forest there is fish port. This is an important area and declared as National park by Department of Environment in 1974. Study was undertaken in this forest in 1967 and in 1982, 1992 and it was found that many changes took place in 25 years. During the visit in 1967, this *K. candel* forest was studied in 1m x 3m quadrates laid on a transect parallel to the river bank. Data of tree height and tree density were collected. Also, number and growth of seedlings were recorded from two sides and from centre of the forest. It was observed that the forest tall but less dense in the centre. In contrast, the trees occurring at the forest edge were dense but of low height.

Results indicated that there was a relation between tree height and tree density. A relation between the seedling number and growth was also established in the centre and two edges of the forest. During next visit in 1992 by author, it was observed that the forest were cleared.

(c) Amami–oshima Island (Fig. 37)

This island is situated in south of Yakushima island. Mangrove plants grow in this island in two north south–blocks close to the confluence of the Sumiyoh river (Su) and the Yakkatsu river (Ya), before they meet the sea. This author visited this forest in 1969 and measured the tree height and plant density in four identified areas in the forest of the south block A, B, C, D.

Interesting observation was that apart from *Kandelia* another species – *Bruguiera gymnorrhiza* also grows in this island. *Bruguiera gymnorrhiza*, however, is present mainly in the forest block occurring in the south. A few plants only occur in the north block in association with *Kandelia candel*. Tree height and density as studied in both these species indicated that in case of *Bruguiera* the density was high but plant height was low. In case of *Kandelia candel* on the contrary only the tall plants occurred in a dense forest. A comparative data of tree height and density for *B. gymnorrhiza* and *K. candel* has been showing. This study was made in all four sites (A, B, C, D) in forest of south block. *Bruguiera* trees in high density occurred in site A when tall trees of *Kandelia* was dense in site D.

My second study of the mangrove plants in this island in 1986 and compared with my first study in 1969. The relative number of crown top of tree of *Bruguiera gymnorrhiza* and *Kandelia candel* as observed during my first and the second visiting times, at the same site. The data were taken from a north–south transect (a–b line). It is indicated that the 17 years gap of the crown top of plant of *B. gymnorrhiza* declined when the crown top of *K. candel* increased.

(d) Katabaru in Okinawa

This island was described in Fig. 37. Patches of mangroves at this place occur in the east and west side of a small river. The eastern side is bounded by a hilly tract. In west there is a road and a garden. Plants growing in the western side receive fresh water supply from the river itself while those in the eastern side get wash out of the hill during rainy season apart from the freshwater from river. This Katabaru forest though small in area is very dense and average height varies from 3~4 m. Different plant species growing there are *Kandelia candel*, *Bruguiera gymnorrhiza* and *Excoecaria agalocha*. *Kandelia candel* is the dominant plant. *Excocaria* plants are relatively few in number. This composition was constant in the western side.. In the eastern side in addition to those plants there are some associates like *Pandanus* and *Hibiscus*. A small plant known as *Clerodendrum inerme* also a mangrove associate grow in the eastern side. Part of this rather a dense forest was cleared and now is being used as a public institution and dumping ground.

(e) Iriomote Island

Iriomote is the last one in this chain and is situated in the extreme south. Mangrove forest is rich comprising of 7 species of plants. Sites of the present study was located along the Nakama–river in south, and the Urauchi–river in north of this round island. Both the sites are now included as the special National natural monument area and declared as national park.

The different plants identified are : *Bruguiera gymnorrhiza, Rhizophora stylosa, Sonneratia alba, Avicennia marina, Kandelia candel, Lumnitzera racemosa* and *Nipa fruticans*. The density of Kandelia and Bruguiera are different. Also plant height of these two plants vary. Important observation was that *Kandelia candel* that was the dominant species in the northern islands are very few in this island in south.

The thickness of leaf, leaf area and dry weight and shape of leaf of young and mature plants of *Bruguiera gymnorrhiza* were compared. It was found that leaf area of young plants are more than those of mature plants. Leaf thickness and the dey weight of leaf in mature plants were more than those of the younger plants. In mature *Bruguiera* plants the knee roots are of large size.

11.3 CONCLUSION OF THE DISTRIBUTION PATTERN OF NATURAL MANGROVE IN JAPAN

It can be concluded that the mangrove distribution and the forest formation in the chain islands in south west Japan is largely determined by the temperature condition. Annual monthly temperature indicates that mean winter temperature (November–March) in northern mangrove growing area like Kiire in Kyushu island is between 7~13°C when temperature during the same period in Iriomote island in south is 18~23°C. Summer temperature, however, in all the islands are closely similar. Southern islands with warmer condition is more suitable for an elaborate mangrove formation. In Iriomote there are 7 different mangrove species as against the presence of only *Kandelia candel* inquire to Yakushima with winter temperature between 7~14°C. Not only the presence of more plant species were observed in the islands in south but also the plant height, and DBH tended to increase with the rise of temperature. It was found that for *Bruguiera* warm climate in south is more suitable. Height of plants and DBH increased in *Bruguiera* growing in southern islands. This species is poorly developed in Amami–Oshima and this island is the northern limit for *Bruguiera*. *Kandelia candel* is a mangrove for colder climate. It is distributed upto Kiire in north, the northern limit is a mangrove in the northern hemisphere though Yakushima climate (12~14°C) is more suitable for this species. In this temperature range in Yakushima plant height of *K. candel* is maximum. With the increase of temperature in southern islands density of Kandelia declines when the presence of *Bruguiera* increase in those islands. Also the DBH of *Bruguiera* increase in Ishigaki and Iriomote where DBH of *K. candel* decrease. *Rhizophora* a less frequent tree in this region, also has temperature preference. *Rhizophora* does not further north of Okinawa. This type of information on the distribution of the mangroves in Japan is important because during implementation of the programme of mangrove plant forestation or in conservation policy of the mangroves in different island, planners will get clues for the selection of appropriate place. Temperature dependent mangrove distribution pattern in Japan is like a field laboratory where students of ecology may be able to do further research. Species preference and the range of distribution of mangrove according to minimum winter temperature is as follows: *Kandelia candel* 7~14°C, *Bruguiera gymnorrhiza* 18–20°C, *Rhizophora* 18–20°C, *Avicennia marina, Sonneratia caseolaris, Lumnitzera racemosa, Nipa fruticans* 20°C and more.

11.4 INTRODUCED MANGROVE IN THE MAIN LAND

Izu-peninsula in central Japan

In southern part of the Izu–Peninsula, in the central Japan, there is an introduced *Kandelia candel* forest. This introduction was made by the *Shizuoka* Prefectual Botanical Garden for preservation of this valuable plant. *Hibiscus hamabo* Seid et Zucc the typical plant for this region with warm temperate climate. It was peninsula before introduction this *Kandelia candel* forest. Following informations were obtained —

The proposed place for the plantation of *Kandelia* was connected with a river and it was the source of fresh water. Water temperature is low during winter but is high during summer. The viviparous seeds were collected from Tanegashima island in 1958 and subsequently transplanted in this peninsula. Initially the plantation was started during winter, but all of trees died. Next plantation was tried in summer, and it was successful. In 1994, author examined a mature forest of *Kandelia candel* in Izu–peninsula and observed that this forest was similar to those at the Tanegashima island. But the portion of the forest in eastern side was changed by construction for fish post in 1996.

11.5 MANGROVES IN SOME OTHER PACIFIC ISLAND

(a) Guam Island

This island is 209 squire mile in area and is located at latitude 130 north and longitude 140 east. Climate is uniformly warm round the year. There are rivers in the southern part of the island and the mangroves grow in the mouth of these rivers. Also, a few mangrove plants occur along the southern coast line. There are total seven mangrove species like *Rhizophora apiculata, R. stylosa, R. mucronata, Lumnitzera racemosa, Xylocarpus moluccensis, X. granatum, Avicennia alba.* All these species occur in mouth of the bay in the western and in the eastern landmass opposite the bay. It was observed that there is a group of *Avicennia* plants which appear intermediate in structure between *A. marina* and *A. alba.* It may be a surviving hybrid of *Avicennia marina* and *Avicennia alba*, and was named as *A. marina* var. *alba* by Stone (1977, 1980). Growth pattern of *A. marina* var. *alba* (named *Avicennia alba* by H. Barth, 1982) vary depending on the place of their occurrence. Plants facing the sea are tall and with high productivity and has been grouped as zone I plants. The plants in zone II grow away from the sea, and are dwarf with poor growth. It was observed that the places where dwarf plants grow are high salinity areas. Also, the under growths in the forest with tall trees were low in comparison to those in the forest with dwarf trees. Both groups of this hybrid *Avicennia* dominate in this bay area. Author visited this forest in 1977, ten months after the Typhoon of 1976. It was observed that the Typhoon caused a severe damage to the mangrove plants though those ultimately survived. The damage was maximum in 20 to 30m stretch of the forest facing the Typhoon wind. There were two patches of tall plants with dwarf plants occurring in between. The tall plants in the front facing the Typhoon were damaged in comparison those in the interior. The tall plants in the front row acted as buffer and so the dwarf plants in the middle were not much damaged. Growth ring count of tall and dwarf trees of *Avicennia alba*, distribution of crown and the density, and plant DBH/D ratio in relation to soil salinity were also studied in *A. alba*.

(b) Palau Island

Palau consists of about 200 islands. There are many other small islands locally known as Rock islands, of which Babelthaup is the major. Size of this island is 43 km by 13 km and is the largest one. There are villages in its coastal plains. The forest is dense and it is difficult to enter. The Gaspan is the longest river in this island. Mangrove forests occur along the side of this river. We investigated this area during the high tide and the pattern of distribution of mangroves was studied in plots along the river at an interval of 300m on a boat. Forest communities tend to change towards the upper part of the stream from the river mouth. The river plain can be divided into lower, middle, and upper zones. The lower zone is dominated by *Sonneratia caseolaris* when in the middle zone *Rhizophora apiculata* is the major tree. In the upper zone *Nipa fruticans*, *Bruguiera gymnorrhiza* and *Scyphiphora hydrophylacea* from a community. The arrangement as found in coastal belts. An endemic species known as *Urandra ammui* (Schellenb.). *Kanehira* was located out at the upper zone along this river. This is a subsidiary species and was named by Watson (1928).

(c) Yap Island

It is a small island with coral reefs and the mangroves. Soil in mangrove region is typical soft mud and difficult to walk into the forest. Major constituent in this forest is *Sonneratia casualaris*. It is a high forest with big tree. The pneumatophores are remarkably big about 2.5m in height and arranged in rows. It is a typical and old *Sonneratia* forest may be used as demonstration.

(d) Palawan Island in Philippines

The island is elongated in shape. Irigan Bay area in south–central part of the island house a dense forest composed of tall trees. A transect shows that *B. gymnorrhiza* and *R. apiculata* with a few *R. sexangra* form a high forest. In climax condition the tree height of *Bruguiera* is very high (40m). At the Tagbilao area in south–central part, typical structured *Sonneratia caseolaris* plant was located. Special character is that the main trunk of the plant was cut. The secondary branch trailed, touched the soil and grew like a runner. The erect portion was cut but the secondary branches trailed to a distance of 35m.

11.6 CONSERVATION AND MANAGEMENT OF MANGROVES

Conservation can near restore an degraded environment only when this work is done with recent informations on the dynamics of this ecosystem–its structure and function. Conserving an ecosystem mean the conserving the biodiversity, the ecological processes, and its integrity. A conserved ecosystem in true sense, is resilient to the outside stresses and is sustainable. In many tropical countries thc mangrove ecosystem, in particular, has been limitlessly exploited for human survival and many of those systems ran beyond their carrying capacity. For the sake of future generation it is the major task of the present ecologists, sociologists, and the politicians to work jointly for the cause of conservation of the mangroves as a self-sustaining system. Biodiversity in an ecosystem is intimately related with its functioning. This relationship is complex and operates in many temporal and spatial scales. Diversity of the dominating trees of the mangrove ecosystems as discussed in the previous chapters is not a complete work. Further work in details on the species diversity and functioning of the ecosystems are yet to be done. With limited informations in hand, strategically we prefer to consider the conservation work to start in certain mangrove areas supposed to be under threat, only with those simple data on the tree species composition, water, and soil salinity. Present status of these factors were identified from different areas through investigations carried out more than once with a reasonable gap of time. It is apparent that trees, soil and water are the dominating ecological characters that determine primary productivity in an mangrove ecosystem; and the secondary or tertiary productivity though never the less important, are dependent on the primary productivity. It is, however, relevant to focus that the species diversity and the dynamics of mangrove ecosystem vary from place to place, and therefore, there can not be a generalized mode for the conservation of the mangrove, and the conservation policy would be based on regional factors. Situation observed in natural mangroves in Kiire village indicate that there was change in the direction of sea current due to human activities in the Kagoshima bay. It resulted in an extension of the shoreline sands towards the mangrove area, and as a result some trees are dying. Measures for the protection taken so far by erection of fencing or by digging canals (to keep the soil moist) are now found to be inadequate to near restore the original habitat of the mangroves there. What more steps can be taken for an effective conservation? It may be suggested that for better results monitoring of water quality (siltation pattern, and presence of chemical effluents in river water) and the soil will be of help. In addition, it will be necessary to identify whether any new construction will obstruct the normal flow of current in the bay.

Loss of small sand bars in Miyara river mangroves area in a span of 25 years may be a natural phenomenon. But the observed die back phenomenon in a *Sonneratia alba* plant is not normal: and it may be due to constraints in the availability of suitable water. Construction of a dam for farming in upper river area seems to be a contributing factor for the deterioration of water quality. Replacement of forests by the orchards and the nursery in hill tracts near the mangrove forest in Gesashi seems to be the sources of agricultural pollutants those are being carried by the river to the mangrove area and cause the death of front line mangrove along the river.

Mangroves have a very delicate habitat and is vulnerable to very minute change in water quality. Monitoring the water quality can provide possible clues for the conservation management. Mangroves growing in natural condition in many places in Japan needs conservation as "as unique natural areas" because the flora, fauna and the landscapes in mangrove growing region are rare in this country. Mangroves growing naturally in the northern hemisphere with rather prolonged low temperature is very uncommon, and so, Japan's mangroves demands tremendous scientific value for further investigation. Biology of the organisms inhabiting in the mangroves and the physical structure of this environment are very specialized. This typical coastal ecosystem is being maintained in nature for thousands of years without any basic change. For short term gain human being can not destroy it. Organisms adapted in mangrove habitat are important gene pool. This salinity tolerant genetic system existing in an mangrove environment can be used for development of trangenic saline resistant crop plants. This landscape is exceedingly beautiful and seems to have potential to earn more money if converted to tourist spots with adequate care instead of further encouraging an industrial growth. Such an areas have aesthetic values too.

Species diversity in tropical mangroves are more than those in mangroves in Japan. Many people in those countries are economically dependent on the mangrove ecosystem. As a result, many large mangrove forests in tropics were cleared. Dinerstein and Wikramanayake (1996) is of opinion that clearing and deforestation cause erosion of species; and they proposed a new approach for conservation and planning known as "CPTI" (conservation potential / threat index). For mangrove species management too, CPTI may be useful because this index forecasts the effects of deforestation on conservation well in advance, and it compares the biological richness with the size of the conserved area, the remaining forest area, and the state of deforestation. Unesco has put special effort to conserve tropical mangroves and this work is in progress in many countries. Both the number of species and the size of the conserved area are important to a manager who will implement the conservation work. In this connection it is relevant to note that conservation is very much dependent on legal, political and social policies of a country. For the success of conservation it may often be necessary to frame new legislation or to amend the older ones depending on the nature of the job.

11.7 MANGROVE ECOSYSTEM

As an evidence, it can be cited that the nature of food chain and wave of the mangrove of Sunderbans, West Bengal, India and that of the Iriomote island at south of Nansei–Shottoh, Japan are different, though both the ecosystems are protected from human interferences. The keystone species in both the Sunderbans and the Iriomote island are represented by crustaceans and the fishes but the umbrella species in two such areas are not similar. While the Sunderbans the umbrella species is the Bengal Tiger, it is the shark in Iriomote in Japan. Oceanic sharks visit the mangrove forest along with sea current at high tide. This differences is due to the fact that mangrove is in the Sunderbans is of tropical origin, and that in Japan is subtropical. Management practices of mangroves will therefore be dependent on the structure and function of ecosystem.

11.8. MANAGEMENT OF MANGROVES IN JAPAN

The mangrove forest is well developed in Iriomote island in south of Nansei–shottoh, Japan. This forest is preserved by the Law and Regulations of Environmental Agency and Cultural Agency of Japan. Before 1970, mangrove products of this forest were used as building material, firewood, charcoal etc. Also the colouring material and tannin used for dying fishnets were derived from mangrove plants. After this, forest was declared as a ``National park'', people are not allowed to cut the trees. Now–a–days such area are used as tourist resort; and as hiding places (in narrow canals) during Typhoon.

Kandelia candel is the only mangrove species represented in Tanegashima island. This island is the northern limit of the mangroves in the northern hemisphere. There are two types of *K. candel* plants in this region. One type is of low height (50~60 cm in height) and grow widely in lagoon area and appear more like a *Kandelia* farm. *Kandelia candel* are represented by tall trees (7~8m in height) along the Minatogawa stream and form a dense community. *Kandelia candel* communities growing in lagoons and those growing along the river need different treatments for protection as the environmental conditions in these two places are not identical.

At the confluence of Gesashi river and the Bay there is a well developed mangrove forest but a distance of only 200~300 meters, at the upper part of this river, there was an older forest (*Alnus formosa* Makino). This natural alder forest is known to function as a purifier of polluted water before it is drained into the mangrove forest. The alder forest was cleared recently and converted into horticultural institutions. This results in the spread of water borne pollutants in the mangrove forest. The author is afraid of this altered environment and invites legal and political control to forest destruction, and also for immediate protection of this buffer zone.

REFERENCES

Abdul Abbas en – Nebaty, 1230, Introduction to 'Ibuel–Bcithat' (Leclery), V. Notices des Manuscripts, 23 (Cited from Bowman, 1917).

Adeqbehin, J.D. and Nwaigbo. 1990. Mangrove Resources in Nigeria : Use and Management Perspectives, *Nature and Resources*, 26 : 2.

Adriani, M.J. 1958. Halophyten. In: W. Ruhland (Ed), Hdb. der Pflanzenphysiology, Bd.4, pp.709–736. Berlin.

Aksornkoae, Sanit. 1975. Structure, regeneration and productivity of mangroves in Thailand. Unpublished Ph.D. thesis. Michigan, Michigan State University, East Lansing, USA.

Aksornkoae, Sanit and Kongsangchi, J. 1981. The mangrove forest structure in the estuarine ecosystem in the Phangnga Bay. Research report submitted to the National Research Council of Thailand (in Thai).

Aksornkoae, Sanit and Saraya, A. 1987. Effects of tin mining on mangrove ecosystems in Ranong, Thailand. In : R.M. Umali (ed) Mangroves of Asia and the Pacific: Status and Management. Technical Report of the UNDP/UNESCO Research and Training Pilot Programme on Mangrove Ecosystems in Asia and the Pacific. Quezon City, Metro Manila, pp.367–381.

Aksornkoae, Sanit. 1987. Mangrove resources. In : Thailand Natural Resources Profile. Ed. National Environment Board, pp.145–163.

Aksornkoae, Sanit. 1987. Thailand. In : R.M. Umali (ed.) : Mangroves of Asia and the Pacific: Status and Management. Technical Report of the UNDP/UNESCO Research and Training Pilot Programme on Mangrove Ecosystems in Asia and the Pacific. Quezon City, Metro Manila, pp.131–150.

Aksornkoae, Sanit and Eiumnoh, A. 1988. Mangrove forests. In : J.N.Paw, S.Bunpapong, A.T. White and M.S.M. Sandorra (eds), The coastal environmental profile of Bank Don Bay and Phagnga Bay, Thailand. Association of Southeast Asian Nations/ United States Coastal Resources Management Project. Technical Publications Series, 2, pp. 21–26.

Aksornkoae, Sanit. 1993. Ecology and Management of Mangroves, Bangkok.

Aksornkoae, Sanit, Paphavasit, N. and Wattayakorn, G. 1993. Mangroves of Thailand : Present status of conservation, use and management. In: B.F. Clough (ed): Technical report of the project, The economic and environmental values of mangrove forests and their present state of conservation in the South–East Asia/Pacific Region.

International Tropical Timber Organization/International Society for Mangrove Ecosystems (ITTO / ISME) Project PD71/89 Rev.1(F), pp.83–134.

Albright, L.C. 1976. In sity aegadation of mangrove tissues. *New Saland Journal of Marine and Freshwater Research*, 10 : 385–389.

Andrews, T.J. and Clough, B.F. 1980. Photosynthesis gas exchange and carbon isotope ratios of some mangroves in North Queensland. *Proc. of Symp. on Prov. and Mgt. of Mangroves*, Papua New Guinea (W. Jank).

APHA 1975. Standard methods for the examination of water and wastewater, 14th edn. Prepared and published by APHA, AWWA and WFCT, 1015, Eighteenth Street NW, Washington DA, USA, pp.1193.

Arisz, W.H., Camphuis, I.J., Heikens, H. and Van Tooren, A.J. 1955. The secretion of the salt glands of *Limonium latifolium* Ktze. *Acta Bot. Neerl.* 5 : 218–246.

Arnold, A. 1955. Die Bedeutung der Chlorionen fur die Pfalnze. *Bot. Stud.*, Vol.2, Jena.

Atkinson, M.R., FIndlay, G.P., Hope, A.B., Pitman, M.G., Saddler, H.D.W. and West, K.R. 1967. Salt regulation in the mangrove *Rhizophora mucronata* Lamk. and *Aegialitis annulata* R.Br. *Aust. J. Biol. Sci.*, 20 : 589–590.

Attims, Y. and Cremer, G. 1967. Les radicelles capillaries des paletuviers dans une mangrove de Cote d'Ivorie, *Adasonia*, Series 2, 7 : 547–551.

Aubreville, A. 1964. Problemes de la mangrove d'hier et d'aujourd'hui. *Adansonia*, N.S. 4 : 19–23.

Austin, H.M. 1971. A survey of the ichthyofauna of the mangroves of western Puerto Rico during December 1967–August 1968, *Carib. J. Sci.*, 11 : 27–39.

Backer, C.A. and Bakhuizen van den Brink, R.C. 1963–68. *Flora of Java.* 3 Vols., Groningen : N.V.P. Noordhoff.

Backer, C.A. and van Steenis, C.G.G. 1954. Sonneratiaceae. *Flora Malesiana* 1, 4 : 280–289.

Ball, M.C. 1980. Patterns of secondary succession in mangrove forest of Southern Florida. *Oecologia*, 44 : 226–35.

Bary, J.G. and Gorham, E. 1964. Litter production in forests of the world. *Advances in Ecological Research*, 2 : 100–157.

Batanouny, K.H. 1973. Kalkdrusen von Limoniastrum monopetalum.Naturwissenschaft. Rundschau, 26 : 213–214.

Batanouny, K.H. and Abu Sitta, Y.M. 1977. Eco-physiological studies on halophytes in arid and semi-arid zones. I. Autecology of the salt secreting halophyte *Limoniastrum monopetalum* (L.) Boiss. *Acta Botanica Acad. Sci. Hung.*, 23 : 13–31.

Batanouny, K.H. and Bataouny, M.H. 1968. Formation of phytogenic hillocks. I. Plants forming phytogenic hillocks. *Acta Bot. Acad. Sci. Hung.*, 14 : 243–252.

Batanouny, K.H. and Batanouny, M.H. 1969. II. Rooting habit of plants forming phytogenic hillocks. *Acta Bot. Acad. Sci. Hung.* 15 : 1–18.

Batanouny, K.H., Hassan, A.A. and Zayed, K.M. 1985. Proline accumulation in plants of different ecological groups as a response to water deficit. *Qatar Univ. Sci. Bull.* 5 : 131–143.

Berry, A.J. 1972. The natural history of west Malaysian mangrove faunas. *Malay Nat. J.* 25 : 135–162.

Bhosale, L.J. and Joshi, G.V. 1973. Ecological studies in mangrove at Ganpatipule. *Proc. 60th Session Indian Science Congress.* III : 373.

Blasco, F. 1977. Outline of ecology, botany and forestry of the mangals of the Indian sub-continent. In V.J.Chapman (Ed.) *Wet Coastal Ecosystems*, Elsevier, Amstardam, pp. 24–260.

Bliss, D.E. 1968. Transition from water to land in Decapod crustaceans. *Am. 2001.* 6: 197–122.

Boto, K.G. and Bunt, J.S. 1981. Tidal export of particulate organic matter from a northern Australian mangrove systems. *Estuarine Coastal and Shelf Science.* 13: 246–255.

Braun–Blanguet, J. 1932. *Plant Sociology : The study of plant communities* (Transl. by G.D. Fuller and H.S.Conrad). Transl. of 1st Ed. of Pflanzenso ziologic (1928), McGraw Hill, New York and London, pp.430.

Brenner, W. 1902. Ueber die Luftwurzeln von *Avicennia tomentosa. Ber d. bot. Gesell.*, 20 : 175–189.

Breteler, F.J. 1969. The Atlantic species of *Rhizophora. Acta Bot. Neerl.* 18 : 434–444. 1977. America's Pacific species of *Rhizophora. Acta Bot. Neerl.* 26 : 225–230.

Briggs, B.G. and Johnson, L.A.S. 1979. Evolution of the Myrtaceae – evidence from inflorescence structure. *Proc. Linn. Soc. N.S.W.* 102(4) : 157–256.

Brinson, M.M. 1977. Decomposition and nutrient exchange of litter in an alluvial swamp forest. *Ecology* 58 : 601–609.

Brinston, M.M., Bradshaw, H.D., Holmes, R.N. and Wilkins, J.R. 1980. Litter fall steamflow and through fall nutrient fluxes in an alluvial swamp forest. *Ecology.* 61: 827–835.

Brown, S., Brinston, M.M. and Lugo, A.E. 1979. *Structure and Function of Riparian Wetlands.* In : R.R. Johnson and J.F. McCormick, technical coordinators, strategies for protection and management of floodplain wetlands and other riparian ecosystems. Forest Service General Tech. Report WO–12, United States Deptt. of Agri., USA.

Brown, F.H.B. 1935. Flora of southeastern Polynesia. III. Dicotyledons. *Bull. Bishop Mus..* 130 : 1–386.

Brown, J.M.A., Outred, H. and Hill, F.C. 1969. Respiratory metabolism in mangrove seedlings. *Plant Physiol.* 44(2) : 287–94.

Brown, W.H. and Fischer, A.F. 1920. Philippine mangrove swamps. In : Minor products of Philippine forests. Ed. W.H. Brown, Vol.1. *Phil. Bur. Forest. Bull.* 22 : 127–248.

Budowski, G. 1965. Distribution of tropical American rain forest species in the light of successional processes. *Turrialba* (Coasta Rica) 15 : 40–2.

Bunt, J.S. and Williams, W.T. 1981. Vegetational relationships in the mangroves of tropical Australia. *Mar. Ecol. Prog. Ser.* 4 : 349–59.

Bunt, J.S., Williams, W.T. and Duke, N.C. 1981. Mangrove distributions in North–East Australia. *J. Biogeogr.* 9 : 111–20.

Burkill, I.H. 1935. *A dictionary of the economic products of the Malay Penninsula.* 2. Vols. Oxford. Oxford University Press.

Byrnes, N.B. 1977. A revision of Combretaceae in Australia. *Contrib. Queenst. Herb.* 20 : 1–88.

Cain, A. and Castro, G.M. De Oliviera. 1959. *Manual of vegetation analysis.* Harper and Brothers Publishers, New York. pp. 325.

Camilleri, J.C. and Ribi, G. 1983. Leaf thickness of mangrove (*Rhizophora mangle*) growing in different salinities. *Biotropica* 15 : 139–141.

Candolle, A.C. de 1841. Second memoire sur la familie des Myrsinaceae. *Ann. Sci. Nat. Series.* 2, 16 : 1–46.

Carey, G. 1934. Further investigations on the embryology of viviparous seeds. *Proc. Linn. Soc. N.S.W.* 59 : 392–410.

Carey, G. and Fraser, L. 1932. The embryology and seedling development of *Aegiceras majus* Gaert. *Proc. Linn. Soc. N.S.W.* 57 : 341–60.

Chai, P.K. 1982. Ecological studies of mangrove forest in Sarawak. Ph.D. thesis. University of Malaysia, Kulala Lumpur.

Chan, E.I. 1977. *Oil pollution and tropical littoral communities.* Biotropical effects of the 1975 Florida Keys oil Spill. (1977) Oil Spill Conference American Petroleum Institute, Washington D.C., USA.

Chapman, V.J. 1940. The Botany of Jamaica shoreline. *Geogr. J. 9C* : 212–23.

Chapman, V.J. 1940. The function of pneumatophores of *Avicennia nitia* Jacq. *Proc.Linn. Soc. Lond.* 153(3) : 228–233.

Chapman, V.J. 1944. 1939 Cambridge University expedition to Jamaica. *J. Linn. Soc. Not.* (London). 52 : 401–535.

Chapman, V.J. 1970. Mangrove phytosociology, *Trop. Ecol.* 11 : 1–19.

Chapman, V.J. and Ronaldson, J.W. 1958. The mangrove and salt marsh flats of the Auckland Isthmus. *N.Z.Dep. Sci. Ind. Res. Bull.* No.125 : 1–79.

Chapman, V.J. 1975. *Mangrove Vegetation*, Carmer, Dehre, pp.425.

Chaudhary, A.B. 1965. A study of the Meliaceae of India with special reference of West Bengal. *Indian For.* 91 : 545–561.

Chengappa, B.S. 1944. Andaman forests and their regenerations. *Indian Forester,* 70 : 297–304.

Chong, V.C. 1979. Mangroves and prawns with special reference to Malaian Environment useful informative paper. Deptt. of Zool. Univ. Malaya.

Christensen, B. 1978. Biomass and primary production of *Rhizophora apiculata* Bl. in a mangrove in Southern Thailand. *Aquatic Botany.* 43 : 52.

Cintron, G., Lugo, A.E., Poal, D.J. and Marris, G. 1978. Mangrove of Arid Environments in Puerto Rico and adjacent Islands. *Biotropica.* 10 : 110–121.

Clarke, G. and Hannon, N.J. 1971. The mangrove swamp and salt marsh communities of the Sydney district. IV. The significance of species interaction. *J. Ecol.*45(3) : 535–53.

Clough, B.F. and Attiwill, P.M. 1975. Nutrient cycling in a community of *Avicennia marina* in a temperate region of Australia. *Proceedings of International Symposium on Biology and Management of Mangrove.* 1 : 137–146.

COE (1983). Global status of mangroves ecosystems. Ed. Saenger Hegeert and Davie Glard, Switzerland.

Conner, W.H. and Day, J.W. 1976. Productivity and composition of a bold cypress water typeto site and bottom and hardware site in a Lausiana swamp. *American Journal of Botany.* 63 : 1354–1364.

Cook, M.T. 1907. The embryology of *Rhizophora mangle. Bull. Torrey Bot. Club.* 34: 271–7.

Corner, E.J.H. 1939. Notes on the systematy and distribution of Malayan phanerogams. III. *Gard. Bull. Straits Settl.* 10 : 239–328.

Cooper, R.E. and Pasha, S.A. 1935. The osmotic and suction pressure of some species of mangrove vegetation. *J. Indian Bot. Soc.* 14 : 100–120.

Crane, J. 1973. Fiddler crabs of the world (Ocypodidae : genus uca) *Princeton Univ. Press. Princeton,* pp. 660.

Crutis, J.T. and McIntosh, R.P. 1951. An Upland forest continuum in the prairie-forest border region of Wisconsin. *Ecol.* 32 : 476–496.

Cruz, A.A. de la. 1964. Mangrove tidal swamps of the tropicas. *Sci. Rev. A.A.* 10(1):9–16.

Cruz de la and Bannag, J.F. 1967. The ecology of a small mangrove path in matabeng bay beach, Botangors Brovinces. *Natural and Applied Science Bulletin.* 60 : 486–494.

Cruz A.A. de la 1976. Functions of coastal wetlands. Association of South East. *Biol. Bull.* 23 : 179–185.

Daniel, M.J. 1975. Preliminary account of the litter production in a New Zealand low land prodocarp–rota broad leaf forest. *N.Z.J. Bot.* 13 : 87.

Dacey, J.W.H. 1980. Internal winds in waterlilies : An adaptation for life in anaerobic sediments. *Science.* 203 : 1253–5.

Dacey, J.W.H. 1981. Pressurized ventilations in the yellow waterlily. *Ecology.* 62 : 1137–47.

Dagar, J.C. 1982. Some ecological aspects of mangrove vegetation of Andaman and Nicobar Island, India. *Sylvetrop. Philipp. For. Res. J.* 7 : 177–216.

Dagar, J.C., Mongia and Bandopadhyay. 1991. Mangroves of Andaman and Nicobar Islands. Oxford & IBSS Publishing Co. Pvt. Ltd., New Delhi.

Dahlgren, R.M.T. 1980. A revised system of classification of the angiosperm. *Bot. J. Linn. Soc.* 80 : 91–124.

Dandy, J.E. and Exell, A.W. 1938. On the nomenclature of three species of *Caesalpinia. J. Bot. Lond.* 76 : 175–83.

Dass, A.K. and Dev Roy M.K. 1980. On the wood-boring mallusks of south Andamans, India. *Rec. Zool. Surv. India.* 77 : 179–187.

Davey, J.E. 1975. Notes on the mechanism of pollen release in *Bruguiera gymnorrhiza. J. S. Afr. Bot.* 41 : 269–72.

Davis, J.H. 1940. The ecology and geologic role of mangroves in Florida. *Pap. Tortugas Lab.* 32 (*Publ. Carn. Inst. No.* 517) : 303–41.

Daubenmire, R.F. 1968. Plant communities : A textbook of plant synecology. Harper and Row, New York, pp. 300.

Davis, J.H. 1960. The ecology and geologic role of mangrove in Florida Carnegic. *Inst. Lash. Puse*, No. 517 : 30–412.

Dehadrai, P.V. and Bhargava, R.M.S. 1972. Seasonal organic production in relation to environmental features in Mardari and Zuri estuaries, Goa. *Indian J. Mar. Sci.* : 52–56.

Diaz-piferrer, M. 1962. The effects of an oil spill on the shore of Guenica Puerto Rico. *Deep Sea Research.* 11 : 855.

Dickinson, R.E. 1928. Distribution of life in Philippines.

Dickinson, C.H. 1974. Decomposition of plant litter in soil. In : C.N. Dickinson and G.J.F. Pugh (Eds.) *Biology of plant litter decomposition*, Vol.2, Academic Press, London and New York, pp. 633–688.

DeVogel, E.F. 1980. *Seedlings of dicotyledons.* Wageningen : Centre for Agricultural Publication and Documentation.

Ding Hou. 1957. A conspectus of the genus *Bruguiera* (Rhizophoraceae). *Nova Guinea n.s.* 8(1) : 163–71.

Ding Hou. 1958. Rhizophoraceae. *Flora Malesiana*, Series 1, 5 : 429–93.

Ding Hou. 1960. A review of the genus *Rhizophora. Blumea* 10(2) : 625–34.

Ding Hou. 1963. Celastraceae. *Flora Malesiana*, Series 1, 6(2) : 227–91.

Ding Hou. 1977. Anacardiaceae. *Flora Malesiana*, Series 1, 8 : 395–548.

Docters van Leeuwen, W. 1911. Ueber die Urasche der weiderholter Versweigung der Stultzwurzeln von *Rhizophora. Ber d. bot. Gesell.* 29 : 476–8.

Duke, N.C. 1984. A mangrove hybrid, *Sonneratia x gulngai* (Sonneratiaceae) from north–eastern Australia. *Austrobaileya.* 2 : 103–5.

Duke, N.C., Birch, W.R. and Williams, W.T. 1981. Growth rings and rainfall correlations in a mangrove tree of the genus *Diospyros* (Ebenaceae). *Aust. J. Bot.* 29 : 135–42.

Duke, N.C. and Bunt, J.S. 1979. The genus *Rhizophora* (Rhizophoraceae) in north–eastern Australia. *Aust. J. Bot.* 27 : 657–78.

Duke, N.C., Bunt, J.S. and Williams, W.T. 1984. Observations on the floral and Vegetative phenologies of north–eastern Australian mangroves. *Aust. J. Bot.* 32: 87–99.

Diude, O. 1896. Deutschtands pflanzengeographic. Handb. Deut. Landesu. Volkskunde. 4 : 502.

Dombois, D.M. and Ellenberg, H. 1974. *Amis and Methods of Vegetation Ecology.* Wiley International, pp. 547.

Du Terte 1967. In : H.H.M.Bowman (Ed.) *Ecology and physiology of Red Mangroves.* Proc. Am. Phil. Soc. 61 : 589–670.

Duck, N.C., Bunt, J.S. and Williams, W.T. 1981. Mangrove litter fall in N.E. Australia. 1 Annual totals by components in selected sp. *Australian Journal of Botany.* 29: 547–553.

Egler, F.E. 1948. The dispersal and establishment of red mangroves, *Rhizophora* in Florida. *Caribb. For.* 9(4) : 299–319.

Egler, F.E. 1952. Southeast saline everglades vegetation, Florida and its management. *Vegetation.* 3(4/5) : 213–165.

Ellis, M.B. 1971. Dematiaceous Hyphomycetes. *Kew Commonwealth Mycological Institute*, pp. 608.

Ellis, M.B. 1979. More Dematiaceous Hyphomycetes. *Kew Commonwealth Mycological Institute*, pp. 507.

Eong, O.J., Khoon, J.S., Hong, W.C. and Dhanraja, G. 1979. Contribution of aquatic productivity in a managed mangrove ecosystems in Malaysia. Paper presented in *Int. Conf. in trends in applied Biol. in South East Asia Univ. Sains Malasia Penang Malasia.*

Essig, F.B. 1971. Pollination in some New Guinea palms. *Principles.* 17 : 75–83.

Ewel, J. 1980. Tropical succession : manifold routes to maturity. *Biotropica* 12 (suppl.): 2–7.

Exell, A.W. 1954. Combretaceae. *Flora Malesiana*, Series 1. 4 : 533–89.

Exell, A.W. and Stace, C. 1966. Revision of Combretaceae. *Bot. Soc. Broteriana.* 40(2) : 5–25.

Faber, C. von 1913. Ueber transpiration und osmotischen druck bei den Mangroven. *Ber. d. bot. Gesell.* 31 : 277–81.

Faber, C. von 1923. Zur Physiologie der Mangroven. Ber. d. bot. Gesell. 41 : 227–34.

Faber, F.C. von 1935. Schimper's Plantgeography auf physiologicher grundlage, 3rd edition, vol.2, Fischer Jena.

Fell, J.W. 1980. The association and potential role of fungi in mangrove detrital systems. *Botanica Marina*, 23 : 257–63.

Fell, J.W., Cefalu, R.C., Master, I.M. and Tallman, A.S. 1975. Microbial activities in the mangrove (*Rhizophora mangle*) leaf detrital system. In C.E. Walsh, S.C. Snedaker, and H.J.Teas (Eds.) *Proceedings International Symposium the Biology & Management of Mangrove*, 8–11 October (9) 4 Honlulu, pp. 661, University of Florida, Gainesville.

Fell, J.W., Master, I.M. and Newell, S.Y. 1980. Laboratory model of the potential role of Fungi (*Phytophthora* spp.) in the decomposition of Red Mangrove (*Rhizophora mangle*) leaf litter, p.359–72. In : K.P. Tenore and B.C. Coull (eds.) *Marine Benthic Dynamic*, Columbia, University of South Carolina Press.

Fell, K.W. 1975. Phycomycetes (*Phytophthora* spp. nov. and *Pythium* sp. nov.) Associated with Degrading Mangrove (*Rhizophora mangle*) leaves. *Can. J. Bot.* 53(24) : 2908–22.

Fell, K.W., Cefalu, R.C., Master, I.M. and Tallman, A.S. 1975. Microbial activities in the Mangrove (*Rhizophora mangle*) leaf detrital system. In : G.Walsh, S.Snedaker and H.Teas (Eds.) Proceedings of International Symposium on the Biology and Management of Mangrove, 8–11 October, Honolulu, Vol. 2, pp.661–79. Gainesville, Institute of Food and Agricultural Sciences, University of Florida.

Fell, J.W. and Hunter, I.L. 1979. Fungi Associated with the decomposition of the black Rush, *Juncus roemarianis*, in South Florida. *Mycologia*, 71(2) : 322– 42.

Fell, J.W. and Master, I.M. 1973. Fungi associated with the degradation of Mangrove (*Rhizophora mangle* L.) Leaves in South Florida. In : H.L.Stevenson and R.R.Colwell (eds.) *Estuarine Microbial Ecology*. Belle W. Baruch Symposium I, pp.455–6, Columbia, University of South Carolina Press.

Fell, J.W. and Newell, S.Y. 1980. Role of Fungi in Carbon Flow and Nitrogen Immobilization in coastal Marine Plant Litter Systems. In : D.T.Wicklow and G.C.Carroll (eds.), *The fungal community : Its organization and role in the ecosystem*, pp.665–78, New York, Marcel Dekker.

Fisher, J.B. 1978. A quantitative study of *Terminalia* branching. Chap. 13 in *Tropical trees as living systems*. eds. P.B. Tomlinson and M.H. Zimmermann, Cambridge, Cambridge University Press.

Fisher, J.B. and Honda, H. 1977. Computer simulation of branching pattern and geometry in *Terminalia* (Combretaceae), a tropical tree. *Bot. Gaz.* 138 : 337– 84.

— 1979a. Branch geometry and effective leaf area : A study of *Terminalia* branching pattern, 1. Theoretical trees. *Am. J. Bot.* 66 : 633–44.

— 1979b. Branch geometry and effective leaf area : A study of *Terminalia* branching pattern, 2. Theoretical trees. *Am. J. Bot.* 66 : 645–55.

Fisher, J.B. and Stevenson. J.W. 1981. Occurrence of reaction wood in branches of dicotyledons and its role in tree architecture. *Bot. Gaz.* 142 : 82–95.

Flowers, T.J., Troke, P.F. and Yeo, A.R. 1977. The mechanism of salt tolerance in halophytes. *Ann. Rev. Plant Physiol.* 28 : 89–121.

Fosberg, F.R. 1939. *Diospyros ferrea* (Ebenaceae) in Hawaii. *Occas. Pap. Bernice Pauahi Bishop Mus.* 15 : 119–31.

— 1975. Phytogeography of Micronesian mangroves. In *Proceedings of the International Symposium on Biology and Management of Mangroves*. eds. G.E. Walsh, S.C. Snedaker and H.J. Teas. Gainesville, Institute of Food and Agricultural Sciences, University of Florida.

Fosberg, F.R. and Sachet, M.–H. 1972. *Thespesia populnea* (L.) Solander ex Correa and *Thespesia populneoides* (Roxburgh) Kosteletsky (Malvaceae). *Smithson. Contrib. Bot.* 7 : 1–13.

Foxworthy, F.W. 1910. Distribution and utilization of mangrove swamps in Malay. *Ann. Jard. Bot. Buitenz.* Series 2, Suppl. 3, part 1 : 319–44.

Garge, A. 1988. Ecological studies of mangroves forests of Andaman Islands. Ph.D. Thesis, Vikram University, Ujjain, pp.293.

Garge, A., Singh, V.P., Pathak, S.M., Donderia, M.S. and Mall, L.P. 1985. Mangroves and chemical characteristics of inundating water in Andaman region. *J. Hydrobiology*, 1(2) : 91–95.

Garge, A., Singh, V.P., Pathak, S.M. and Mall, L.P. 1987. Mangals hydrology of Ritches archipelago of Andaman Island. *Proc. Nat. Symp. on Perspectives in Hydrobiology*, Vikram University, Ujjain (Awarded) Sec. V (430) : 221–224.

Giesey, J.P. 1980. Microcosms in Ecological Research. Springfield Va. Technical Information Centre, *U.S. Department of Energy*, 1, pp.110.

Gill, A.M. 1971. Endogenous control of growth-ring development in *Avicennia For. Sci.*, 17 : 462–5.

Gill, A.M. and Tomlinson, P.B. 1969. Studies on the growth of red mangrove (*Rhizophora mangle* L.) I, Habit and general Morphology. *Biotropica* 1(1) : 1–9.

— 1971a. Studies on the growth of red mangrove (*Rhizophora mangle* L.), 2, Growth and differentiation of aerial roots. *Biotropica*, 3(1) : 63–77.

— 1971b. Studies on the growth of red mangrove (*Rhizophora mangle* L.), 3, Phenology of the shoot. *Biotropica*, 3(2) : 109–24.

— 1975. Aerial roots : an array of forms and functions. Chap.12 in *The development and function of roots*, eds. J.G.Torrey and D.T.Clarkson. London : Academic Press.

— 1977. Studies on the growth of red mangrove (*Rhizophora mangle* L.), 4, The adult root system. *Biotropica*, 9(3) : 145–55.

Golley, F.B., McGinnis, J.T., Clements, R.G., Child, G.I. and Duever, M.J. 1975. *Mineral Cycling in a tropical moist forest Ecosystems*. University of Georgia Press Athans Ga, 248 pp.

Golley, F.B., Odum, H.T. and Wilson, R.F. 1962. The structure and metabolism of a Puerto Rico red mangrove forests in May. *Ecology* 43 : 9–19.

Golley, P.M., Golley, F.B. 1972. Tropical Ecology with an emphasis on organic productivity. Athans. *Ga Inst. Soc. Trop. Ecol. Int. Assoc. Ecol. & Indian Nat. Sci. Acad.* p.418.

Gosselink, J.C. and Turner, R.E. 1878. The role of hydrology in freshwater wetland ecosystems. 63–78 In: R.E. Good, D.F. Whigham and R.L. Simpson (Eds.) *Freshwater Wetlands : Ecological Processes and Management Potential*. Academic Press, New York, U.S.A.

Gosz, J.R., Likens, G.E., Bormann, F.H. 1973. Nutrient Release from Decomposing Leaf and Branch litter in the Hubbard Brook Forest, New Hampshire. *Ecol. Monogr.* 43 : 173–191.

Goulter, P.F.E. and Allaway, W.G. 1979. Litter fall and decomposition in a mangrove stand *Avicennia marina* (Forsk.). Vich. in Middle Harbour, Sydney. *Aust. J. Marine and Freshwater Res.*, 30 : 541–546.

Graham, A. 1977. New records of *Pelliceria* (Theaceae / Pelliceriaceae) in the tertiary of the Caribbean. *Biotropica*, 9 : 48–52.

Greenway, H. and Munns, R. 1980. Mechanisms of salt tolerance in nonhalophytes. *Ann. Rev. Plant Physiol.*, 31 : 149–90.

Guppy, H.B. 1906. *Observations of a naturalist in the Pacific between 1896 & 1899, Plant Dispersal*, Vol.2 pp. 627. London : McMillan.

— 1917. *Plant seeds and currents in the West Indies and Azores*, pp. 531. London: Williams & Norgate.

Hamball, G.C. and Yata Zalpu, M. 1984. Nitrogen fixing bacteria from warty lenticellate. Bark of Mangrove Tree – *Bruguiera gymnorhiza* (L.) Lamk. *Applied and Environmental Microbiology*, Vol.47(1) : 44–48.

Harber, W. 1958. Okologische Untersuchungen der Bodenatmung. *Flora*, 146 : 109– 157.

Harley, H.L. 1971. Fungi in Ecosystems. *J. Appl. Ecol.*, 8(3) : 627–42.

Haberlandt, G. 1910. Eine botanische Tropenreise. Leipzig : Wilhelm Engelmann. 1928. *Physiological Plant Anatomy*. London : Macmillan.

Halle, F., Oldeman, R.A.A. and Tomlinson, P.B. 1978. *Tropical trees and forests and architectural analysis*. Berlin : Springer–Verlag.

Hartog, C. den. 1970. The sea–grasses of the World. *Verhandl. Kon. Ned. Akad. Wetensch. Nat.*, 59(1) : 1–275.

Hattink, T.A. 1974. A revision of Malesian *Caesalpinia*, including *Mezoneuron* (Leguminose–Caesalpiniaceae). *Reinwardtia*, 9(1) : 1–69.

Hart, M.G.R. 1959. Sulphur Oxidation in tidal mangrove swamps of Sierra Leone. *Plant and Soil*, 11 : 215–236.

Hart, M.G.R. 1962. Observations on the source of acid in empoldered mangrove soils. I. Formation of elements sulpher. *Plant and Soil*, 17 : 87–89.

Heald, E.J. 1971. The production of organic detritus in a south Florida estuary. *Univ. Miami Sea Grant Tech. Bull.*, 6 : 162.

Heald, E.J. 1969. The production of organic detritus in a South Florida estuary. Unpublished Ph.D. thesis. University of Miami, Florida, USA.

Hesse, P.R. 1961a. Some difference between the soils of *Rhizophora* and *Avicennia* swamps in Sierra Leone. *Plant and Soil*, 14 : 335–46.

Hesse, P.R. 1961b. The decomposition of organic matter in a mangrove swamp soil. *Plant and Soil.*, 14 : 249–263.

Hemsley, W.B. 1879–88. *Botany in Biologia Centrali–Americana*. 5 vols. London. Porter and Dulau.

Holbrook, N.M. and Putz, F.E. 1982. Vegetative seaward expansion of *Sonneratia alba* trees in a Malaysian mangrove forest. *Malay. For.* 45 : 278–81.

Hosakawa, T., Tagawa, H. and Chapman, V.J. 1977. Mangals in Micronesia, Taiwan, Japan, the Philippines and Oceania. Chap.14 in *Ecosystems of the world*, Vol.1, *Wet coastal ecosystems*. Amsterdam : Elsevier Scientific.

Howard, R.A. 1981. Three experiences with manchineel (*Hippomane* spp., Euphorbiaceae. *Biotropica* 13 : 224–7.

Howe, M.A. 1911. A little known mangrove of Panama. *J.N.Y. Bot. Guard.* 12 : 61–72.

Holdridge, L.R. 1967. *Life Zone Ecology*. Tropical Science Conference, Sen Jose Costa Rica, p.206.

Holthuis, L.B. 1956. Three species of Crustacea decapoda macrura from Southern Brasil, including a new species of *Upogebia*. *Zool. Meded.* 34 : 173–181.

Jackson, M.L. 1958. *Soil Analysis*. Prentice Hall Englwood Cliffs. N.J. USA, pp.498.

Jansonnius, H.H. 1950. The vessels in the wood of Javan mangrove trees. *Blumea*, 6 : 465–9.

Jernelow, A.O., Linden, O. and Rosenblum, I. 1976. The St. Peter Oil Spill – An ecological and socio–economic study of effects. Columbia and Ecuador. May– June 1976. *Swedish water and air pollution Research Inst. Stockholm*, Sweden, pp.34

Jenik, J. 1970. Root system of tropical trees, 5. The peg roots and the pneumathodes of *Laguncularia racemosa* Gaertn. *Preslia* 42 : 105–13.

— 1978. Roots and root systems in tropical trees : morphologic and ecologic aspects. Chap.14 in *Tropical trees as living systems*, eds. P.B. Tomlinson and M.H. Zimmerman. Cambridge : Cambrdige University Press.

Jonker, F.P. 1959. The genus *Rhizophora* in Suriname. *Acta Bot. Neerl.* 8 : 58–60.

Joshi, G.V., Pimplaskar, M. and Bhosale, L.J. 1972. Physiological studies in germination of mangroves. *Bot. Mar.* 15 : 91–5.

Joshi, G.V. 1973. Soil plant relationship in the plants of the saline soils of the Deccan. *Proc. Symp. on Deccan Trap Country*, held under the auspices of the Indian National Science Academy, 45 : 30–38.

Joshi, G.V., Sontakke, S.D. and Bhosale, L.J. 1980. Photosynthesis in mangroves. Proceeding of Symposium on Biology and Management of Mangrove Papua New Guinea (D.W.Jank Publications).

Juncosa, A.M. 1982a. Embryo and seedling development in the Rhizophoraceae. Ph.D. thesis, Duke University, Durham, North Carolina.

— 1982b. Developmental morphology of the embryo and seedling of *Rhizophora mangle* L. (Rhizophoraceae). *Am. J. Bot.* 69 : 1599–611.

— 1984a. Embryogenesis and seedling development in *Cassipourea elliptica* (Sw.) Poir (Rhizophoraceae). *Am. J. Bot.* 71 : 70–9.

— 1984b. Embryogenesis and seedling developmental morphology of the seedling in *Bruguiera exaristata* Ding Hou (Rhizophoraceae). *Am. J. Bot.* 71 : 180–91.

Karsten, G. 1890. Ueber die Mangrove–Vegetation in Malayischen Archipel. *Ber d. bot. Gesell.* 8 : 49–55.

— 1891. Ueber die Mangrove–Vegetation in Malayischen Archipel. *Eine morphologi schbiologische Studie. Bibl. Bot.* 22 : 71.

Kearney, T.H. 1954. Notes on Malavaceae V. *Leaf. West Bot.* 7 : 118–19.

Keay, R.W.J. 1953. *Rhizophora* in West Africa. *Kew Bull.* 1953 : 121–7.

Kenneally, K.F., Wilson, P.G. and Semeniuk, J. 1978. A new character for distinguishing vegetative material of the mangrove genera *Bruguiera* and *Rhizophora* (Rhizophoraceae). *Nuytsia* 4 : 178–80.

Kipp–Goller, A. 1940. Ueber Bau und Entwicklung der viviparen Mangrovekeimlinge. *Z. Bot.* 35 : 1–40.

Kloss, C.B. 1902. The Andaman and Nicobar. Vivek Publishing House Kamala Nagar, New Delhi, pp.373.

Knapp–van Meeuwen, M.S. 1970. A revision of four genera of the tribe–Leguminosae Caesalpinoideae–Cynometreae in Indomalesia and the Pacific. *Blumea* 18(1) : 1–52.

Kobuski, C.E. 1951. Studies in the Theaceae XXII, The genus *Pelliciera*. *J. Arnold. Arbor.*, 32 : 256–62.

Kostermans, A.J.G.J. 1959. Monograph of the genus *Heritiera aiton* (Sterculiaceae). *Reinwardtia*, 4 : 465–83.

Kolehmainer, S.E., Martin, F.D. and Scholander, P.B. 1975. Environmental effect of cooling system at nuclear power plant Inst. *Atome. Eng. Agerey Viema Italy*, pp.409–422.

Krishnamurthy, K. 1971. Phytoplankton pigments in Porto Novo waters (India). *Int. Rev. Ges. Hydrobiol.*, 56 : 373–282.

Krishnamurthy, K. Prince Jeyaseelan, M.J. and Sultan Ali, M.A. 1984. The distribution of fishes in the global mangroves. *Curr. Sci.* 53(5) : 901–906.

Krishnamurthy, K. 1979. Mangroves. *Sci. Reporter*, 16 : 481–483.

Krishnamurthy, K. and Sunderraj, V. 1973. A survey of environmental features in a sections of Valler–Coleroon estuarine system, South India. *Mar. Biol.* 23(3) : 229–237.

Lacerda, L.D. de 1981. Mangrove wood pulp, an alternative food source for the tree crab *Aratus pisonii. Biotropica* 13 : 317.

Lacerda, L.D. de and May, D.V. 1982. Evolution of a new community type during the degradation of a mangrove ecosystem. *Biotropica*, 14 : 238–9.

Lam, H.J. 1934. Materials towards a study of the flora of the island of New Guinea. *Blume*, 1 : 115–159.

Larue, C.D. and Muzik, T.J. 1951. Does the mangrove really plant its seedlings ? *Science*, 114 : 661–2.

— 1954. Growth regeneration and precocious rooting in *Rhizophora mangle. Pap. Mich. Acad. Sci. Arts. Lett.* (Part I, Botany and Forestry), 39 : 9–29.

Lawrence, D.B. 1949. Self–erecting habit of seedling red mangroves (*Rhizophora mangle* L.) *Am. J. Bot.*, 36(5) : 426–7.

Lee, D. 1980. *The sinking ark : environmental problems in Malaysia and southeast Asia.* Kula Lumpur : Heinemann.

Lersten, N.R. and Curtis, J.D. 1974. Colleter anatomy in red mangrove, *Rhizophora mangle* (Rhizophoraceae). *Can. J. Bot.* 52 : 2277–80.

Lewis, D. and Rao, A.N. 1971. Evolution of dimorphism and population polymorphism in *Pemphis acidula* Frost. *Proc. R. Soc. Lon. B.*, 178 : 79–94.

Lewis, R.R. 1983. Impact of oil spills on mangrove forests. Chap.19 in *Biology and Ecology of Mangroves*, Ed.H.J.Teas. *Tasks for Vegetation Science*, 8. The Hague : Junk.

Lear, R. and Turner, T. 1977. *Mangrove of Australia.* Univ. Queensland Press, Brisbane, pp.84.

Lewis, A.D. and Ellway, C.P. 1971. Fishes of Taliebudgera Creek, South Queensland. *Operculum* 3 : 60–63.

Liebau, O. 1914. Beitrage zur Anatomie und Morphologie der Mangrove–Pflanzen, insbesondere ihres Wurzelsystems. *Beit. Biol. Pflanz.* 12 : 181–213.

Linden, O. and Jernal, O.R. Al 1979. The mangrove swamp – An environment in danger. UNESCO Publication 1979. Regional Dacca (1978).

Lugo, A.E. 1980. Mangrove ecosystems : successional or steady state. *Biotropica*, 12 (suppl.) : 65–72.

Lugo, A.E. and Snedaker, S.C. 1974. The ecology of mangroves. *Annual Review of Ecology* and *Systematics*, 5 : 39–64.

Lugo, A.E. and Snedaker, S.C. 1975. Properties of mangrove forest in Southern Florida. *Proc. on Intern. Sympo. on biology and Management of Mangroves*, 1 : 170– 212.

Lundegradh, H. 1927. Carbon dioxide evolution of soil and crop growth. *Soil Sci.*, 23 : 4117–453.

Lush, D.C. and Hynesh, H.B.N. 1973. The formation of particles in freshwater Leachates of dead leaves. *Limnol. Oceanogr.* 18(6) : 968–77.

MacArthur, R.H. and Wilson, E.O. 1967. *The theory of island biogeography.* Princeton : Princeton University Press.

MacNae, W. 1968. A general account of the fauna and flora of mangrove swamps and forests in the Indo–West Pacific region. *Adv. Mar. Biol.*, 6 : 73–270.

MacNae, W. and Fosberg, F.R. 1981. Sonneratiaceae. In : M.D.Dassanayake (Ed.) *A Revised Handbook to the Flora of Ceylon*, Vol.3, American Publishing, New Delhi.

MacNae, W. 1968. A general account of the flora and fauna of a mangrove swamps forests in the Indo-West Pacific region. *Adv. Mar. Biol.*, 6 : 73–270.

MacNae, W. and Kalk, M. 1962. The ecology of the mangrove swamps of Inhaca Island, Mozambique, *J. Ecol.*, 50 : 19–34.

Mall, L.P., Singh, V.P., Garge, A. and Pathak, S.M. 1985. A new approach towards the mangrove forest flora of Andaman islands. *Indian For.*, 111(5) : 290–300.

Mall, L.P., Singh, V.P., Garge, A. and Pathak, S.M. 1987. Ecological studies of mangroves at Ritchies Archipelago. *Trop. Ecol.* 28 : 182–192.

Mall, L.P., Singh, V.P., Garge, A. and Pathak, S.M. 1985. Mangrove vegetation it's structure, ecology and importance. *Trends in plant Research*, pp.457–472.

Mall, L.P., Singh, V.P., Garge, A. and Pathak, S.M. 1985. Mangrove forests of Andaman and some aspects of its ecology. *Proc. Nat. Symp. on Biol. Utilization and Conservation of Mangroves*. Shivaji University, Kolhapur, pp.25–38.

Mall, L.P., Singh, V.P and Garge, A.1991. Study of biomass. Litterfall, litter decomposition and soil respiration in monogeneric mangrove forests and mixed mangrove forests. *Tropical Ecology*, 32(1) : 144–152.

Malley, D.F. 1978. Degradation of Mangrove Leaf litter by Tropical Seasarmid Crab *Chiromanthes onychophrum. Mar. Biol.*, 4 : 377–86.

Markgraf, F. 1927. Die Apocynaceen von Neu-Gurinea, 117. In : C.Lauterbach (ed.) *Beitrage zur Flora von Papuasien*, 14. Leipzig : Max Weg. (Sonderabdruck aw Engler, *Bot. Jahrbucher* 61 : 1927).

McVeen, D.B. 1976. Report on land use in the Andaman and Nicobar Island, IUCN, Merges Switzerland.

Moldenke, H.N. 1960. Materials towards a monograph of the genus *Avicennia* L. I and II. *Phytologia* 7 : 123–68, 179–252, 259–63.

Moore, H.E. 1973. The major groups of palms and their distribution. *Gentes Herbarum*, 11(2) : 27-141.

Mukherjee, J. and Chanda, S. 1973. Biosynthesis of *Avicennia* L. in relation to taxonomy, *Geophytology*, 3 : 85–8.

Mullan, D.P. 1931a. Observations on the water-storing devices in the leaves of some Indian Halophytes. *J. Ind. Bot. Soc.* 10 : 126–32.

— 1931b. On the occurrence of glandular hairs (salt glands) on the leaves of some Indian Halophytes. *J. Ind. Bot. Soc.* 10 : 184–9.

Muller, J. 1981. Fossil pollen records of extant angiosperms. *Bot. Rev.*, 47 : 1–142.

Muller, J. and Hou Liu, S.Y. 1966. Hybrids and chromosomes in the genus *Sonneratia blumea*, 14(2) : 337–43.

Muller, J. and van Steenis, C.G.G. 1968. The genus *Sonneratia* in Australia. *N. Queensl. Nat.*, 34 : 6–8.

Nawalkar, B.S. 1961. Importance of mangroves. *Trop. Ecol.* 2 : 89–93.

Nawalkar, B.S. 1959. Studies in the ecology of the mangroves. *J. Univ. Bombay*, N.S., 28(3) : 6–10.

Nawalkar, B.S. 1973. Ecological study and geographical distribution survey of the halophytes of Bombay and Maharashtra, India. *Proc. Indian Sci. Congr.*, 59(3) : 394–395.

Nawalkar, B.S. and Bharucha, F.R. 1949. Studies in the ecology of mangroves. *J. Univ. Bombay N.S.*, 18(3) : 17–35.

Nawalkar, B.S. 1940. Studies in the ecology of mangroves. I. Determination of the osmotic pressure of *Avicennia alba* Blume. *J. Univ. Bombay*, 8 : 58–74.

Nawalkar, B.S. 1941. Studies in the ecology of mangroves. II. Physical factors of the mangrove soil. *J. Univ. Bombay*, 9 : 78–92.

Nawalkar, B.S. 1951. Succession of mangrove vegetation in Bombay and Islands. *J. Bombay Nat. Hist. Soc.*, 50 : 157–160.

Nearchus, 235 B.C. Arr Anab VI (cited from Bowman, 1917).

Newell, S.Y. 1981. Fungi and Bacteria in or on Leaves of Eelgrass (*Zostera marina* L.) from Chesapcake Bay. *Applied and Environmental Microbiology*, 41(55) : 1919–24.

Newell, S.Y. and Cooksey, K.E., Fell, J.W., Master, I.M., Miller, C. and Walter, M.A. 1981. Acute Impact of an Organophosphorus Insecticide on Microbes and Small Invertebrates of a Mangrove estuary. *Arch. Environm. Contam. Toxicol.*, 10 : 227–35.

Newell, S.Y. and Fell, J.W. 1980. Mycoflora of Turtlegrass (*Thalassia testudinum* Konig) as Recorded after Seater Incubation. *Botanica Marina*, 23 : 257–63.

Newell, S.Y. 1982. Surface sterilization and the active mycoflora of a seagrass. *Botanica Marina*, 25 : 339–46.

Nie, N.H., Hull, C.H., Jenkins, J.G. and Steinbrener, K., Bent, D.H. 1975. *SPSS Statistical Package for the Social Sciences*. McGraw Hill, New York, pp. 675.

Nakanischi, S. 1964. An epiphytic community on the mangrove tree, *Kandelia candel. Hikobia*, 4 : 124.

Ng, F.S.P. 1978. Strategies of establishment in Malayan forest trees. Chap.5 In : P.B.Tomlinson and M.H.Zimmermann (Eds.) *Tropical trees as living systems*, Cambridge University Press, Cambridge.

Noamesi, G.K. 1958. A revision of the Xylocarpeae (Meliaceae). Thesis. University of Wisconsin, Madison.

Odum, E.P. 1969. The strategy of ecosystem development. *Science*, 164 : 262–270.

Odum, E.P. 1971. *Fundamentals of Ecology*. Philadelphia, W.B.Saunders Co., pp.574.

Odum, E.P. 1980. In : V.S. Kennedy (Ed.) The status of three ecosystem level hypothesis regarding salt marsh estuaristidal subsidy, outwelling and detritus- based food chains, *Estuarine Perspective* Academic Press, New York, pp. 483–495.

Odum, W.E. and Heald, E.J. 1972. Trophic analysis of an estuarine mangrove community. *Bulletin of Marine Science*, 22 : 671–738.

Odum, W.E. 1970. Pathways of energy flow in a south Florida estuary. Ph.D. Thesis, University of Miami, Coral Gables, Florida.

Odum, W.E. 1972. Pathways of energy flow in a south Florida estuary. *Univ. of Miami, Sea Grant Tech. Bull No.7*, pp.162.

Ogura, Y. 1940. On the types of abnormal roots in mangrove and swamp plants. *Bot. Mag. Tokyo*, 54 : 389–404.

Olexa, M.T. and Freeman, T.E. 1978. A gall disease of red mangrove caused by *Cylindrocarpon didynum Plant Dis. Rep.*, 62 :´283–5.

Oostting, H.J. 1956. The study of plant communities. W.H.Freeman and company, San Fransisco and London, pp.440.

Osborne, D.J. and Berjak, P. 1997. The making of mangrove : The remarkable pioneering role played by seeds of *Avicennia marina. Endeavour*, 21(4) : 145–147.

Oviedo, G.F. 1926. Primera parie de la historia natural general de las Indias (cited from Bowman, 1917).

Onell, R.V. 1976. Ecosystem Persistence and Heterotrophic Regulation. *Ecology* VI, 57(6) : 1244–53.

Paivokee, A.E.A. 1984. Tapping patterns in the Nipa palm (*Nypa fruticans* Wurmb.) *Principes*, 28 : 132–7.

Pannier, F. 1959. El efecto de distintas concentraciones salinas sobre el desarrollo de *Rhizophora mangle* L. *Acta Cient. Venezolana*, 10(3) : 68–78.

— 1962. Estudio fisiologico sobre la vivipara de *Rhizophora mangle* L. *Acta Cient. Venezolana*, 13(6) : 184–97.

— 1965. Problemas de translocacion en *Rhizophora mangle*. 16th Congr. das. B.B. Doc., 15, Itabuna, Bahia.

Pannier, F. and Pannier, R.F. 1975. Physiology of vivipary in *Rhizophora mangle* L. *Proc. Int. Symp. Biol. Mgt. Mangroves*, 2 : 632–9.

Panshin, A.J. 1932. An anatomical study of the woods of the Philippine mangrove swamps. *Philipp. J. Sci.*, 48(2) : 143–205.

Parkinson, C.E. 1923. A forest flora of the Andaman islands. Bishen Singh, Dehradun, pp.325.

Parkinson, C.E. 1934. The Indian species of *Xylocarpus*. *Ind. For.*, 60 : 136–8.

Pax, F. and Hoffman, K. 1912. Euphorbiaceae. In Engler's *Das Pflanzenfamilien*, 52 (4.147.5 Hippomaneae). Leipzig : Wilhelm Engelman.

— 1931. Euphorbiaceae. In Engler and Prantl, *Die naturlichen Pflanzenfamilien*, ed.2, Bd 19c, 11–233, Leipzig, Wilhelm Engelmann.

Paynes, P.D.W. 1967. A monograph of the genus *Barringtonia* (Lecythidaceae). *Blumea*, 15 : 157–265.

Piper, C.S. 1944. Soil and Plant Analysis : A monograph from waite Agricultural Research Institute University of Adelaide Press, Adelaide, South Australia.

Pool, D.J., Snedaker, S.C. and Lugo, A.E. 1975. Litter production in mangrove forests of south Florida and Puerto–Rico. In : G.Walsh, S.C.Snedaker and H.Teas (eds.) Proc. of Intern symposium on Biology and management of mangrove I pp.213–237.

Pool, D.J., Snedaker, S.C. and Lugo, A.E. 1977. Structure of mangrove forest in Florida, Puerto Rico, Mexico and Costa Rica.

Por, F.D. 1984. Editor's note on tides and water levels in the mangal. *Hydrobiology of mangal*, W.Junk Publishers, pp.25–26.

Por, F.D., Dar, I. and Amir, A. 1977. The mangal of Sinai limits of an ecosystem. Helgal. Wiss. Meeresunters, 30 : 295–314.

Polhill, R.M. 1971. Some observations on generic limits in Dalbergieae– Lonchocarpineae Benth. (Leguminosae). *Kew Bull.* 25 : 259–73.

Prance, G.T. 1975. Revisao taxonomica das especies amazonicas de Rhizophoraceae. *Acta Amazonica* 5(1) : 5–22.

Primack, R.B., Duke, N.C. and Tomlinson, P.B. 1981. Floral morphology in relation to pollination ecology in five Queensland coastal plants. *Austrobaileya*, 4 : 346–55.

Primack, R.B. and Tomlinson, P.B. 1978. Sugar secretions from the buds of *Rhizophora*. *Biotropica* 10(1) : 74–5.

— 1980. Variation in tropical forest breeding systems. *Biotropica*, 12 : 229–31.

Qasim, S.Z. and Reddy, C.V.G. 19667. The estimation of plant pigments of the Cochin backwater during the monsoon months. *Bull. Mar. Sci.*, 17 : 95–110.

Rabinowitz, D. 1978a. Dispersal properties of mangrove propagules. *Biotropica*, 10:47–57.

— 1978b. Early growth of mangrove seedlings in Panama, and an hypothesis concerning the relationship of dispersal and zonation. *J. Biogeogr.*, 5 : 113–33.

Radlkofer, L. 1932. Sapindaceae. In : A.Engler, *Das Pflanzenreich*, Heft 98b, 4, 165, Leipzig : Wilhelm Engelmann.

Rains, D.W. and Epstein, E. 1967. Preferential absorption of potassium by leaf tissue of the mangrove *Avicennia marina* : an aspect of halophytic competence in coping with salt. *Aust. J. Biol. Sci.*, 20 : 847–57.

Rao, A.N. 1971. Morphology and morphogenesis of foliar sclereids in *Aegiceras corniculatum. Isr. J. Bot.*, 20 : 124–32.

Rao, T.A. 1973. An ecological approach towards classification of coastal vegetation in India. *Indian Sci. Congr. Assoc.*, 60th Session, pp.396.

Rao, T.A. and Sastry, A.R.K. 1972. An ecological approach towards classification of coastal vegetation of India. *Indian For.*, 98 : 594–607.

Raymond, A. and Phillips, T.L. 1983. Evidence for an upper carboniferous mangrove community. Chap. 2 in *The biology and ecology of mangroves*. Ed.H.J.Teas. *Tasks for vegetation science* 8. The Hague : Junk.

Reichie, D.E., McBrayer, J.F. and Ausmus, B.S. 1975. Ecological energetics of decomposer invertebrates in deciduous forest and total soil respiration budget. In : J.Vanek (ed.) *Progress in soil zoology*. Academia Publishing House of Cseko–Slovak Academy of Sciences. Prague, pp.282–292.

Rehm, A. and Humm, H.J. 1973. *Sphaeroma terebrans* : a threat to the mangroves of southwestern Florida. *Science*, 182 : 173–4.

Reimold, R.J. 1977. Mangals and salt marshes of eastern United States. Chap.7 in *Ecosystems of the World*, Vol.1 *Wet coastal ecosystems*, Ed. V.J.Chapman, Amsterdam : Elsevier Scientific.

Retallack, G. and Dilcher, D.L. 1981. A coastal hypothesis for the dispersal and rise to dominance of flowering plants. In : K.J.Niklas (ed.) *Paleobotany, paleoecology and evolution*. Praeger, New York, pp.27–77.

Rheede (tot Drakenstein), H. 1678–1703. *Hortus indicus malabaricus*. Amsterdam : Amstelodami.

Ribi, G. 1982. Does the wood boring isopod *Sphaeroma terebrans* benefit red mangroves (*Rhizophora mangle*) ? *Bull. Mar. Sci. Miami*, 31 : 925–8.

Ridley, H.N. 1930. *The dispersal of plants throughout the world*. Kent : Ashford 1938. Notes on *Xylocarpus, Kew Bull.* 1938 : 288–92.

Rollet, B. 1981. *Bibliography on mangrove research*, 1600, 1975. UNESCO, Paris.

Roth, I. 1965. Histogenese der Lentizellen am Hypokotyl von *Rhizophora mangle* L. *Ost. Bot. Z.* 112 : 640–53.

Rodin, L.E. and Bazilevich, N.I. 1967. *Production and Mineral Cycling in Terrestrial Vegetation*. London, Oliver & Boyd, pp.288.

Saenger, P., Specht, M.M., Specht, R.L. and Chapman, V.J. 1977. Mangal and coastal salt–marsh communities in Australia. In : Chapman, V.J. (ed.) *Ecosystems of the World*, Vol.1, Elsevier, Amsterdam, pp.293–346.

Saenger, P. 1982. Morphological, anatomical and reproductive adaptations of Australian mangroves. In : Clough, B.F. (ed.) *Mangrove ecosystems in Australia – Structure Function and Management*, Australian Institute of Marine Science and Australian National University Press, Canberra, pp.153–191.

Sahni, K.C. 1958. Mangrove forests in the Andaman and Nicobar islands. *Indian Forester*, 84(9) : 554–562.

Schimper, A.F.W. 1903. Plant geography upon a physiological basis. Englemenn and Wheldon and Westley Ltd., pp.395–410.

Schimper, A.F.W. 1891. *Botanische Mittheilungen aus den Tropen*. Heft 3 : *Die Indo Malayische Strandflora*. Jena : Gustav Fischer.

Schmid, R. 1980. Comparative anatomy and morphology of *Psiloxylon* and *Heteropyxis* and the subfamilial and tribal classification of Myrtaceae. *Taxon.* 29(5/6) : 559–95.

Schmidt, J. 1903. *Bidrag till kundskab om skuddene hos den gamle verdens mangrove-traeer*, Copenhagen.

Schnetter, M.L. 1978. Der Einfluss von Ausserfaktoren auf die Struktur des Blattes von *Avicennia germinans* (L.) unter naturlichen Bedingungen. *Beitr. Biol. Pflanz.* 54 : 13-28.

Scholander, P.F. 1968. How mangroves desalinate water. *Physiol. Plant* 21 : 251–61.

Scholander, P.F., Hammel, H.T., Hemmingsen, A.E. and Bradstreet, E.D. 1965. Sap pressure in vascular plants. *Science* 148 : 339–40.

Scholander, P.F., Hammel, H.T., Hemmingsen, E.A. and Bradstreet, E.D. 1964. Hydrostatic pressure and osmotic potential in leaves of mangroves and some other plants. *Proc. Nat. Acad. Sci. U.S.A.* 52 : 119–25.

Scholander, P.F., Hammel, H.T., Hemmingsen, E.A. and Garay, W. 1962. Salt balance in mangroves. *Plant Physiol.* 37 : 722–9.

Scholander, P.F., Van Dam, L. and Scholander, S.I. 1955. Gas exchange in the roots of mangroves. *Am. J. Bot.*, 42 : 92–8.

Schumann, K. 1891. Rubiaceae. In : Engler and Prantl (eds.) *Die naturlichen Pflanzenfamilien*, IV, Leipzig, Wilhelm Engelman, pp.1–156.

Semeniuk, K. 1980. Mangrove zonation along an eroding coastline in King Sound, north-western Australia. *J. Ecol.*, 68 : 789–812.

— 1983. Mangrove distribution in north-western Australia in relationship to regional and local freshwater seepage. *Vegetatio*, 53 : 11–31.

Semeniuk, V., Kenneally, K.F. and Wilson, P.G. 1978. *Mangroves of western Australia*, Handbook No.12, Perth : Western Australian Naturalists Club.

Semple, J.C. 1970. The distribution of pubescent leaved individuals of *Conocarpus erectus* (Combretaceae). *Rhodora*, 72 : 544–6.

Shimony, C., Fahn, A. and Reinhold, L. 1973. Ultrastructure and ion gradients in the salt glands of *Avicennia marina* (Forssk.). *Vierh. New Phytol.*, 72 : 27–36.

Sidhu, S.S. 1968. Further studies on the cytology of mangrove species of India. *Caroylogia*, 21 : 353–7.

Simberloff, D., Brown, B.J. and Lowrie, S. 1978. Isopod and insect root borers may benefit Florida mangroves. *Science*, 210 : 630–2.

Sinclair, J. 1968. Florae Malesianae precursores, 17. The genus *Myristica* in Malesia and outside Malesia. *Gard. Bull. Singapore*, 23 : 1–540.

Singh, J.S. and Gupta, S.R. 1977. Plant decomposition and soil respiration in terrestrial ecosystems. *Bot. Rev.*, 43 : 449–528.

Singh, V.P. 1989. Mangrove forests of Andaman Island – Their structure, Status and Future Prospects. In : Ram Prakash (ed.) *Advances in forestry research in India*, Vol. 5.

Singh, V.P. 2000. A new approach of Nitrogen fixation and uptake in mangroves ecosystem. *Indian Journal of Forestry*, 23(4) : 405–410.

Singh, V.P., Garge, A., Pathak, S.M. and Mall, L.P. 1986. Mangrove forests of Andaman islands in relation to human interference. *Environmental Conservation*, 13(2):169-172.

Singh, V.P., Garge, A., Pathak, S.M. and Mall, L.P. 1987. Pattern and process in mangrove forests of the Andaman islands. *Vegetatio*, 21(3) : 185–188.

Singh, V.P. and Mall, L.P. 1980. Some ecological aspects to mangrove vegetation of the south Andaman islands in India. *Sylvatrop. Philipp. for Res.*, 5 : 243–253.

Singh, V.P., Mall, L.P., Garge, A. and Pathak, S.M. 1986. Some ecological aspects of mangrove forest of Andaman islands. *Jour. Bombay Natu. Hist. Society*, 83(3) : 525–537.

Singh, V.P., Mall, L.P., Garge, A. and Pathak, S.M. 1987. A new record of some mangrove species from Andaman islands and their distribution. *Indian For.*, 113(3) : 214–217.

Sioli, Harald, 1954. Beitrage zur regionalen Limnologie des Amazonasgebretes. In : *Arch. Hydrobiol.* 49 : 441–518.

Sioli, Harald, 1954. Betrachutungen uber den Begriff der Fruchtbarkeit eines Gebietes anhand der Verhaltnisse in Boden und Gewassern Amazoniens. In : *Forsch und Fortsch*, 28 : 65–72.

Sioli, Harald, 1965. Bemerkungen zur Typologie amazonischer Flusse. In : *Amazoniana*, 1 : 267–277.

Sioli, Harald, 1968. Zur Okologie des Amazonas–Gebietes. In : Fittkau, Ernst Josef / Illies, J. / Klonge, H./ Schwabe, G.H. and H. Sioli (eds.) : Biogeography and Ecology in South America. Vol.1. Monographiae Biologicae 18 : 137–170.

Sioli, Harald, 1984. The Amazon Limnology and landscape ecology of a mighty tropical river and its basin – Dodrecht, Boston, Lancaster : Junk Publ. 1984 – 763p.

Slobodkin, L.B. 1964. Experimental populations of Hydra Jubilee Symposium. In : *Suppl. to J. Ecol.*, 52 : 131–148.

Snedaker, S.C. 1978. Mangroves their value and preparation. *Nature and Resources*, 14 : 6–13.

Snadaker, S.C. and Snedaker, J.G. 1984. *The mangrove ecosystem : research methods.* A monographs on oceanographic methodology–8, UNESCO, Paris.

Snedaker, S.C. and Brown, M.S. 1981. Primary productivity of mangroves. In : C.C.Black and Mitsui (eds.) *Handbook Series of Biosolar Resources*, Vol. 1, Basic Principle, pp. 477–86.

Snedaker, S.C. and Lugo, A. 1973. The role of mangrove ecosystem in the maintenance of environmental quality and a high productivity of desirable fisheries. *Atlanta Ga. US Bureau of Sports, Fisheries and Wildlife*, pp.263.

Soto, R. and Jimenez, J.A. 1982. Analisis fisonomico estructural del manglar de Puerto Soley, La Cruz, Guanacaste, Costa Rica, *Rev. Biol. Trop.* 30 : 161–8.

Sperry, J.S. 1983. Observations on the structure and function of hydathodes in *Blechnum lehmannii. Am. Fern. J.* 73 : 65–72.

Stace, C.A. 1965a. Cuticular studies as an aid to plant taxonomy. *Bull. Br. Mus. Nat. Hist.* 4 : 3–78.

— 1965b. The significance of the leaf epidermis in the taxonomy of Combretaceae, 1. A general review of tribal generic and specific characters. *J. Linn. Soc. Bot.*, 59 : 229–52.

Start, A.N. and Marshall, A.G. 1976. Nectarivorous bats as pollinators of trees in West Malaysia. In : J.Burley and B.T.Styles (eds.) *Tropical trees, variation breeding and conservation*, Academic Press, New York, pp.141–50.

Stearn, W.T. 1958. A key to West Indian mangroves. *Kew Bull.*, 1985 : 33–7.

Steenis, C.G.G.J. van. 1928. The Bignoniaceae of the Netherlands Indies. *Bull. Jard. Bot. Buitenz.* Series 3, 10 : 173–290.

— 1936. *Osbornia octodonia*, een weinig bekende mangrove–boom. *Trop. Nat.* 25:194-6.

-------------— 1937. De soorten van het geschlacht *Acanthus* in Nederlandsch–Indie. *Trop. Nat.* 26 : 202–7.

———— 1948. Plumbaginaceae. *Flora Malesiana* 1, 4 : 107–12.

———— 1949. Vicarism in the Malaysian flora. *Flora Malesiana* 1, 4 : 59.

———— 1958. Ecology of mangroves. *Flora Malesiana.*, 1, 5 : 431–41.

———— 1962. The distribution of mangrove plant genera and its significance for palaeogeography. *Proc. Kon. Net. Amsterdam*, Series C, 65 : 164–9.

———— 1977. Bignoniaceae. *Flora Malesiana* 1(8) : 114–86.

Steenis–Kruseman, M.V. van 1950. Malaysian plant collectors and collections. *Flora Malesiana* 1(1) : 1–63a.

Stehli, F.G. and Wells, J.W. 1971. Diversity and age patterns in hermatypic corals. *Syst. Zool.* 20 : 115–26.

Stern, W.L. and Voigt, G.K. 1959. Effect of salt concentration on growth of red mangrove in culture. *Bot. Gaz.*, 121(1) : 36–9.

Stevens, P.F. 1980. A revision of the Old World species of *Calophyllum* (Guttiferae). *J. Arnold Arbor.*, 61 : 117–699.

Studholme, W.P. and Philipson, W.R. 1966. A comparison of the cambium in two woods with included phloem *Heimerliodendron brunonianum* and *Avicennia resinifera. N.Z.J. Bot.*, 4(4) : 355–65.

Subba Rao, N.V., 1970. On the collection of strombidae (Molluska Gastropoda) from Bay of Bengal, Arabian Sea and Western Indian Ocean with some new records. I. Genus. *Strombus. J. Mar. Biol. Ass. India*, 12 : 109–124.

Sukriotigono, S. Kusurata Kartawinata and Isamu Yamada, 1980. The mangrove forest in Bungin river, Banyusin South Sumatra. *Asian Symp. on mangrove and environment. Res. and Management*, Univ. of Malaya, Kulalalumpur Maleysia. 25–29 Aug.

Sandaraj, V. and Krishnamurthy, K. 1972. Nutrients and plankton; back water and adjoining mangrove environment. *Symp. Estuarine Biol. Centre Adv. Stud. Mar. Biol.* Porto Novo India, 20–24 Jan.

Suzuki, E. and Tagawa, H. 1983. Biomass of mangrove forest and a sedgemarsh on Ishigaki island, South Japan. *Jap. J. Ecol.*, 33 : 231–234.

Sussex, I. 1975. Growth and metabolism of the embryo and attached seedling of the viviparous mangrove, *Rhizophora mangle. Am. J. Bot.*, 62 : 948–53.

Swingle, W.T. 1943. The botany of citrus and the wold relatives of its orange subfamily. In : H.J.Webber and L.D.Batchelor (eds.) *The citrus industry*, Vol.1, Burkley : Los Angeles, Division of Agricultural Sciences.

Tan, H. and Rao, A.N. 1981. Vivipary in *Opiorrhiza tomentosa* Jacq. (Rubiaceae). *Biotropica* 13 : 232–3.

Tattersfield, F., Martin, J.T. and Howes, F.N. 1940. Some fish poison plants and their insecticidal properties. *Kew Bull.* 1940 : 169–80.

Teas, H.J. 1982. An epidemic dieback gall disease of *Rhizophora* mangroves in the Gambia, West Africa. *Plant Dis. Rep.*, 66 : 522–3.

Thom, B.G. 1967. Mangrove ecology and deltaic geomorphology : Tobasco, Mexico. *J. Ecol.*, 55 : 301–43.

———— 1975. Mangrove ecology from a geomorphological viewpoint. In : G. Walsh, S. Snedaker and H. Teas (eds.) *Proc. Int. Symp. Biol. Manag. Mangroves*, Gainesville : University of Florida Press, pp.469–81.

Tieghem, Ph. van 1898. Avicenniacees et Symphoremacees. Place de ces deux nouvelles families dans la classification. *J. Bot. Paris*, 12 : 345–65.

Tobe, H. and Raven, P.H. 1983. An embryological analysis of Myrtales : its definition and characteristics. *Ann. Missouri Bot. Gard.*, 70 : 71–94.

Tomlinson, P.B. 1961. Palmae. In : C.R.Metcalfe (ed.) *Anatomy of the Monocotyledons*, Vol. 2, Clerendon Press, Oxford.

---------- 1971. The shoot apex and its dichotomous branching in the *Nypa* palm. *Ann. Bot. Lond.* 35 : 865–79.

---------- 1978. *Rhizophora* in Australia – some clarification of taxonomy and distribution. *J. Arnold Arbor.*, 59 : 156–69.

---------- 1980. *The biology of trees native to tropical Florida.* Petersham, Mass : published privately.

---------- 1982a. Field collection and study of Old World (Indo–Pacific) mangroves. *Nat. Geog. Soc. Res. Rep.*, 14 : 669–77.

---------- 1982b. *Helobiae (Alismatidae)*, In : C.R. Metcalfe (ed.) *Anatomy of the monocotyledons*, Vol. 7, Clarendon Press, Oxford.

Tomlinson, P.B., Bunt, J.S., Primack, R.B. and Duke, N.C. 1978. *Lumnitzera rosea* (Combretaceae) – its status and floral morphology. *J. Arnold Arbor.*, 59 : 342– 51.

Tomlinson, P.B., Primack, R.B. and Bunt, J.S. 1979. Preliminary observations in floral biology in mangrove *Rhizophoraceae. Biotropica*, 11 : 256–77.

Tomlinson, P.B. and Wheat, D.W. 1979. Bijugate phyllotaxis in Rhizophoreae (Rhizophoraceae). *Bot. J. Linn. Soc.*, 78 : 317–21.

Tomlinson, P.B. and Womersley, J.S. 1976. A species of *Rhizophora* new to New Guinea and Queensland, with notes relevant to the genus. *Contrib. Herb. Aust.*, 19 : 1–10.

Tralau, H. 1964. The genus *Nypa* van Wurmb. *K. svensk Vetensk. Akad. Handl.*, Series 5, 10(1) : 5–29.

Trochain, J. and Dulau, L. 1942. Quelques particularites anatomiques d'*Avicennia nitida* (Verbenaceae) de la mangrove ouest africaine. *Bull. Soc. Hist. Nat. Toulouse*, 77 : 271–81.

Troll, W. 1933a. *Camptostemon schultzii* Mast und *Camptostemon philippensis* (vid.). Becc. als neue Vertreter der austral–asiatischen Mangrove vegetation. *Flora n.f.* 128 : 348–60.

------------ 1933b. Ueber *Acrostichum aureum* L., *Acrostichum speciosum* Willd. und neotene Formen des lezteren. *Flora n.f.* 128 : 301–28.

------------ 1943. *Vergleichende Morphologie den hoheren Pflanzen* bd. 1 t 3. Vegetationsorgane. Berlin : Gebruder, Borntraeger.

Thothathri, K. 1962. Contribution to the flora of the Andaman and Nicobar Islands. *Bull. Bot. Surv. India*, 4 : 281–296.

Tomlinson, T.E. 1957. Relationship between mangrove vegetation, soil texture and reaction of surface soil after empolding saline swamps in Sierra Leone : *Trop. Agric.*, 34 : 41–50.

Twilley, R.R. 1985. Exchange of organic carbon in basin mangrove forests in a southwest Florida estuary. *Estuarine, Coastal and Shelf Science* 20 :

Uhl, N.W. 1972. Inflorescence and flower structure in *Nypa fruticans* (Palmae). *Am. J. Bot.* 59 : 729–43.

Ulken, A. 1983. Distribution of Phycomycetes in mangrove swamps with brackish waters and waters of high salinity. In : H.Teas. (ed.) Chap.12 in *Biology and ecology of mangroves. Tasks for vegetation science*, 8. The Hague : Junk.

United Nations Educational Scientific and Cultural Organization. 1978. Secondary successions. Chap.9 in *Tropical forest ecosystems*. A state–of–knowledge report prepared by UNESCO/UNEP/FAO. Paris : UNESCO.

Untawale, A.G. 1980. Bibliography of Indian mangroves. National Institute of Oceanography, Dona Pola Goa.

Untawale, A.G., Dwivedi, S.N. and Singhbal, S.Y.S. 1973. Ecology of mangroves in Mandovi and Zuari estuary and interconnecting cumburjua canal of Goa. *Ind. J. Marine Sci.*, 2 : 47–53.

Valeton, T. 1895. Les *Cerebera* du Jardin Botanique de Buitenzorg. *Ann. Jard. Bot. Buitenz.* 12 : 238–48.

Van der Pijl, L. 1936. Fledermause und Blumen. *Flora* 31 : 1–40.

van Vliet, G.J.C.M. 1976. Wood anatomy of Rhizophoraceae. *Leiden Bot. Ser.*, 3 : 20–75.

Van Steenis, C.G.G.J. 1958. Ecology (mangrove species) flora. *Malesiana Ser.*, 15:431-441.

Van Steenis C.G.G.J. 1962. The distribution of mangroves plant genera and its significance for palaeo–geography. *Verh. K. Ned. Akad. Wet. Ser. C.*, 65(2) : 164–169.

Waisel, Y. 1972. *Biology of halophytes*. Academic Press, New York.

Walsh, G.E. 1974. Mangroves : A Review. In : *Ecology of Halophytes*, Academic Press, New York, pp.51–174.

———— 1977. Exploitation of mangal. In : V.J.Chapman (ed.) *Ecosystems of the world, 1. Wet coastal ecosystems*, Chapter 16, Elsevier Scientific, Amsterdam.

Walsh, G.E., Ainsworth, K.A. and Rigby, R. 1979. Resistance of red mangrove (*Rhizophora mangle* L.) to lead, cadmium and mercury. *Biotropica*, 11 : 22–7.

Walsh, G.E., Snedaker, S.C. and Teas, H.J. (eds.) 1975. *Proceedings of the International Symposium on Biology and Management of Mangroves.* 2 Vols. Gainesville : Institute of Food and Agricultural Sciences, University of Florida.

Walter, H. and Steiner, M. 1936. Oekologie der ost–afrikanschen Mangroven. *Z. Bot.*, 30 : 65–93.

Ward, C.J. and Steinke, T.D. 1982. A note on the distribution and approximate areas of mangroves in South Africa. *S. Afr. J. Bot.*, 1 : 51–3.

Watson, J.G. 1929. Mangrove forests of the Malay Peninsula. *Malay For. Rec. No.6.*

Weber–El Ghobary, M.O. 1984. The systematic relationships of *Aegialitis* (Plumbaginaceae) as revealed by pollen morphology. *Plant. Syst. Evol.*, 144 : 53–8.

Webster, G.L. 1967. The genera of Euphorbiaceae in the southeastern United States. *Arnold Arbor.*, 48 : 303–430.

———— 1975. Conspectus of a new classification of the Euphorbiaceae. *Taxon.*, 24 : 593–601.

Wells, A.G. 1983. Distribution of mangrove species in Australia. In : Teas, H.J. (ed.) *Biology and Ecology of Mangroves.* Chap. 6 *Tasks for vegetation Science*, Junk, The Hague.

West, R.C. 1977. Tidal salt–marsh and mangal formations of middle and South America. In : *Ecosystems of the world,* 1, *Wet coastal ecosystems.* Elsevier Scientific, Amsterdam.

Westermaier, M. 1900. Zur Kenntnis der Pneumatophoren. *Botanische Untersuchungen im Amschluss an eine Tropenreise*, Heft 1. Freiberg.

White, C.T. 1926. A variety of *Ceriops tagal* C.B. Rob. (= *C.candollean* W.&A.), *J. Bot. Lond.*, 64 : 220–1.

Weinstein, M.P., Courtney, C.M. and Kinch, J.C. 1977. The macro island estuary a summary of the physiochemical and biological parameters. *Fla. Sci.* 40 : 97– 124.

Wharton, C.H. and Brinson, M.M. 1979. Characteristics of southeastern river systems. In : R.R. Johnson and J.F. Mccornic, Technical coordinators. Strategies for production and management of floodplain wet islands and other riparian ecosystems. *Forest Service General Technical Report* WO, 12 Deptt. of Agric., USA.

Wisweswara Rao, V. 1975. Bleniid fishes from Godavari estuary. *J. Bombay Nat. Hist. Soc.*, 70: 480–487.

Woodroffe, C.D. 1982. Litter production and decomposition in the New Zealand mangrove *Avicennia marina* var. *resinifera*. *New Zealand Journal of Marine and Freshwater Research*, 16 : 179–188.

Woodroffe, C.D. 1983. Development of mangrove forests from a geological perspective. In : H.Teas (Ed.) *Biology and ecology of mangroves. Tasks for vegetation science*, 8. Junk, The Hague.

Wright, D.F. 1977. A North Queensland mangrove pollen flora. B.Sc. Hons. Botany thesis. James Cook University, Townsville, Queensland.

Wyatt–Smith, J. 1953a. The Malayan species of *Bruguiera*. *Malay For.*, 16(3) : 156–61.

———— 1953b. The Malayan species of *Sonneratia*. *Malay*, 16 : 213–16.

———— 1954. The Malayan species of *Avicennia*. *Malay For.*, 17 : 21–5.

———— 1960. Field key to the trees of mangrove forests in Malaya. *Malay For.*, 23 : 126–32.

Yaldwin, J.C. and Wodzicki, K. 1979. Systematics and ecology of the land crabs of the Tokelau islands. *Central Pacific Act a Res. Bull.*, 235 : 1–53.

Yamashiro, M. 1961. Ecological study on *Kandelia candel* (L.) Druce, with special reference to the structure and falling of the seedlings. *Hikobia*, 2(3) : 209–14.

Zahran, M.A. 1977. Africa A, Wet formations of the African Red Sea coast. In : V.J.Chapman (Ed.) *Ecosystems of the World*, 1, Chap. 10 *Wet coastal ecosystems*, Elsevier Scientific, Amsterdam.

Zamski, E. 1979. The mode of secondary growth and the three dimensional structure of the phloem in *Avicennia Bot. Gaz.*, 140 : 67–76.

———— 1981. Does successive cambia differentiation in *Avicennia* depend on leaf and branch initiation? *Isr. J. Bot.*, 30 : 57–64.

Zimmermann, M.H. 1983. *Xylem structure and the ascent of sap.* Springer–Verlag, Berlin.

INDEX